Dynamics of Open Quantum Systems: Quantum Fluctuations, Decoherence and Emergent Phenomena

Guest Editors

Fernando C. Lombardo
Paula I. Villar

Basel • Beijing • Wuhan • Barcelona • Belgrade • Novi Sad • Cluj • Manchester

Guest Editors

Fernando C. Lombardo
Departamento de Fisica
Juan Jose Giambiagi, and
IFIBA CONICET
Buenos Aires
Argentina

Paula I. Villar
Departamento de Fisica
Juan Jose Giambiagi, and
IFIBA CONICET
Buenos Aires
Argentina

Editorial Office
MDPI AG
Grosspeteranlage 5
4052 Basel, Switzerland

This is a reprint of the Special Issue, published open access by the journal *Entropy* (ISSN 1099-4300), freely accessible at: www.mdpi.com/journal/entropy/special_issues/094C0KPR45.

For citation purposes, cite each article independently as indicated on the article page online and using the guide below:

Lastname, A.A.; Lastname, B.B. Article Title. *Journal Name* **Year**, *Volume Number*, Page Range.

ISBN 978-3-7258-2650-6 (Hbk)
ISBN 978-3-7258-2649-0 (PDF)
https://doi.org/10.3390/books978-3-7258-2649-0

© 2024 by the authors. Articles in this book are Open Access and distributed under the Creative Commons Attribution (CC BY) license. The book as a whole is distributed by MDPI under the terms and conditions of the Creative Commons Attribution-NonCommercial-NoDerivs (CC BY-NC-ND) license (https://creativecommons.org/licenses/by-nc-nd/4.0/).

Contents

About the Editors vii

Preface ix

Ludmila Viotti, Fernando C. Lombardo and Paula I. Villar
Geometric Phase of a Transmon in a Dissipative Quantum Circuit
Reprinted from: *Entropy* **2024**, *26*, 89, https://doi.org/10.3390/e26010089 1

Aroaldo S. Santos, Pedro H. Pereira, Patrícia P. Abrantes, Carlos Farina, Paulo A. Maia Neto and Reinaldo de Melo e Souza
Time-Dependent Effective Hamiltonians for Light–Matter Interactions
Reprinted from: *Entropy* **2024**, *26*, 527, https://doi.org/10.3390/e26060527 17

Xiangjia Meng, Yaxin Sun, Qinglong Wang, Jing Ren, Xiangji Cai and Artur Czerwinski
Dephasing Dynamics in a Non-Equilibrium Fluctuating Environment
Reprinted from: *Entropy* **2023**, *25*, 634, https://doi.org/10.3390/e25040634 40

Rahma Abdelmagid, Khadija Alshehhi and Gehad Sadiek
Entanglement Degradation in Two Interacting Qubits Coupled to Dephasing Environments
Reprinted from: *Entropy* **2023**, *25*, 1458, https://doi.org/10.3390/e25101458 54

Arzu Kurt
Interplay between Non-Markovianity of Noise and Dynamics in Quantum Systems
Reprinted from: *Entropy* **2023**, *25*, 501, https://doi.org/10.3390/e25030501 69

Eloi Flament, François Impens and David Guéry-Odelin
Emulating Non-Hermitian Dynamics in a Finite Non-Dissipative Quantum System
Reprinted from: *Entropy* **2023**, *25*, 1256, https://doi.org/10.3390/e25091256 81

Oleg V. Morzhin and Alexander N. Pechen
Control of the von Neumann Entropy for an Open Two-Qubit System Using Coherent and Incoherent Drives [†]
Reprinted from: *Entropy* **2023**, *26*, 36, https://doi.org/10.3390/e26010036 98

Feng Tian, Jian Zou, Hai Li, Liping Han and Bin Shao
Relationship between Information Scrambling and Quantum Darwinism
Reprinted from: *Entropy* **2023**, *26*, 19, https://doi.org/10.3390/e26010019 120

Federico Cerisola, Franco Mayo and Augusto J. Roncaglia
A Wigner Quasiprobability Distribution of Work
Reprinted from: *Entropy* **2023**, *25*, 1439, https://doi.org/10.3390/e25101439 137

Nicolás F. Del Grosso, Fernando C. Lombardo, Francisco D. Mazzitelli and Paula I. Villar
Adiabatic Shortcuts Completion in Quantum Field Theory: Annihilation of Created Particles
Reprinted from: *Entropy* **2023**, *25*, 1249, https://doi.org/10.3390/e25091249 151

About the Editors

Fernando C. Lombardo

Fernando Lombardo is a theoretical physicist with a Ph.D. from the University of Buenos Aires (UBA) and postdoctoral experience at Imperial College London. He specializes in the physics of the Casimir effect—both static and dynamic—as well as in the study of decoherence processes and the quantum-to-classical transition in open systems. He was a former director of the Department of Physics at UBA. Fernando has made significant contributions to understanding quantum phenomena, particularly in applications related to light-matter interaction. Recently, he has extended his research interests to explore these phenomena within the framework of circuit quantum electrodynamics (cQED).

Paula I. Villar

Paula I. Villar is a leading physicist specializing in the study of geometric phases in open quantum systems. She is a Ph.D. graduate from the University of Buenos Aires (UBA) and a researcher at CONICET. Paula's work delves deeply into understanding geometric effects and their applications within quantum mechanics, especially in the context of dissipative and open systems. Her expertise extends to the physics of superconducting circuits, where she investigates quantum gate design, photon generation, and controlled-squeezed states in circuit quantum electrodynamics (cQED). Her contributions are marked by a commitment to both scientific rigor and interdisciplinary collaboration, making her a valuable asset to the field of quantum physics.

Article

Geometric Phase of a Transmon in a Dissipative Quantum Circuit

Ludmila Viotti [1,†], **Fernando C. Lombardo** [2,3,*,†] **and Paula I. Villar** [2,3,†]

1. The Abdus Salam International Center for Theoretical Physics, Strada Costiera 11, 34151 Trieste, Italy
2. Departamento de Física, Facultad de Ciencias Exactas y Naturales, Universidad de Buenos Aires, Buenos Aires 1428, Argentina
3. Instituto de Física de Buenos Aires (IFIBA), CONICET—Universidad de Buenos Aires, Buenos Aires 1428, Argentina
* Correspondence: lombardo@df.uba.ar
† These authors contributed equally to this work.

Abstract: Superconducting circuits reveal themselves as promising physical devices with multiple uses. Within those uses, the fundamental concept of the geometric phase accumulated by the state of a system shows up recurrently, as, for example, in the construction of geometric gates. Given this framework, we study the geometric phases acquired by a paradigmatic setup: a transmon coupled to a superconductor resonating cavity. We do so both for the case in which the evolution is unitary and when it is subjected to dissipative effects. These models offer a comprehensive quantum description of an anharmonic system interacting with a single mode of the electromagnetic field within a perfect or dissipative cavity, respectively. In the dissipative model, the non-unitary effects arise from dephasing, relaxation, and decay of the transmon coupled to its environment. Our approach enables a comparison of the geometric phases obtained in these models, leading to a thorough understanding of the corrections introduced by the presence of the environment.

Keywords: geometric phases; circuit QED; Kerr coupling

Citation: Viotti, L.; Lombardo, F.C.; Villar, P.I. Geometric Phase of a Transmon in a Dissipative Quantum Circuit. *Entropy* **2024**, *26*, 89. https://doi.org/10.3390/e26010089

Academic Editor: Giuliano Benenti

Received: 25 December 2023
Revised: 15 January 2024
Accepted: 16 January 2024
Published: 22 January 2024

Copyright: © 2024 by the authors. Licensee MDPI, Basel, Switzerland. This article is an open access article distributed under the terms and conditions of the Creative Commons Attribution (CC BY) license (https://creativecommons.org/licenses/by/4.0/).

1. Introduction

Significant advancements in coherent superconducting circuits have enabled the development of a diverse range of qubit designs, encompassing the classic flux [1–4], phase [5–8], and charge [9–11] qubits, as well as more contemporary designs like transmon [12] and fluxonium [13] circuits. The mathematical representation and quantum dynamics of these qubits, when transversely coupled to resonators, are described by circuit quantum electrodynamics (cQED) [14–16]. Although these qubits are designed to behave as a two-level system within a superconducting circuit, they inherently possess multiple additional energy levels that can influence their interactions with other components. Focusing on the transmon qubit, one of its main advantages is its longer coherence timescale when compared to other circuits, which is essential for performing quantum computations and quantum error correction operations [17–19]. This increased coherence can be attributed to the non-harmonic energy level structure, which helps suppress certain types of decohering effects [20].

Within the landscape of physical systems provided by cQED setups, the objects known as geometric phases (GPs) have lately played an important role, mainly for implementing measurements and other quantum operations and therefore allowing GPs to be harnessed for various applications, such as geometric gates.

The idea that the phase acquired by the state of a quantum system can be decomposed into a dynamical and a geometrical component originated with Berry's theoretical work, where it was constrained to the context of adiabatic, cyclic, unitary evolution [21]. Subsequently, the concept of GP has been extended to non-adiabatic cyclic, non-cyclic, and even non-unitary evolutions [22–29]. These generalizations consistently reduce to less comprehensive results as additional conditions are met. The GP has also been elucidated as a consequence of quantum kinematics, interpreted in terms of a parallel transport condition dependent solely on the geometry of the Hilbert space, from which its name is derived [30].

As evident in the extensive literature, GPs have evolved into not only a fruitful avenue for exploring fundamental aspects of quantum systems but also a topic of technological interest. For instance, due to its resilience to fluctuations in a coupled bath, the GP has been proposed as a significant resource for constructing phase gates [31–33] for quantum information processing. Superconducting circuits have been extensively investigated as the physical system allowing this aim [34–38]. In a more fundamental approach, Berry phase has also been theoretically studied [39] and measured [40,41] in several circuit architectures. The vacuum GP accumulated by a superconducting artificial atom that was interacting with a single mode of a microwave cavity was measured [42] as well, while the corrections on the GP introduced by transitions to higher exited levels of the transmon were examined in [43].

When dealing with non-unitary dynamics leading to mixed states, the GP needs further generalization from its pure-state definition. A well defined proposal that applies under these conditions was presented in [28]. Thereafter, this definition has been applied to measure the corrections induced on the GP in non-unitary evolutions [44] and to explain the noise effects observed in the GP of a superconducting qubit [41,45]. Particularly, the GP of a two-level system under the influence of an external environment has been studied in a wide variety of scenarios [46,47]. Even though the GP is not an immediate reflection of the dynamics and can therefore remain robust to the effect of the environment, it differs, in the general case, from that accumulated by the associated closed system, as the evolution is now affected by non-unitary effects such as decoherence and dissipation. Under suitable conditions, the non-unitary GP can be measured through interferometric (atomic interference) [48,49], spin echo [41], and NRM [44,50] experiments.

In a previous study [51], we thoroughly examined the GP accumulated by a two-level system (TLS) that was interacting with a single mode of the quantized electromagnetic field within a dissipative cavity, a physical setup known as a dissipative Jaynes–Cummings (JC) model. Addressing the scenario frequently encountered in semiconductor cavity quantum electrodynamics (QED) [52], the interaction between the atom-mode system and its environment was characterized by the flow of photons through the cavity mirrors and the continuous, incoherent pumping spontaneously exciting the TLS.

In the present paper, we extend the work on the JC model to encompass the scenario of a nonlinear transmon coupled to a transmission line or resonator. Additionally, both the transmon and resonator are coupled to two semi-infinite waveguides serving as the surrounding environments. We will investigate the dynamics of the composite system, both in the qubit sector and in the two-excitation sector. Restricting to the one-excitation sector will allow for direct comparison of the results previously obtained, which implies the comparison of two different architectures in which atom–cavity dynamics emerge. It is worth highlighting a major difference between both studies even at this early stage, which is the non-monotonic behavior encountered in the GP under certain environmental conditions. Thereafter, to further explore the richer nature of the transmon atom, we also examine the GP and its environmentally induced corrections when the two-excitation levels are involved in the dynamics.

In the next section, we will introduce the Hamiltonian describing the non-harmonic transmon-field system under investigation and the coupling of the composite system to the environment. In this section, we will also present some insights about the dynamic evolution of the system and the definition of the geometric phase. In Section 3, we will describe the one-excitation subspace dynamics and the correspondence with previous results. The two-excitation space and the role of charging energy and non-harmonicity is discussed in Section 4. Section 5 summarizes our main conclusions.

2. Transmon with Atomic–Kerr Interaction

Due to their substantial dimensions, stemming from the necessity of maintaining low charging energy (via large capacitance), transmon qubits inherently lend themselves to capacitive coupling with microwave resonators. This coupling is reflected in the transmon Hamiltonian $\hat{H} = 4E_c(\hat{n} - n_g)^2 - E_J cos\hat{\varphi}$ by the substitution of the classical voltage source V_g with the resonator, a quantized gate voltage $n_g \to -\hat{n}_r$, representing the charge bias of the transmon due to the resonator. The Hamiltonian of the combined system is therefore [12]

$$\hat{H} = 4E_c(\hat{n} + n_r)^2 - E_J cos\hat{\varphi} - \sum_m \hbar\omega_m \hat{a}_m^\dagger \hat{a}_m, \tag{1}$$

where $\hat{n} = \hat{Q}/2e$ is the charge number operator, and $\hat{\varphi} = (2\pi/\Phi_0)\hat{\Phi}$ (mod 2π) is the phase operator, defined by the charge and phase operators, respectively, of the quantum circuit. The charging energy is $E_c = e^2/2C_\Sigma$, with $C_\Sigma = C_J + C_S$ being the sum of the junction's capacitance C_J and the shunt capacitance C_S. The operator $\hat{n}_r$ can be written as $\hat{n}_r = \sum_m \hat{n}_m$, with $\hat{n}_m = (C_g/C_m)\hat{Q}_m/2e$ being the contribution to the charge bias to the mth resonator mode. In this expression, C_g is the capacitance of the gate and C_m is the corresponding associated resonator mode capacitance. It is usual to consider $C_g \ll C_\Sigma, C_m$. When assuming that the transmon frequency is much closer to one of the resonator modes than all the other modes, it is possible to truncate the sum over m to a single term. In this single-mode approximation for the resonator or cavity, the Hamiltonian reduces to a single oscillator of frequency denoted by ω_r coupled to a transmon.

Expressed in terms of creation and annihilation operators, in the single-mode approximation, the Hamiltonian for the transmon–resonator cavity reduces to

$$\hat{H} = \hbar\omega_r \hat{a}^\dagger \hat{a} + \hbar\omega_q \hat{b}^\dagger \hat{b} - \frac{E_c}{2}\hat{b}^\dagger \hat{b}^\dagger \hat{b}\hat{b} - \hbar g(\hat{b}^\dagger - \hat{b})(\hat{a}^\dagger - \hat{a}), \tag{2}$$

where the term $\hat{n}_r^2$ has been absorbed in the charging energy term of the resonator mode and therefore leads to a renormalization of the resonator frequency, which we omit for simplicity. The frequency of the mode of interest is ω_r, and the coupling constant between the artificial atom and the resonator is given by the relation $g = \omega_r C_g/C_\Sigma (E_J/2E_c)^{1/4} (\pi Z_r/R_K)^{1/2}$, where Z_r is the characteristic impedance of the resonator mode and $R_K = h/2e^2$ is the resistance quantum. The above Hamiltonian can be simplified further in the experimentally relevant situation where the coupling constant is much smaller than the system frequencies, $|g| \ll \omega_r, \omega_q$. After rotating-wave approximation, the Hamiltonian reads

$$\hat{H}_K \approx \hbar\omega_r \hat{a}^\dagger \hat{a} + \hbar\omega_q \hat{b}^\dagger \hat{b} - \frac{E_c}{2}\hat{b}^\dagger \hat{b}^\dagger \hat{b}\hat{b} + \hbar g(\hat{b}^\dagger \hat{a} + \hat{b}\hat{a}^\dagger). \tag{3}$$

This is the Hamiltonian we shall consider to study the interaction between the artificial atom (in the transmon regime $E_J/E_c \gg 1$) and the electromagnetic field mode of the resonator. In this Hamiltonian, the term proportional to E_c is the so-called Kerr-like interaction term, or non-harmonic term.

2.1. Coupling to the Environment

Hitherto, our focus has been on quantum systems completely isolated from the surrounding environment. Nevertheless, a comprehensive portrayal of quantum circuits necessitates consideration of the manner in which these systems engage with their environment, encompassing both measurement apparatus and control circuitry. Indeed, the environment assumes a dual function in quantum technology; portraying quantum systems as entirely isolated is not only impossible due to inevitable coupling with undesirable environmental degrees of freedom but also renders a perfectly isolated system impractical for manipulation. Such a system would lack utility since we would be devoid of the means to control or observe it. Given these considerations, in this section, we investigate our quantum system coupled to external semi-infinite transmission lines that represent the measurement and control mechanisms while constituting the primary environment leading to photon loss and spontaneous decay of the transmon as their main environmental effect, respectively. We also consider the existence of a flux line for tunability of the transmon that leads to dephasing due to flux noise.

To study the open system comprising the transmon–resonator and transmission lines (reservoir), we will employ the conventional formalism provided by Lindblad master equations. In this context, we can model the transmission lines as a set of harmonic oscillators, similar to the approach taken in quantum Brownian motion models, which are paradigmatic examples of open quantum systems. Assuming, therefore, that the transmon and the resonator are coupled to independent baths (this system is illustrated in Figure 1), the master equation for the composite system can be expressed as follows:

$$\dot{\rho} = -i[\hat{H}_K, \rho] + \kappa \mathcal{D}[\hat{a}]\rho + \gamma \mathcal{D}[\hat{b}]\rho + \gamma_\varphi \mathcal{D}[\hat{b}^\dagger \hat{b}^\dagger]\rho, \qquad (4)$$

where ρ is the density matrix of the composite system (transmon–resonator), and $\hat{H}_K$ is the Hamiltonian of Equation (3). While this equation may suggest that dissipative processes independently impact the transmon and the resonator, the entanglement introduced by $\hat{H}_K$ implies that events such as the loss of a resonator photon can result in qubit relaxation. In this master equation, the coefficient κ is the photon decay rate; γ represents the relaxation rate of the artificial atom, which is related to the qubit-environment coupling strength evaluated at the qubit frequency; and γ_φ is the pure dephasing rate that superconducting quantum circuits can also suffer, caused, for example, by fluctuations of parameters controlling their transition frequency and by dispersive coupling to other degrees of freedom in their environment. The symbol $\mathcal{D}[\hat{\mathcal{O}}]\rho$ in Equation (4) represents the dissipator

$$\mathcal{D}[\hat{\mathcal{O}}]\rho = \mathcal{O}\rho\mathcal{O}^\dagger - \frac{1}{2}\{\mathcal{O}^\dagger\mathcal{O}; \rho\}, \qquad (5)$$

where $\{.;.\}$, is the anticommutator.

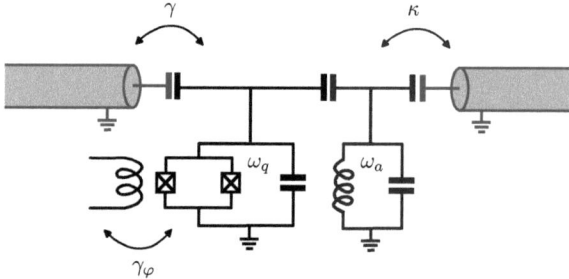

Figure 1. Transmon capacitively coupled to the resonator and both of them capacitively coupled to transmission lines. The coefficient κ is associated with the coupling between the resonator and the readout transmission line and represents the photon decay rate. In addition, γ is related to the qubit-environment coupling strength introduced by the control mechanism and represents the relaxation rate. Finally, γ_φ is the pure dephasing rate emerging due to the flux noise taking place in the flux line that allows for tunability of the transmon frequency.

We will work on the basis $\mathcal{B} = \{|m\,n\rangle\}$, where m refers to the mth energy eigenstate of the transmon and n is the nth Fock state of the mode field in the resonator.

2.2. Dynamic Evolution of the Open System

To address the dynamics of the transmon–field composite system, we numerically solve the master Equation (4), constrained to the subspace with two or less excitations, so that the base $\mathcal{B}$ reduces to $\{|00\rangle, |10\rangle, |01\rangle, |20\rangle, |11\rangle, |02\rangle\}$. At a given instant t, the state of the system is described by a 6×6 density matrix

$$\rho(t) = \begin{bmatrix} \rho_{00} & \rho_{01} & \rho_{02} & \rho_{03} & \rho_{04} & \rho_{05} \\ \rho_{10} & \rho_{11} & \rho_{12} & \rho_{13} & \rho_{14} & \rho_{15} \\ \rho_{20} & \rho_{21} & \rho_{22} & \rho_{23} & \rho_{24} & \rho_{25} \\ \rho_{30} & \rho_{31} & \rho_{32} & \rho_{33} & \rho_{34} & \rho_{35} \\ \rho_{40} & \rho_{41} & \rho_{42} & \rho_{43} & \rho_{44} & \rho_{45} \\ \rho_{50} & \rho_{51} & \rho_{52} & \rho_{53} & \rho_{54} & \rho_{55} \end{bmatrix}, \quad (6)$$

which can be decomposed into blocks as in Equation (6). Elements belonging to different blocks satisfy differential equations that are decoupled so, by taking an initial condition with vanishing ρ_{ij} elements for the off-diagonal blocks, the state of the system remains block diagonal along the whole evolution.

In this way, when the system is prepared in a one-excitation state $|\Psi(0)\rangle \in \{|10\rangle, |01\rangle\}$, the evolution remains restricted to the subspace form by the 2×2 block and ρ_{00}. In that case, the dynamics renders independent on the value of the capacitance energy E_c generating the anharmonicity. On the other hand, by setting the initial state as a pure state within the 3×3 block, all the matrix elements in the block-diagonal subspace are involved in the evolution.

2.3. Geometric Phase in the Open System

As already noted in the introductory Section 1, a generalized definition of a GP that is suitable to be computed for a mixed state under non-unitary evolution was proposed in [27]. It reads

$$\phi_g[\rho] = \arg \left\{ \sum_{k=1}^{N} \sqrt{\epsilon_k(0)\epsilon_k(T)} \, \langle \psi_k(0)|\psi_k(T)\rangle \, e^{-\int_0^T dt \, \langle \psi_k(t)|\dot{\psi}_k(t)\rangle} \right\}, \quad (7)$$

where $\psi_k(t)$ are the instantaneous eigenvectors of the density matrix, and $\epsilon_k(t)$ are the corresponding eigenvalues. This formula provides a well defined GP that, although defined for non-degenerate but otherwise general mixed states, when computed over pure states under unitary evolution, reduces to the expression

$$\Phi_g(t) = \arg \langle \Psi(0)|\Psi(t)\rangle - \mathrm{Im}\left\{ \int_0^t dt' \, \langle \Psi(t')|\dot{\Psi}(t')\rangle \right\}, \quad (8)$$

defined over the most general unitary evolution of a pure state $|\Psi(t)\rangle$ [23,53]. It is also manifestly gauge invariant and therefore depends solely on the path traced by the state in the ray space. When dealing with pure initial states, $\epsilon_k(0) = 1$ for the specific k labeling the initial state and vanishes for all $k' \neq k$. Therefore, Equation (7) reduces to a simpler form

$$\Phi_g(t) = \arg \langle \psi_+(0)|\psi_+(t)\rangle - \mathrm{Im}\left\{ \int_0^t dt' \, \langle \psi_+(t')|\dot{\psi}_+(t')\rangle \right\} \quad (9)$$

where $|\psi_+(t)\rangle$ is the eigenvector of $\rho(t)$ that coincides, at $t = 0$, with the initial state. This is the eigenvector such that $\epsilon_+(0) = 1$. Equation (9) has the exact same functional form of the GP defined over unitary evolutions, for which the only difference is that the pure state involved is the eigenstate $|\psi_+(t)\rangle$ of the density matrix and not the state of the system itself, which is now a mixed state.

We will restrict our analysis to pure initial states, so that the usual analysis applied to pure-state GPs can be immediately extrapolated by observing the behavior of the density matrix eigenstate $|\psi_+(t)\rangle$.

3. One-Excitation State: Dissipative Jaynes–Cummings Model

It will be useful to start by exploring some aspects of the better-known subspace accessed when preparing the system in a state with only one excitation. In this case, as stated in Section 2.2, the state of the system remains constrained to the 2×2 block spanned by $\{|10\rangle, |01\rangle\}$, with the only exception of showing population exchange to ρ_{00}.

On general grounds, the populations ρ_{11} and ρ_{22} of energy levels with one excitation oscillate while decaying, whereas the population of the vacuum state ρ_{00} grows. Depending on the parameters, the asymptotic state can be a pure $|00\rangle$ state or a mixed state with non-vanishing but suppressed ρ_{11} and ρ_{22} populations. The only non-zero coherences $\rho_{12/21}(t)$ increase in absolute value up to a maximum value and then vanish asymptotically. Figure 2 shows the explicit situation in which the state is prepared in a state $|\Psi(0)\rangle = |01\rangle$ with one field excitation.

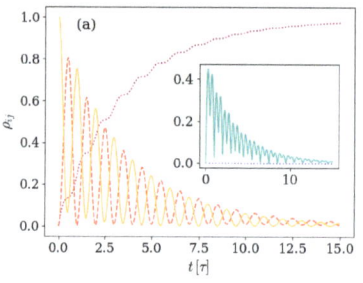 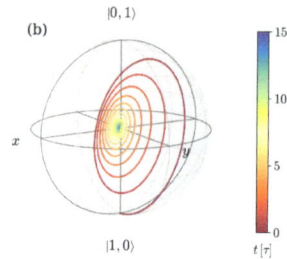

Figure 2. Dynamics of a system prepared in an initial $|\Psi(0)\rangle = |01\rangle$ and characterized by a detuning $\Delta = 0.017 \omega_q$ and an atom-field coupling $g = 0.028 \omega_q$. The environment is described by a photon loss rate $\kappa = 0.005 \omega_q$ and negligible atom relaxation γ and dephasing γ_φ rates. Panel (**a**) displays the density matrix elements evolution, and panel (**b**) shows the Bloch sphere representation of the 2×2 block. In panel (**a**), the main plot displays the evolution of populations ρ_{00}, ρ_{11}, and ρ_{22} with dotted purple, dashed orange, and solid yellow lines, whereas the inset displays the absolute value $|\rho_{12}|$.

It is useful to notice already at this point that the relation between the parameters defining the unitary evolution and the parameters defining the environmental effects results in a dynamic that exhibits specific characteristics. This is noticeable in Figure 2, where the absolute value of the only non-vanishing coherences reaches a minimum $|\rho_{12}| \sim 0$ at $t \sim 7.5\,\tau$, where it is smaller than the asymptotic value. This fact will be shown in different ways when observing the evolution of the eigenstate $|\psi_+(t)\rangle$ and the GP. In this archetypal example, we use $\omega_q = 2\pi \times 6$ GHz, $g = 2\pi \times 166.85$ MHz, and $\kappa = 2\pi \times 30$ MHz, which constitute up-to-date typical values [54–57], while we consider a non-dispersive detuning $\Delta = \omega_q - \omega_r$ originating from a resonator frequency $\omega_r = 2\pi \times 5.97$ GHz. In what follows, we will keep the typical values for the artificial atom frequency ω_q and the atom-mode coupling g while inspecting the dynamics arising in different conditions defined by different parameter values. In this sense, the atom-mode detuning Δ will be modified from closer-to-resonance values $\Delta \sim \mathcal{O}(10)$ MHz to values within the dispersive regime $\Delta \sim \mathcal{O}(1)$ GHz. The artificial atom decay rate will also be increased to $\gamma = 2\pi \times 30$ MHz. On the other hand, along the entire work, time is measured in units of $\tau = 2\pi/\Omega$, with the JC–Rabi frequency $\Omega = \sqrt{\Delta^2 + 4g^2}$.

The eigenvector $|\psi_+(t)\rangle$ of the density matrix that is involved in the expression for the geometrical phase belongs, in this case, to the 2×2 block and can thus be observed on the Bloch sphere. In order to inspect the dependence of the dynamics with the detuning Δ, Figure 3 shows the evolution of $|\psi_+(t)\rangle$ on the Bloch sphere for a system that is prepared in a state $|\Psi(0)\rangle = |01\rangle$ with one field excitation for three different relations of the detuning Δ to the frequency ω_q associated with the artificial atom. The remaining conditions are taken to be equal in all three cases. Panel (**a**) exhibits the case with $\Delta = 0.0017 \omega_q$, panel (**b**) displays the case where $\Delta = 0.017\omega_q$, and panel (c) shows the case with $\Delta = 0.17\omega_q$. In all plots, the time is given by the color according to the color bar on the very right of the figure.

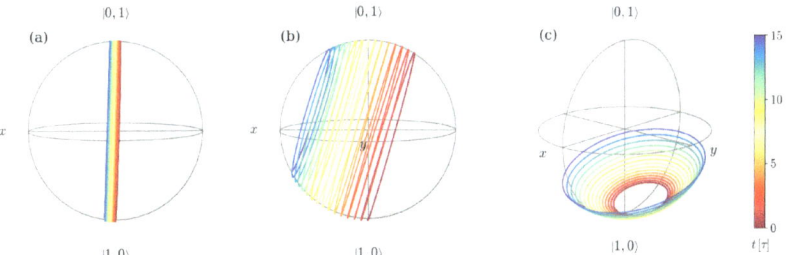

Figure 3. Trajectory displayed on the Bloch sphere by the eigenstate $|\psi_+(t)\rangle$ of the density matrix for the non-unitary evolution of a system prepared in an initial state $|\Psi(0)\rangle = |01\rangle$ for different ratios of the detuning Δ to the artificial atom frequency ω_q. Panels (**a**–**c**) show the cases with $\Delta = 0.0017\,\omega_q$, $\Delta = 0.017\,\omega_q$, and $\Delta = 0.17\,\omega_q$, respectively. The environment and remaining features of the system are the same in all three panels. The environment is characterized by a photon loss rate $\kappa = 0.005\,\omega_q$ and negligible atom decay and dephasing rates, and the atom-field coupling considered satisfies the relation $g = 0.028\,\omega_q$.

Figure 3 shows that the path described by $|\psi_+(t)\rangle$ on the Bloch sphere is in all three cases a spiral that starts in the south pole of the sphere. The axis along which the spiral winds and moves differs in all three cases. Under the conditions in panel (a) of Figure 3, in which the system is closer to resonance, the spiral axis is almost the *x*-axis and the curve moves little along it. Thus, the state traces a path that slightly deviates from vertical rings. When increasing the detuning, the axis of the spiral tilts and the turns separate from each other, as visible in panel (b). By further increasing the detuning, the axis of the spiral gets closer and closer to the *z*-axis and the initial turns get again closer while spreading for longer times.

In some cases, as displayed in panels (a) and (b) of Figure 3, this behavior implies the exploration of different hemispheres of the sphere, in which the winding starts on one side of a certain (different in each case) great circle and crosses to the other side at some point. When this happens, the GP accumulated changes sign. In order to see this, Figure 4 shows both the GP accumulated as a function of time (a), and the corresponding path traced by the $|\psi_+(t)\rangle$ eigenstate on the Bloch sphere (b), for the characteristic example depicted in Figures 2 and 3b.

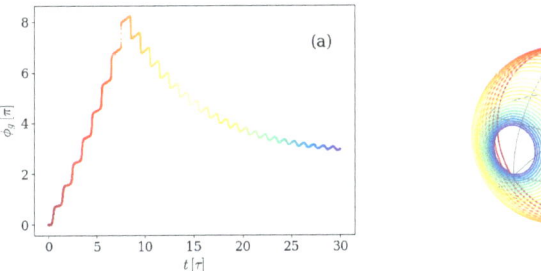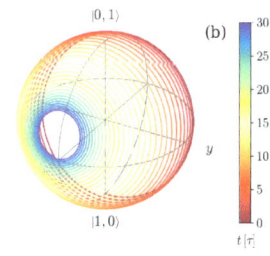

Figure 4. (**a**) Geometric phase accumulated over time by an initial state $|\Psi(0)\rangle = |01\rangle$ and (**b**) Bloch sphere depicting the path traced by the density matrix eigenvalue $|\Psi_+(t)\rangle$ for a system with detuning $\Delta = 0.017\,\omega_q$ and atom-field coupling $g = 0.028\,\omega_q$. The environment is characterized by a photon loss rate $\kappa = 0.005\,\omega_q$ for the artificial atom and negligible atom decay γ and dephasing γ_φ rates. The color depicts, in both panels, the time instant as indicated by the color bar on the right.

As expected, the accumulated GP has a sign while the state is winding the sphere on one hemisphere, and it changes sign as soon as the winding takes place on the other hemisphere. It can be seen from Figure 2 that this occurs at the specific time instant in which the coherences reach their minimum value. Therefore, we can classify the dynamics of the system initialized with one excitation as two kinds. One kind of evolution is that in which the coherences reach a minimum lower than the asymptotic value, the winding of the eigenstate $|\psi_+(t)\rangle$ changes hemisphere, and the GP changes sign, whereas the other is that in which none of these things happen.

To better explore under which physical circumstances the GP accumulation is non-monotone, we study the dependence in three characteristics of the system and environment. These are: the initial state of the system, which we take to be either $|10\rangle$ or $|01\rangle$; the main source of decoherence and dissipative effects, which we consider to be either the photon loss κ or the atom spontaneous decay γ; and the detuning Δ. The results of this examination are displayed in Figures 5 and 6, which show the GP accumulated in time for different combinations of the parameters characterizing the system. In both figures, solid lines represent the GP accumulated by the dissipative system, and the unitary results are introduced as dotted lines for reference.

Figure 5 corresponds to the previously explored case in which the photon loss process is the main source of dissipation. Three different ratios of the detuning to the frequency associated with the artificial atom $\Delta = 0.0017\,\omega_q$, $\Delta = 0.017\,\omega_q$, and $\Delta = 0.034\,\omega_q$ are displayed. Panel (a) shows the GP accumulated by a system prepared in the first excited level of the transmon atom $|\Psi(0)\rangle = |10\rangle$, and, in panel (b), the initial state has a single field excitation $|\Psi(0)\rangle = |01\rangle$.

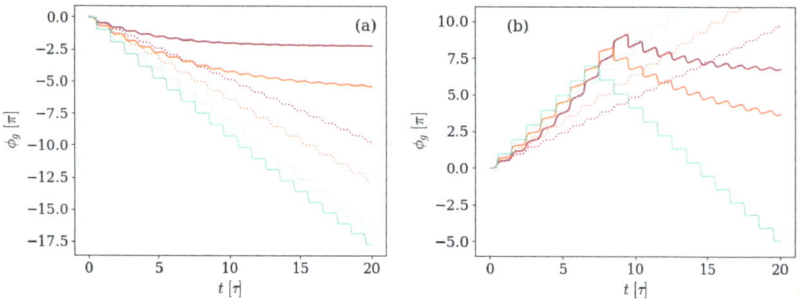

Figure 5. Geometric phase accumulated by systems with different values of detuning Δ, but otherwise equal. In panel (**a**) the system is prepared in an initial state $|\Psi(0)\rangle = |10\rangle$, whereas panel (**b**) shows the case with $|\Psi(0)\rangle = |01\rangle$. The detuning values $\Delta/\omega_q = 0.0017$, 0.017, and 0.34 are depicted by the light blue, red, and purple solid lines, respectively. The atom-field coupling satisfies $g = 0.028\,\omega_q$, and the environment is characterized by photon loss rate $\kappa = 0.005\,\omega_q$ and negligible atom-decay γ and dephasing γ_φ rates. The unitary results are included as dotted lines for reference, following the same Δ/ω_q-to-color code.

The two kinds of evolution, giving rise to monotonic or non-monotonic GPs, are clearly observed in Figure 5. In panel (a), the GP accumulated by an initial $|10\rangle$ state is softer than the unitary result due to the environmental effects on the dynamics. When the state reaches the steady state and therefore stops moving on the ray space, the GP settles. On the other hand, in panel (b), the GP accumulated by an initial $|01\rangle$ is non-monotonic, with the change of direction found sooner for smaller Δ/ω_q ratios. Therefore, the dynamics leading to non-monotonic GPs are found when the main environmental effects are those affecting the initial excitation of the system. It is worth noticing that the results in Figure 5 are in full agreement with the results obtained in [51], in which the explored situation was that of a system prepared in an initial $|10\rangle$ state and afterwards evolving in an imperfect cavity, corresponding to the case displayed in panel (a).

Likewise, Figure 6, where the GP accumulated in time is shown for the same three ratios of the detuning to the frequency of the transmon $\Delta = 0.0017\,\omega_q$, $\Delta = 0.017\,\omega_q$, and $\Delta = 0.034\,\omega_q$, but the main environmental effect is the atom spontaneous decay. Once again, panel (a) shows the case in which the system is prepared in the first excited level of the transmon atom $|\Psi(0)\rangle = |10\rangle$, whereas in panel (b) the initial state has a single field excitation $|\Psi(0)\rangle = |01\rangle$.

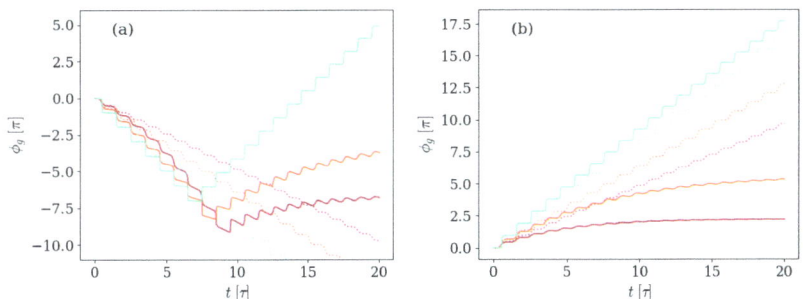

Figure 6. Geometric phase accumulated by systems with different values of detuning Δ, but otherwise equal. In panel (**a**) the system is prepared in an initial state $|\Psi(0)\rangle = |10\rangle$, whereas panel (**b**) shows the case with $|\Psi(0)\rangle = |01\rangle$. The detuning values $\Delta/\omega_q = 0.0017$, 0.017, and 0.34 are depicted by the light blue, red, and purple solid lines, respectively. The atom-field coupling satisfies $g = 0.028\,\omega_q$, and the environment is characterized by a decay rate $\gamma = 0.005\,\omega_q$ and negligible photon loss κ and dephasing γ_φ rates. The unitary results are included as dotted lines for reference, following the same Δ/ω_q-to-color code.

Consistent with the statement that non-monotonic GPs are found when the main environmental effects are those affecting the initial excitation of the system, in Figure 6 a change in the GP direction is only observed in panel (a), in which the initial state is the first excited level of the transmon atom and vacuum field $|\Psi(0)\rangle = |10\rangle$. As was also found in Figure 5, the strongest the atom decay rate in relation to the frequencies associated with unitary evolution, the sooner the GP accumulation changes sign.

Comparing to previous results in [51], the absence of non-monotonic behavior found there can be explained as a combination of both the initial state in which the atom–photon system was prepared and the main environmental phenomena affecting it when the physical architecture is semiconductor cavities. In that case, the effect considered in Figure 6 was absent.

The evolution in Figure 6 can be re-observed, giving emphasis to the time instants by displaying it in the same manner as Figure 4. The analog plots compose Figure 7, which thus shows the GP accumulated as a function of time (a) and the corresponding path traced by the $|\psi_+(t)\rangle$ eigenstate on the Bloch sphere (b) for a system prepared in the first excited level of the transmon atom and vacuum field, with detuning-to-frequency rate $\Delta = 0.017\,\omega_q$, and atom decay rate $\gamma = 0.005\,\omega_q$.

Once again, the change in the direction of the GP coincides in time with the moment in which the path traced by $|\psi_+(t)\rangle$ crosses from one side to the opposite of a great circle.

In order to re-state the description in terms of the behavior of the coherence, we go back to the initially considered case in which the initial state of the system has a single field excitation $|\Psi(t)\rangle = |01\rangle$. The analysis performed indicates there are two main situations in which the GP accumulated by this state will be monotonic: (a) if the relation between the environmental effects is such that the photon loss results are negligible in comparison with the atom decay rate, and (b) if, even though the main source of decoherence were the photon loss, the unitary parameters are strong enough to prevent the hemisphere crossing until the steady state is achieved.

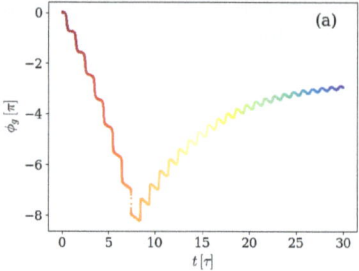 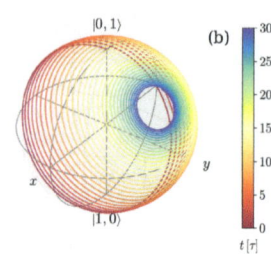

Figure 7. (**a**) Geometric phase accumulated over time by an initial state $|\Psi(0)\rangle = |10\rangle$ and (**b**) Bloch sphere depicting the path traced by the density matrix eigenvalue $|\psi_+(t)\rangle$ for a system with detuning $\Delta = 0.017\,\omega_q$ and spin-field coupling $g = 0.028\,\omega_q$. The environment is characterized by a relaxation rate $\gamma = 0.005\,\omega_q$ for the artificial atom and negligible photon loss κ and dephasing γ_φ rates. The color depicts, in both panels, the time instant, as indicated by the color bar on the right.

Figure 8 shows the density matrix elements evolution for both these situations in panels (a) and (b), respectively.

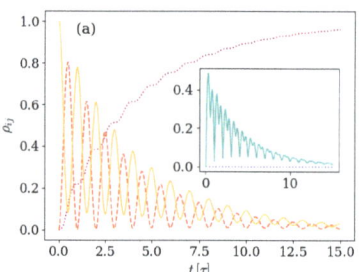 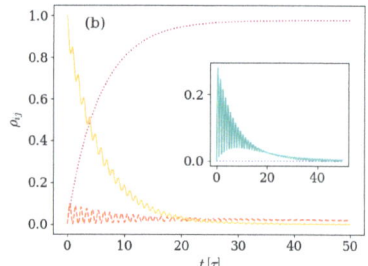

Figure 8. Density matrix elements evolution in time for two different conditions under which the GP does not change monotony. In both cases, the main plot displays the evolution of populations ρ_{00}, ρ_{11}, and ρ_{22} with dotted purple, dashed orange, and solid yellow lines, and the insert displays the absolute value $|\rho_{12}|$ (solid light-blue line) and the vanishing of the remaining coherences $|\rho_{i0}|$ (blue dotted line). Also in both panels, the system is prepared in an initial $|\Psi(0)\rangle = |01\rangle$ with atom-field coupling $g = 0.028\,\omega_q$. Panel (**a**) shows the case with $\Delta = 0.0017\,\omega_q$, $\gamma = 0.005\,\omega_q$, and negligible photon loss κ rate, whereas panel (**b**) addresses the case with $\Delta = 0.017\,\omega_q$, $\kappa = 0.005\,\omega_q$, and negligible atom decay γ rate. Dephasing is considered a subleading process on both plots.

It is immediately noticed that, in none of those cases, the absolute value of the coherences goes below its asymptotic value. In panel (a), where the photon loss is negligible in comparison with the atom decay rate, the excited populations ρ_{11} and ρ_{22} decay oscillating as the ground state gets populated. Different from what was observed in Figure 2, the absolute value of the non-zero coherences $|\rho_{12}|$ never reached a minimum below its asymptotic value. A similar behavior is observed in panel (b), where the increase in the detuning ratio Δ/ω_q results in a relatively less strong environment that requires observation over longer timescales. As in panel (a), in this case the coherences absolute value is never below the asymptotic value.

4. Two-Excitation Space: Role of Charging Energy and Non-Harmonicity

When turning to higher excited initial states, the evolution of the system takes place in the full six-dimensional Hilbert space described in Section 2.2. As stated there, if the system is prepared in a state with a defined number of excitations, the originally vanishing coherences remain zero at all following times.

On general grounds, the excited populations decay as lower-energy states get populated and, in the same way as described in Section 3, the system evolves to an asymptotic state that can be either the pure ground state $|00\rangle$ or a mixed state with non-vanishing but suppressed populations on exited states. The non-zero coherences increase in absolute value up to a maximum value, to vanish asymptotically afterwards.

In order to make the most simple generalization possible of the one-excitation case, we consider in this section a system that is prepared in a two-excitation state $|\Psi(0)\rangle = |11\rangle$, with excitations of a different nature. Figure 9 shows an specific example depicting the above described behavior.

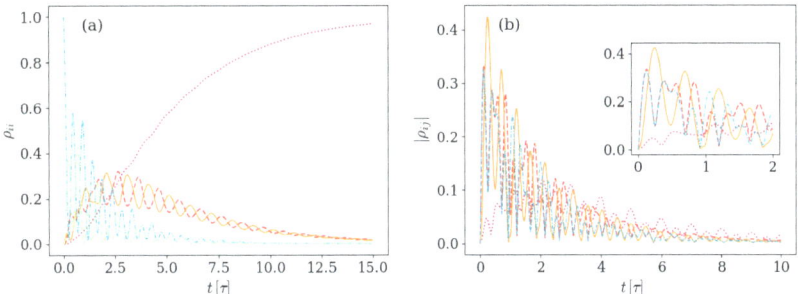

Figure 9. Dynamics of the system, depicted by the time evolution of the density matrix elements. Panel (**a**) shows the evolution of the populations $\rho_{ii}(t)$, with line stiles associated with matrix elements as follows: the dotted purple line shows ρ_{00}, the dashed red line shows ρ_{11}, the solid orange line shows ρ_{22}, the dot-dashed green line shows ρ_{33}, and the double dot-dashed light blue line shows ρ_{44}. Panel (**b**) displays the evolution, in absolute value, of the non-zero coherences ρ_{ij}, $i \neq j$. In this panel, the dotted purple line, dashed red line, solid orange line, dot-dashed green line, and double dot-dashed light blue line correspond to the matrix elements $\rho_{12}, \rho_{34}, \rho_{35}$, and ρ_{45}. The considered system is prepared in an initial state $|\Psi(0)\rangle = |11\rangle$ in the first excited level of the artificial atom and one field excitation. It is further characterized by a detuning value $\Delta = 0.0017\,\omega_q$, atom-field coupling rate of $g = 0.028\,\omega_q$, and anharmonicity $E_c = 0.035\,\omega_q$. The environment is characterized by photon loss $\kappa = 0.005\,\omega_q$ rate and negligible atom decay γ and dephasing γ_φ rates. The unitary results are included as dotted lines for reference.

It can be observed in panel (a) of Figure 9 that the population $\rho_{44}(t)$ associated with the $|11\rangle$ state decreases and shows non-harmonic oscillations. Along with this decrement, an immediate increment of the $\rho_{33}(t)$ and $\rho_{55}(t)$ populations, associated with the remaining two-excitation states $|20\rangle$ and $|02\rangle$, takes place. These are suppressed within a few Rabi periods. Due to the effect of the environment, the $\rho_{11}(t)$ and $\rho_{22}(t)$ elements associated with one-excitation states also show an increment and afterwards decrement while oscillate harmonically. The timescale of the increment–decrement of these elements is longer than the timescale shown by the two-excitation populations. Finally, the $\rho_{00}(t)$ population associated with the ground state monotonically increases along the whole evolution up to a steady value ~ 1. The non-zero coherences are now four (and their corresponding complex conjugates). Panel (b) of Figure 9 shows the evolutions of these elements. In all cases, the absolute value of these elements show an initial increment and asymptotic vanishing. Noticeably, the timescale associated with the $\rho_{12}(t)$ coherence belonging to the one-excitation subspace is larger than the timescale in which the coherences ρ_{34}, ρ_{35}, and ρ_{45}, belonging to the two-excitation block of $\rho(t)$, evolve.

In order to reproduce the analysis performed for evolutions constrained to the one and zero-excitation subspace, we turn now to the observation of the GP accumulated by the state of the system in time. With this purpose, Figure 10 shows the GP as a function of time for systems characterized by different ratios of the detuning Δ to the frequency ω_q associated with the transmon first transition, but otherwise equal. On each panel, the main environmental effect is of a different kind: in panel (a) the main environmental effect is the spontaneous decay of the transmon, whereas in panel (b) the main environmental effect is the photon loss.

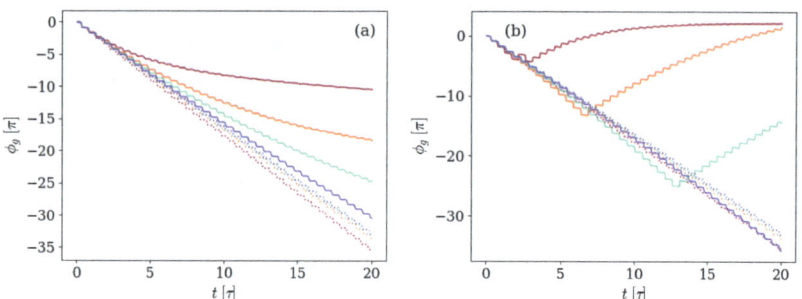

Figure 10. Geometric phase accumulated by a system prepared in an initial state with one excitation of each kind $|\Psi(0)\rangle = |11\rangle$ and characterized by different values of detuning Δ, but otherwise equal. The detuning values $\Delta/\omega_q = 0.0017$, 0.013, 0.015, and 0.017 are depicted by the purple, red, light blue, and blue solid lines, respectively, the atom-field coupling satisfies $g = 0.028\,\omega_q$, and the anharmonicity is $E_c = 0.035\,\omega_q$. In panel (**a**), the environment is characterized by an atom decay rate $\gamma = 0.005\,\omega_q$ and negligible photon loss κ and dephasing γ_φ rates. On the other hand, in panel (**b**), the environment is characterized by photon loss rate $\kappa = 0.005\,\omega_q$ and negligible atom decay γ and dephasing γ_φ ratios. The unitary results are included as dotted lines for reference, following the same Δ/ω_q-to-color code.

In panel (a) of Figure 10, where the initial $|11\rangle$ state evolves subjected to the spontaneous decay of the artificial atom, the GP accumulated along the evolution differs from the unitary GP, in that the non-unitary increment is slower and softer up to a point in which it is completely stopped by the meeting of a steady state that does not move on the ray space any more. The bigger the atom-decay rate relative to the unitary parameters, the strongest the effect of the environment on the dynamics, which is reflected in the same way by the GP.

In addition, panel (b) shows the evolution of the same system for the case in which the main environmental effect is the photon loss. Under these circumstances, the non-monotonic behavior of the GP already observed in Section 3 is recovered. If the detuning is small enough, the environment has the effect of changing the sign in the GP accumulated, which afterwards tends to an asymptotic state as the system reaches a steady state that does not move in the ray space. However, the differences observed when comparing to Figure 5 are not only quantitative but also qualitative: the GP acquired for $\Delta = 0.0017\,\omega_q$ not only changes direction but also does so more than once. Instead of observing a single minimum step, there are two.

With the anharmonicity between levels being one of the main differences between these state subspaces, we further inspect the effect of the ratio E_c/ω_q with focus on this difference in the behaviors observed in Figures 5 and 10.

Effect of the Anharmonicity

In order to examine the effect of the anhamonicity E_c on the GP accumulated by the non-unitary system, in panel (a) of Figure 11 we reproduce Figure 10b for a different value of the anharmonicity E_c. In doing so, the behavior of both the unitary and non-unitary GP accumulated in time is qualitatively modified depending on the relation between parameters.

With regard to the unitary results, displayed in the figure in dotted lines, the GP accumulated when decreasing the anharmonicity remains qualitatively equal to the previous situation for the smallest $\Delta = 0.0017\omega_q$ ratio, as depicted by the purple dotted line. However, when increasing the detuning, the unitary GP reaches a regime in which it changes sign periodically, leading to a step-like oscillation around a fixed value.

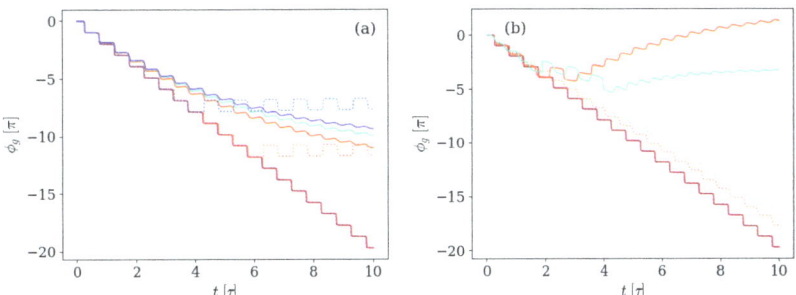

Figure 11. Geometric phase accumulated by a system prepared in an initial state with one excitation of each kind $|\Psi(0)\rangle = |11\rangle$. Panel (**a**) reproduces (**b**), changing the values of the anharmonicity to $E_c = 0.003\,\omega_q$. The considered detuning values are $\Delta/\omega_q = 0.0017$, 0.013, 0.015, and 0.017, depicted by the purple, red, light blue, and blue solid lines, respectively, and the atom-field coupling satisfies $g = 0.028\,\omega_q$. The environment is characterized by a photon loss rate $\kappa = 0.005\,\omega_q$ and negligible atom decay γ and dephasing γ_φ rates. On the other hand, panel (**b**) shows the GP accumulated for systems with different values of the anharmonicity $E_c/\omega_q = 0.003$, 0.035, and 0.067, depicted by the purple, red, and light blue solid lines, respectively. The system and environment are further characterized by a detuning $\Delta = 0.0017\,\omega_q$, an atom-field coupling $g = 0.028\,\omega_q$, photon loss rate $\kappa = 0.005\,\omega_q$, and negligible atom decay γ and dephasing γ_φ rates. On both panels, the unitary results are included as dotted lines for reference, following the same (**a**) Δ/ω_q-to-color and (**b**) E_c/ω_q-to-color codes.

Regarding the GP accumulated in the presence of an environment, the non-monotonic behavior observed in Figure 10b completely disappears, and the GP performs steps that get softer and smaller as time goes by up to the stationary regime in which the state stops moving in ray space, so no further GP is accumulated.

To further inspect this effect, panel (b) of Figure 11 shows the GP accumulated in time for a system that is prepared in the same $|\Psi(0)\rangle = |11\rangle$ state with fixed detuning ratio $\Delta = 0.0017\omega_q$ and environment conditions described by $\kappa = 0.005\,\omega_q$, but different values $E_c/\omega_q = 0.003$, 0.035, and 0.067 of the anharmonicity E_c. With increasing anharmonicity, the non-monotonic behavior is recovered, and more changes are observed in the GP sign for greater E_c values.

As a whole, the GP accumulated by an initial state with one photon and the transmon in its first excited level shows the non-monotonic behavior only when the main effect of the environment is the photon loss, observed when both the anharmonicity splitting the transmon levels and the effect of the environment are big enough, leading to a system that is closer to a two-level system. Reducing the anharmonicity leads to a degeneracy in the artificial atom states that prevents the environmental non-monotonic behavior while introducing qualitative changes in the unitary GP.

5. Discussion

In this paper we examined the dynamics beyond the two-level approximation of a transmon. Particularly, we studied the open dynamics of a nonlinear transmon coupled to a one-mode resonator and a transmission line. We have shown that the density matrix can be decomposed into blocks that satisfy differential equations that decoupled under particular initial conditions. Therefore, we were able to separately analyze one-excitation and two-excitation subspaces. We further explored the geometric phase accumulated by the state in order to have a better insight into the richer nature of the transmon artificial atom.

In the case of a one-excitation state, we retrieved results of the dissipative Jaynes–Cummings model. However, as we contemplated complex Linblad equations (with three different noise channels), we can complete already existing results. For example, we found the existence of a non-monotonic behavior in the accumulated geometric phase due to a combination of both the initial state and the main leading noise ruling the dynamics.

In the case of the two-excitation state, we studied the dissipative dynamics of the system in a regime where E_c cannot be neglected. The open dynamics are more complex, but we were able to present studies on the numerically accumulated geometric phase on a 3×3 subspace and study its behavior as a function of the anharmonicity rates. Again, the GP accumulated by an initial state with one photon and the transmon in its first excited level shows the non-monotonic behavior only when the main effect of the environment is the photon loss. This can be better understood with the help of the complete analysis on the 2×2 subspace (where we can interpret results with the help of the Bloch sphere).

It is important to remark that in the near-resonance case, the accumulated geometric phase seems to remain robust, as previous results stated.

Author Contributions: Conceptualization, Methodology, Formal analysis, and Writing—original draft, by L.V., F.C.L. and P.I.V. All authors contributed equally to this work. All authors have read and agreed to the published version of the manuscript.

Funding: This research was funded by Agencia Nacional de Promoción Científica y Tecnológica (ANPCyT), Consejo Nacional de Investigaciones Científicas y Técnicas (CONICET), and Universidad de 728 Buenos Aires (UBA).

Institutional Review Board Statement: Not applicable.

Informed Consent Statement: Not applicable.

Data Availability Statement: No new data were created or analyzed in this study. Data sharing is not applicable to this article.

Conflicts of Interest: The authors declare no conflicts of interest.

References

1. Friedman, J.R.; Patel, V.; Chen, W.; Tolpygo, S.; Lukens, J.E. Quantum superposition of distinct macroscopic states. *Nature* **2000**, *406*, 43–46. [CrossRef] [PubMed]
2. Van Der Wal, C.H.; Ter Haar, A.; Wilhelm, F.; Schouten, R.; Harmans, C.; Orlando, T.; Lloyd, S.; Mooij, J. Quantum superposition of macroscopic persistent-current states. *Science* **2000**, *290*, 773–777. [CrossRef] [PubMed]
3. Yan, F.; Gustavsson, S.; Kamal, A.; Birenbaum, J.; Sears, A.P.; Hover, D.; Gudmundsen, T.J.; Rosenberg, D.; Samach, G.; Weber, S.; et al. The flux qubit revisited to enhance coherence and reproducibility. *Nat. Commun.* **2016**, *7*, 12964. [CrossRef]
4. Steffen, M.; Brito, F.; DiVincenzo, D.; Farinelli, M.; Keefe, G.; Ketchen, M.; Kumar, S.; Milliken, F.; Rothwell, M.B.; Rozen, J.; et al. Quantum information storage using tunable flux qubits. *J. Phys. Condens. Matter* **2010**, *22*, 053201. [CrossRef] [PubMed]
5. Martinis, J.M.; Nam, S.; Aumentado, J.; Urbina, C. Rabi oscillations in a large Josephson-junction qubit. *Phys. Rev. Lett.* **2002**, *89*, 117901. [CrossRef]
6. Ansmann, M.; Wang, H.; Bialczak, R.C.; Hofheinz, M.; Lucero, E.; Neeley, M.; O'Connell, A.D.; Sank, D.; Weides, M.; Wenner, J.; et al. Violation of Bell's inequality in Josephson phase qubits. *Nature* **2009**, *461*, 504–506. [CrossRef]
7. Strauch, F.W.; Johnson, P.R.; Dragt, A.J.; Lobb, C.; Anderson, J.; Wellstood, F. Quantum logic gates for coupled superconducting phase qubits. *Phys. Rev. Lett.* **2003**, *91*, 167005. [CrossRef]
8. Chen, Y.; Sank, D.; O'Malley, P.; White, T.; Barends, R.; Chiaro, B.; Kelly, J.; Lucero, E.; Mariantoni, M.; Megrant, A.; et al. Multiplexed dispersive readout of superconducting phase qubits. *Appl. Phys. Lett.* **2012**. [CrossRef]
9. Nakamura, Y.; Pashkin, Y.A.; Tsai, J. Coherent control of macroscopic quantum states in a single-Cooper-pair box. *Nature* **1999**, *398*, 786–788. [CrossRef]
10. Bouchiat, V.; Vion, D.; Joyez, P.; Esteve, D.; Devoret, M. Quantum coherence with a single Cooper pair. *Phys. Scr.* **1998**, *1998*, 165. [CrossRef]
11. Pashkin, Y.A.; Astafiev, O.; Yamamoto, T.; Nakamura, Y.; Tsai, J. Josephson charge qubits: A brief review. *Quantum Inf. Process.* **2009**, *8*, 55–80. [CrossRef]
12. Koch, J.; Yu, T.M.; Gambetta, J.; Houck, A.A.; Schuster, D.I.; Majer, J.; Blais, A.; Devoret, M.H.; Girvin, S.M.; Schoelkopf, R.J. Charge-insensitive qubit design derived from the Cooper pair box. *Phys. Rev. A* **2007**, *76*, 042319. [CrossRef]
13. Manucharyan, V.E.; Koch, J.; Glazman, L.I.; Devoret, M.H. Fluxonium: Single cooper-pair circuit free of charge offsets. *Science* **2009**, *326*, 113–116. [CrossRef] [PubMed]
14. Blais, A.; Grimsmo, A.L.; Girvin, S.M.; Wallraff, A. Circuit quantum electrodynamics. *Rev. Mod. Phys.* **2021**, *93*, 025005. [CrossRef]

15. Girvin, S.M. Circuit QED: Superconducting qubits coupled to microwave photons. In *Quantum Machines: Measurement and Control of Engineered Quantum Systems: Lecture Notes of the Les Houches Summer School: Volume 96, July 2011*; Oxford University Press: Oxford, UK, 2014.
16. Makhlin, Y.; Schön, G.; Shnirman, A. Quantum-state engineering with Josephson-junction devices. *Rev. Mod. Phys.* **2001**, *73*, 357. [CrossRef]
17. Marques, J.; Ali, H.; Varbanov, B.; Finkel, M.; Veen, H.; van der Meer, S.; Valles-Sanclemente, S.; Muthusubramanian, N.; Beekman, M.; Haider, N.; et al. All-microwave leakage reduction units for quantum error correction with superconducting transmon qubits. *Phys. Rev. Lett.* **2023**, *130*, 250602. [CrossRef]
18. Battistel, F.; Varbanov, B.M.; Terhal, B.M. Hardware-efficient leakage-reduction scheme for quantum error correction with superconducting transmon qubits. *PRX Quantum* **2021**, *2*, 030314. [CrossRef]
19. Kim, Y.; Kang, J.; Kwon, Y. Design of quantum error correcting code for biased error on heavy-hexagon structure. *Quantum Inf. Process.* **2023**, *22*, 230. [CrossRef]
20. Chen, L.; Li, H.X.; Lu, Y.; Warren, C.W.; Križan, C.J.; Kosen, S.; Rommel, M.; Ahmed, S.; Osman, A.; Biznárová, J.; et al. Transmon qubit readout fidelity at the threshold for quantum error correction without a quantum-limited amplifier. *NPJ Quantum Inf.* **2023**, *9*, 26. [CrossRef]
21. Berry, M.V. Quantal Phase Factors Accompanying Adiabatic Changes. *Proc. R. Soc. Lond.* **1984**, *392*, 45–57.
22. Aharonov, Y.; Anandan, J. Phase change during a cyclic quantum evolution. *Phys. Rev. Lett.* **1987**, *58*, 1593. [CrossRef] [PubMed]
23. Samuel, J.; Bhandari, R. General setting for Berry's phase. *Phys. Rev. Lett.* **1988**, *60*, 2339. [CrossRef] [PubMed]
24. Sjöqvist, E.; Pati, A.K.; Ekert, A.; Anandan, J.S.; Ericsson, M.; Oi, D.K.; Vedral, V. Geometric Phases for Mixed States in Interferometry. *Phys. Rev. Lett.* **2000**, *85*, 2845. [CrossRef] [PubMed]
25. Wilczek, F.; Zee, A. Appearance of gauge structure in simple dynamical systems. *Phys. Rev. Lett.* **1984**, *52*, 2111. [CrossRef]
26. Anandan, J. Non-adiabatic non-abelian geometric phase. *Phys. Lett. A* **1988**, *133*, 171–175. [CrossRef]
27. Singh, K.; Tong, D.; Basu, K.; Chen, J.; Du, J. Geometric phases for nondegenerate and degenerate mixed states. *Phys. Rev. A* **2003**, *67*, 032106. [CrossRef]
28. Tong, D.; Sjöqvist, E.; Kwek, L.C.; Oh, C.H. Kinematic approach to the mixed state geometric phase in nonunitary evolution. *Phys. Rev. Lett.* **2004**, *93*, 080405. [CrossRef]
29. Wu, S.; Huang, X.; Wang, L.; Yi, X. Information flow, non-Markovianity, and geometric phases. *Phys. Rev. A* **2010**, *82*, 052111. [CrossRef]
30. Simon, B. Holonomy, the Quantum Adiabatic Theorem, and Berry's Phase. *Phys. Rev. Lett.* **1983**, *51*, 2167–2170. [CrossRef]
31. Jones, J.A.; Vedral, V.; Ekert, A.; Castagnoli, G. Geometric quantum computation using nuclear magnetic resonance. *Nature* **2000**, *403*, 869–871. [CrossRef]
32. Sjöqvist, E.; Tong, D.M.; Andersson, L.M.; Hessmo, B.; Johansson, M.; Singh, K. Non-adiabatic holonomic quantum computation. *New J. Phys.* **2012**, *14*, 103035. [CrossRef]
33. Xu, G.; Zhang, J.; Tong, D.; Sjöqvist, E.; Kwek, L. Nonadiabatic holonomic quantum computation in decoherence-free subspaces. *Phys. Rev. Lett.* **2012**, *109*, 170501. [CrossRef]
34. Kamleitner, I.; Solinas, P.; Müller, C.; Shnirman, A.; Möttönen, M. Geometric quantum gates with superconducting qubits. *Phys. Rev. B* **2011**, *83*, 214518. [CrossRef]
35. Song, C.; Zheng, S.B.; Zhang, P.; Xu, K.; Zhang, L.; Guo, Q.; Liu, W.; Xu, D.; Deng, H.; Huang, K.; et al. Continuous-variable geometric phase and its manipulation for quantum computation in a superconducting circuit. *Nat. Commun.* **2017**, *8*, 1061. [CrossRef]
36. Abdumalikov, A.A., Jr.; Fink, J.M.; Juliusson, K.; Pechal, M.; Berger, S.; Wallraff, A.; Filipp, S. Experimental realization of non-Abelian non-adiabatic geometric gates. *Nature* **2013**, *496*, 482–485. [CrossRef]
37. Leibfried, D.; DeMarco, B.; Meyer, V.; Lucas, D.; Barrett, M.; Britton, J.; Itano, W.M.; Jelenković, B.; Langer, C.; Rosenband, T.; et al. Experimental demonstration of a robust, high-fidelity geometric two ion-qubit phase gate. *Nature* **2003**, *422*, 412–415. [CrossRef]
38. Xu, Y.; Hua, Z.; Chen, T.; Pan, X.; Li, X.; Han, J.; Cai, W.; Ma, Y.; Wang, H.; Song, Y.; et al. Experimental implementation of universal nonadiabatic geometric quantum gates in a superconducting circuit. *Phys. Rev. Lett.* **2020**, *124*, 230503. [CrossRef]
39. Aunola, M.; Toppari, J. Connecting Berry's phase and the pumped charge in a Cooper pair pump. *Phys. Rev. B* **2003**, *68*, 020502. [CrossRef]
40. Möttönen, M.; Pekola, J.P.; Vartiainen, J.J.; Brosco, V.; Hekking, F.W. Measurement scheme of the Berry phase in superconducting circuits. *Phys. Rev. B* **2006**, *73*, 214523. [CrossRef]
41. Leek, P.J.; Fink, J.; Blais, A.; Bianchetti, R.; Goppl, M.; Gambetta, J.M.; Schuster, D.I.; Frunzio, L.; Schoelkopf, R.J.; Wallraff, A. Observation of Berry's phase in a solid-state qubit. *Science* **2007**, *318*, 1889–1892. [CrossRef]
42. Gasparinetti, S.; Berger, S.; Abdumalikov, A.A.; Pechal, M.; Filipp, S.; Wallraff, A.J. Measurement of a vacuum-induced geometric phase. *Sci. Adv.* **2016**, *2*, e1501732. [CrossRef]
43. Berger, S.; Pechal, M.; Pugnetti, S.; Abdumalikov Jr, A.; Steffen, L.; Fedorov, A.; Wallraff, A.; Filipp, S. Geometric phases in superconducting qubits beyond the two-level approximation. *Phys. Rev. B* **2012**, *85*, 220502. [CrossRef]
44. Cucchietti, F.M.; Zhang, J.F.; Lombardo, F.C.; Villar, P.I.; Laflamme, R. Geometric phase with nonunitary evolution in the presence of a quantum critical bath. *Phys. Rev. Lett.* **2010**, *105*, 240406. [CrossRef]

45. Lombardo, F.C.; Villar, P.I. Corrections to the Berry phase in a solid-state qubit due to low-frequency noise. *Phys. Rev. A* **2014**, *89*, 012110. [CrossRef]
46. Lombardo, F.C.; Villar, P.I. Environmentally induced corrections to the geometric phase in a two-level system. *Int. J. Quantum Inf.* **2008**, *06*, 707–713. [CrossRef]
47. Lombardo, F.C.; Villar, P.I. Nonunitary geometric phases: A qubit coupled to an environment with random noise. *Phys. Rev. A* **2013**, *87*, 032338. [CrossRef]
48. Fuentes-Guridi, I.; Carollo, A.; Bose, S.; Vedral, V. Vacuum induced spin-1/2 Berry's phase. *Phys. Rev. Lett.* **2002**, *89*, 220404. [CrossRef]
49. Moore, D.J.; Stedman, G. Non-adiabatic Berry phase for periodic Hamiltonians. *J. Phys. Math. Gen.* **1990**, *23*, 2049. [CrossRef]
50. Du, J.; Zou, P.; Shi, M.; Kwek, L.C.; Pan, J.W.; Oh, C.H.; Ekert, A.; Oi, D.K.; Ericsson, M. Observation of geometric phases for mixed states using NMR interferometry. *Phys. Rev. Lett.* **2003**, *91*, 100403. [CrossRef]
51. Viotti, L.; Lombardo, F.C.; Villar, P.I. Geometric phase in a dissipative Jaynes-Cummings model: Theoretical explanation for resonance robustness. *Phys. Rev. A* **2022**, *105*, 022218. [CrossRef]
52. Khitrova, G.; Gibbs, H.; Kira, M.; Koch, S.W.; Scherer, A. Vacuum Rabi splitting in semiconductors. *Nat. Phys.* **2006**, *2*, 81–90. [CrossRef]
53. Mukunda, N.; Simon, R. Quantum kinematic approach to the geometric phase. I. General formalism. *Ann. Phys.* **1993**, *228*, 205–268. [CrossRef]
54. Ahmad, H.G.; Jordan, C.; van den Boogaart, R.; Waardenburg, D.; Zachariadis, C.; Mastrovito, P.; Georgiev, A.L.; Montemurro, D.; Pepe, G.P.; Arthers, M.; et al. Investigating the Individual Performances of Coupled Superconducting Transmon Qubits. *Condens. Matter* **2023**, *8*, 29. [CrossRef]
55. Lledó, C.; Dassonneville, R.; Moulinas, A.; Cohen, J.; Shillito, R.; Bienfait, A.; Huard, B.; Blais, A. Cloaking a qubit in a cavity. *Nat. Commun.* **2023**, *14*, 6313. [CrossRef]
56. Swiadek, F.; Shillito, R.; Magnard, P.; Remm, A.; Hellings, C.; Lacroix, N.; Ficheux, Q.; Zanuz, D.C.; Norris, G.J.; Blais, A.; et al. Enhancing dispersive readout of superconducting qubits through dynamic control of the dispersive shift: Experiment and theory. *arXiv* **2023**, arXiv:2307.07765.
57. Cohen, J.; Petrescu, A.; Shillito, R.; Blais, A. Reminiscence of classical chaos in driven transmons. *PRX Quantum* **2023**, *4*, 020312. [CrossRef]

Disclaimer/Publisher's Note: The statements, opinions and data contained in all publications are solely those of the individual author(s) and contributor(s) and not of MDPI and/or the editor(s). MDPI and/or the editor(s) disclaim responsibility for any injury to people or property resulting from any ideas, methods, instructions or products referred to in the content.

Article

Time-Dependent Effective Hamiltonians for Light–Matter Interactions

Aroaldo S. Santos [1,2], Pedro H. Pereira [1], Patrícia P. Abrantes [3], Carlos Farina [3], Paulo A. Maia Neto [3,*] and Reinaldo de Melo e Souza [1]

1. Instituto de Física, Universidade Federal Fluminense, Niterói 24210-346, Rio de Janeiro, Brazil; arofisica@gmail.com (A.S.S.); pedro_h@id.uff.br (P.H.P.); reinaldos@id.uff.br (R.d.M.e.S.)
2. Instituto Federal do Paraná, Telêmaco Borba 84269-090, Paraná, Brazil
3. Instituto de Física, Universidade Federal do Rio de Janeiro, Rio de Janeiro 21941-972, Rio de Janeiro, Brazil; ppabrantes91@gmail.com (P.P.A.); farina@if.ufrj.br (C.F.)

* Correspondence: pamn@if.ufrj.br

Abstract: In this paper, we present a systematic approach to building useful time-dependent effective Hamiltonians in molecular quantum electrodynamics. The method is based on considering part of the system as an open quantum system and choosing a convenient unitary transformation based on the evolution operator. We illustrate our formalism by obtaining four Hamiltonians, each suitable to a different class of applications. We show that we may treat several effects of molecular quantum electrodynamics with a direct first-order perturbation theory. In addition, our effective Hamiltonians shed light on interesting physical aspects that are not explicit when employing more standard approaches. As applications, we discuss three examples: two-photon spontaneous emission, resonance energy transfer, and dispersion interactions.

Keywords: effective Hamiltonians; time-dependent Hamiltonians; quantum fluctuations; molecular quantum electrodynamics; light–matter interactions

Citation: Santos, A.S.; Pereira, P.H.; Abrantes, P.P.; Farina, C.; Maia Neto, P.A.; de Melo e Souza, R. Time-Dependent Effective Hamiltonians for Light–Matter Interactions. *Entropy* 2024, 26, 527. https://doi.org/10.3390/e26060527

Academic Editors: Fernando C. Lombardo and Paula I. Villar

Received: 27 May 2024
Revised: 12 June 2024
Accepted: 14 June 2024
Published: 19 June 2024

Copyright: © 2024 by the authors. Licensee MDPI, Basel, Switzerland. This article is an open access article distributed under the terms and conditions of the Creative Commons Attribution (CC BY) license (https://creativecommons.org/licenses/by/4.0/).

1. Introduction

In molecular quantum electrodynamics, atoms and molecules are treated within non-relativistic quantum mechanics, and the electromagnetic field mediating the interactions is quantized. This approach, born with Dirac's seminal treatment of spontaneous emission [1], is still an ongoing and intense research field, especially with the unprecedented control of light–matter interactions at the atomic scale reached in the last decades.

All phenomena in this topic can be fully understood by starting with the classical minimal coupling Hamiltonian and quantizing it. For molecular quantum electrodynamics, the most convenient approach is to work in the Coulomb gauge. Throughout this work, we shall deal with neutral molecules, in which case we can make a unitary transformation on the minimal coupling Hamiltonian and work with the equivalent multipolar Hamiltonian [2–6]. Nonetheless, the generality of these Hamiltonians is also their main weakness, since we must perform extensive calculations to obtain the quantities describing most of the effects. For instance, the interaction between two nonpolar molecules in their ground state results from a tedious fourth-order perturbative calculation.

Here comes the convenience of working with effective Hamiltonians, which are tailored for each specific application, bringing several physical insights and shortening the technical calculation to a much simpler and lower perturbative order analysis. An insightful example is the dynamical polarizability (DP) Hamiltonian, obtained by R. Passante and collaborators [7,8], which is built directly on the molecular dynamical polarizability instead of its electric dipole operator, capturing better the physics governing the interaction. Indeed, nonpolar molecules do not possess permanent electric dipole moments, and their interaction is possible only due to virtual internal transitions that are automatically taken into account by the dynamical polarizability. This is the main message of effective Hamiltonians: building the relevant physical mechanism into the Hamiltonian lowers the perturbative order required in calculations. With the DP Hamiltonian, intermolecular interactions are determined by means of a second-order calculation. Effective Hamiltonians and actions are also useful in describing non-stationary systems and have been employed to develop a multipolar approach to the dynamical Casimir effect [9] and to understand its microscopic origin [10–14].

Effective Hamiltonians are easily constructed from unitary transformations [15], but there is no general recipe to generate useful ones. In this paper, we fill this gap by presenting a systematic method to obtain convenient effective Hamiltonians and by discussing their physical implications. To illustrate our approach, we derive four general effective Hamiltonians allowing us to extend the scope of applications to important phenomena in molecular quantum electrodynamics.

Our method is based on choosing a unitary transformation inspired by (but not equal to) the Hermitian conjugate of the evolution operator for the system. A key concept for our formalism is that the linear susceptibility χ of a quantum system is built from the unequal time commutator of an appropriate operator $\mathcal{O}$ describing the system. For the context explored in this paper, we shall take $\mathcal{O}$ as being (i) the molecular dipole operator, in which case χ is related to the molecular polarizability, and (ii) the electric field operator, with χ representing the electric field generated by a point dipole (Green function), as discussed in Appendix A.

The importance of the unequal time commutators of the electric field operators in connection to the measurability of the fields was stressed in the literature [16,17]. Here, we show that the unequal time commutators can be taken as the basis to generate convenient effective Hamiltonians by allowing a given degree of freedom to effectively dress another one. Physically, this is equivalent to considering part of our system (a molecule or the electromagnetic field) as an open quantum system that is effectively dressed by an appropriate unitary evolution operator, thus yielding an effective time-dependent Hamiltonian for the system.

In Section 2, we employ our formalism to set up the Hamiltonian H_M^eff, in which the molecular degrees of freedom are dressed by the field. This Hamiltonian generalizes the DP one in two aspects: (i) it naturally accounts for internal dissipation in the molecules, and (ii) it does not require the molecules to remain in the same internal state (usually the ground state) during the process to be described. The latter aspect is a key element and enables us to evaluate the two-photon spontaneous emission (TPSE) in Section 3 within first-order perturbation theory—a much simpler route than the one commonly followed in the literature.

From Section 4 on, we work with situations involving more than one molecule. We employ our method to build the second effective Hamiltonian H_F^eff, where one molecule, say molecule B, dresses the field. With this dressing, the field acting on the other molecules is given by the superposition of the vacuum electric field with the electric dipole field generated by B. In Section 5, we demonstrate the convenience of H_F^eff by directly computing the resonance energy transfer (RET) between two quantum emitters in first order.

Then, in Section 6, we analyze dipole–dipole correlation effects and show that different effective Hamiltonians are convenient depending on the distance separating the molecules. In the asymptotic long-distance limit, we demonstrate the Hamiltonian H_MF^eff, in which the field dresses the molecules, and, in turn, one of the dressed molecules dresses back the field. This Hamiltonian is similar to H_F^eff, but now the electric field generated by molecule B does not depend on the molecular dipole operator. Instead, it is produced by the dipole induced by the vacuum itself. In the particular case where we assume the molecules to be in the ground state, in the long-distance limit, we recover the Hamiltonian originally proposed by P.W. Milonni [18].

Finally, for the short-distance limit (the non-retarded regime), we follow a complementary route: first, one molecule dresses the field, and then the dressed field dresses back the molecules. This leads us to a new effective Hamiltonian $H_{\text{FM}}^{\text{eff}}$. We demonstrate its convenience by applying it in Section 7 to obtain the London interaction energy in first-order perturbation theory and show that this route provides some relevant physical insights. Indeed, our approach can quantify the contributions to the interaction energy coming from the dipole fluctuations of each molecule. The above examples surely do not exhaust the list of useful effective Hamiltonians, and our method should be valuable for several additional applications.

2. The Field Dresses the Molecules

We begin with a single neutral molecule at position R in the presence of the quantized electromagnetic field. In the dipole approximation, the Hamiltonian describing the system is

$$H_{\text{total}} = H_0 - \boldsymbol{d} \cdot \boldsymbol{E}(\boldsymbol{R}), \tag{1}$$

where $\boldsymbol{d}$ is the molecular dipole operator, and $\boldsymbol{E}(\boldsymbol{R})$ is the quantized electric field evaluated at the molecule's center-of-mass position $\boldsymbol{R}$. Note that in the dipole approximation, the electric field can be taken as uniform over the scale of the molecule. H_0 stands for the free Hamiltonians and is given by

$$H_0 = H_{0m} + \sum_{k\sigma} \hbar \omega_k \left(a_{k\sigma}^\dagger a_{k\sigma} + \frac{1}{2} \right). \tag{2}$$

In this expression $a_{k\sigma}$ ($a_{k\sigma}^\dagger$) stands for the annihilation (creation) operator for a photon with wavevector $\boldsymbol{k}$ and polarization σ, whose electromagnetic field oscillates with a frequency $\omega_k = |\boldsymbol{k}|c$. H_{0m} is the free molecular Hamiltonian whose eigenstates and eigenenergies are assumed to be known. The coupling between the molecule and field, given by the second term on the right-hand side of Equation (1), can be treated as a perturbation. Therefore, it is convenient to work in the interaction picture with the interaction Hamiltonian

$$H(t) = -\boldsymbol{d}(t) \cdot \boldsymbol{E}(\boldsymbol{R}, t). \tag{3}$$

The time dependence is obtained by evolving the operators with the free Hamiltonian H_0. A nonpolar molecule is characterized by not having a permanent electric dipole in its ground state $|g\rangle$, i.e., $\langle g|\boldsymbol{d}|g\rangle = 0$. Thus, any process during which the molecule does not excite, such as the Stark shift, must be obtained at least through second-order perturbation theory. If $|\psi(t)\rangle$ symbolizes the state of the molecule–field system in the interaction picture, then its evolution can be written as

$$i\hbar \frac{d}{dt} |\psi(t)\rangle = H(t) |\psi(t)\rangle. \tag{4}$$

An equivalent description is generated once we apply a unitary transformation to the state. We choose it as

$$U_{\text{M}}(t) = e^{\frac{i}{\hbar} \int_{-\infty}^{t} dt' H(t')} \tag{5}$$

Note that transformation (5) implements the Heisenberg picture to first order in H, thus canceling, at this order, the time evolution of $|\psi(t)\rangle$, which is precisely our goal. If $[H(t), H(t')] = 0$, then the transformation given in Equation (5) would implement the Heisenberg picture exactly. Therefore, in the representation defined by U_{M}, the time evolution of $|\psi(t)\rangle$ results from the non-vanishing value of the commutator $[H(t), H(t')]$, which is consistent with the discussion on linear susceptibilities outlined in Section 1. As will become clear below, this transformation effectively dresses the molecular degree of freedom indicated by the subscript M. Next, we derive the equation satisfied by $|\psi_{\text{M}}(t)\rangle = U_{\text{M}}|\psi(t)\rangle$. From Equation (4), we find

$$i\hbar \frac{d}{dt} |\psi_{\text{M}}(t)\rangle = H_{\text{M}} |\psi_{\text{M}}(t)\rangle, \tag{6}$$

with
$$H_M(t) = U_M(t)H(t)U_M(t)^{-1} + i\hbar \frac{dU_M(t)}{dt}U_M^{-1}(t). \quad (7)$$

Here enters the fact that we are not looking for an equivalent Hamiltonian but rather an effective one. We desire an equivalent Hamiltonian only up to quadratic order in the dipole operator, and thus, we are allowed to expand U_M and collect results up to the second order in H, obtaining

$$U_M(t) \approx 1 + \frac{i}{\hbar}\int_{-\infty}^t dt' H(t') - \frac{1}{2\hbar^2}\left[\int_{-\infty}^t dt' H(t')\right]^2, \quad (8)$$

$$\frac{d}{dt}U_M(t) \approx \frac{i}{\hbar}H(t) - \frac{1}{2\hbar^2}\int_{-\infty}^t dt'\{H(t), H(t')\}. \quad (9)$$

Note that the expansion in Equation (8) does not correspond to a Dyson series, since the unitary transformation given in Equation (5) is not an evolution operator (it lacks a time-ordering operator). These expansions differ in the second-order term, and we stress that the third term on the right side of Equation (8) is proportional to the square of the second term, which is crucial for the results we will obtain. We define the effective Hamiltonian $H_M^{\text{eff}}(t)$ as the second-order approximation of H_M, which is obtained by substituting the previous relations into Equation (7):

$$H_M^{\text{eff}}(t) = -\frac{i}{2\hbar}\int_{-\infty}^t dt'[H(t), H(t')], \quad (10)$$

where we used the identity $2H(t)H(t') = [H(t), H(t')] + \{H(t), H(t')\}$. Notice that the linear term in the dipole vanished. From Equation (3) and since electric field operators at the same spatial point commute at all times (see Appendix A), we are left with

$$H_M^{\text{eff}}(t) = -\frac{i}{2\hbar}\int_{-\infty}^t dt'[d_j(t), d_l(t')]E_j(\mathbf{R}, t)E_l(\mathbf{R}, t'), \quad (11)$$

where we employed Einstein notation and denoted by $j, l = 1, 2, 3$ the Cartesian components of the operators. The great convenience of this Hamiltonian is that it is quadratic in the operators, thus halving the required perturbation order in comparison to the Hamiltonian given by Equation (3). We point out that our demonstration remains the same whether the electric field is quantized or not. As an example, with this effective Hamiltonian, the Stark effect can be obtained from a first-order perturbative calculation. We emphasize that this Hamiltonian is valid only within first-order perturbation theory, but improvements can be made if one keeps extra terms in Equations (8) and (9). H_M^{eff} mixes both the materials' and fields' degrees of freedom. When the atom is assumed to remain in the ground state, we may take the expectation value of H_M^{eff} in the molecular's subspace defined by the ground state through the evaluation of $\langle g|H_M^{\text{eff}}(t)|g\rangle$. We stress here that we are not acting on the field subspace, and thus, this average is still an operator in the field variables, which we denote by

$$H_M^{\text{eff}(gg)}(t) = -\frac{1}{2}\int_{-\infty}^\infty dt'\alpha_{jl}(t-t')E_j(\mathbf{R}, t)E_l(\mathbf{R}, t'), \quad (12)$$

with
$$\alpha_{jl}(t-t') = \frac{i}{\hbar}\theta(t-t')\langle g|[d_j(t), d_l(t')]|g\rangle \quad (13)$$

being the molecular dynamical polarizability tensor for the ground state describing its linear response to an applied electric field—see Appendix A for details. For practical applications and some physical interpretations, it is generally more suitable to work with the dynamical polarizability in the Fourier representation rather than in the time domain. To do so, we write the free electromagnetic field in the usual form:

$$E(R, t') = \sum_{k\sigma} E_{k\sigma}(R, t') = \sum_{k\sigma} \left[E_{k\sigma}^{(+)}(R) e^{-i\omega_k t'} + E_{k\sigma}^{(-)}(R) e^{i\omega_k t'} \right], \quad (14)$$

where $\omega_k = c|k|$, σ is the polarization degree of freedom, and the superscript $+$ $(-)$ refers to positive (negative) frequencies of the field. Substituting Equation (14) into (12), we arrive at

$$H_M^{\text{eff}(gg)}(t) = -\frac{1}{2} d^{\text{ind}}(t) \cdot E(R, t), \quad (15)$$

where

$$d_j^{\text{ind}}(t) = \sum_{k\sigma} \left[\alpha_{jl}(\omega_k) E_{k\sigma}^{l(+)}(R, t) + \alpha_{jl}(-\omega_k) E_{k\sigma}^{l(-)}(R, t) \right] \quad (16)$$

stands for the vacuum-induced dipole operator, and $E_{k\sigma}^{l(\pm)}(R, t) = E_{k\sigma}^{l(\pm)}(R) e^{\mp i\omega_k t}$ is the l-th Cartesian component of $E_{k\sigma}^{(\pm)}(R, t)$. Notice that $d^{\text{ind}}(t)$ acts on the field's Hilbert space. Due to the reality of $\alpha(t - t')$, $\alpha_{jl}(-\omega_k) = \alpha_{jl}^*(\omega_k)$, and, thus, $d^{\text{ind}}(t)$ is an Hermitian operator. If dissipation is negligible, we re-obtain as a particular case the DP Hamiltonian [7]

$$H_M^{\text{eff}(DP)}(t) = -\frac{1}{2} \sum_{k\sigma} \alpha_{jl}(\omega_k) E_{k\sigma}^l(R, t) E^j(R, t). \quad (17)$$

Physically, this is the quantum counterpart of the interaction energy of a polarizable system without permanent electric dipole moments in the presence of an external electric field. In the case without dissipation, the dynamical polarizability in the Fourier space is given by (see Appendix A)

$$\alpha_{jl}(\omega) = -\frac{1}{\hbar} \sum_{r \neq g} d_j^{gr} d_l^{rg} \left(\frac{1}{\omega - \omega_{rg}} - \frac{1}{\omega + \omega_{rg}} \right), \quad (18)$$

where r denotes the excited internal molecular states, and $d^{gr} = \langle g|d|r\rangle = d^{rg*}$ is the transition dipole moment between states g and r, while ω_{rg} is the corresponding transition frequency.

There are some subtleties concerning the unitary transformation employed in this section. One could argue that once the integration present in Equation (5) starts from $-\infty$, our truncation in Equation (9) is not rigorous. The point is that the molecule has a finite memory, characterized by a time scale τ. This means that $\alpha_{jl}(t - t')$ vanishes for $t - t' \gg \tau$ in Equation (13), enabling the lower limit in (5) to be replaced by $t - \tau$. The validity of this truncation is then tantamount to the validity of the perturbative method in the molecule–field interaction, justifying our approach. This argument also underlies the convenience of working in the Fourier space. Convergence of the time integration in Equation (12) requires that we account for dissipation in the polarizability. Nevertheless, in many cases of interest, the most relevant Fourier modes are far from molecular resonances, and we may neglect dissipation when using Equations (15) and (16).

Another important aspect is that the effective Hamiltonian (11) is convenient only when first-order perturbation theory in Hamiltonian (3) vanishes, even though regularization techniques may render it applicable if this is not the case [8]. We may separate the main applications of the effective Hamiltonian H_M^{eff} into two groups: (i) the molecule remains in the same internal state during the entire process, and the expectation value of the electric dipole operator in this state is zero; (ii) the molecule undergoes a transition between two internal states, but the electric dipole operator is unable to connect these two states. Examples involving (i) have already been discussed in the literature [7,8], in contrast with case (ii). One fascinating example of this second group is the two-photon spontaneous emission, with selection rules forbidding the one-photon transition. In the next section, we explore this example from the perspective of the effective Hamiltonian derived in this section.

3. Application to the Two-Photon Spontaneous Emission

An excited molecule may decay to its ground state through the emission of two photons in the so-called two-photon spontaneous emission (TPSE). This phenomenon is particularly interesting when the one-photon transition is forbidden. The TPSE makes the vacuum unstable and is responsible for the initial buildup of the intracavity field in two-photon micromasers [19–21]. More recently, it was shown that the simultaneously emitted photons can be indistinguishable and entangled in time and frequency [22–24], renewing the interest in this phenomenon [25–29]. This section aims to obtain the TPSE rate directly from first-order perturbation theory in Hamiltonian (11). Let us consider that a molecule in an internal state $|e\rangle$ decays in vacuum to its ground state $|g\rangle$ through the emission of two photons with wavevectors k and k' and polarizations σ and σ'. To this end, it suffices to analyze the matrix element of H_M^{eff} connecting the initial and the final states, given by

$$\langle g; 1_{k\sigma} 1_{k'\sigma'} | H_M^{\text{eff}}(t) | e; 0\rangle = -\frac{1}{2} \int_{-\infty}^{\infty} dt' D_{jl}^{ge}(t,t') \langle 1_{k\sigma} 1_{k'\sigma'} | E_j(0,t) E_l(0,t') | 0\rangle, \qquad (19)$$

where we chose the origin of our coordinate system at the position of the molecule. Here, $|0\rangle$ denotes the vacuum state of the electromagnetic field. We also define

$$D_{jl}^{ge}(t,t') = \frac{i}{\hbar} \theta(t-t') \langle g | [d_j(t), d_l(t')] | e\rangle, \qquad (20)$$

which involves only the molecular degrees of freedom and quantifies the linear response of the molecule to an applied field connecting internal states $|e\rangle$ and $|g\rangle$. Note that, when taking $|e\rangle = |g\rangle$ in Equation (20), the tensor $\overleftrightarrow{D}$ yields as a particular case the polarizability of the molecule, which is given by Equation (13). The TPSE rate is immediately obtained in the long-time limit by substituting Equation (19) into Fermi's golden rule. In general, it is more convenient to represent $\overleftrightarrow{D}$ in Fourier space. We begin by writing

$$d(t) = e^{iH_{0m}(t-t_0)} d e^{-iH_{0m}(t-t_0)} \qquad (21)$$

and, at the end, we take $t_0 \to -\infty$. In this expression, H_{0m} denotes the free molecular Hamiltonian, with eigenstates satisfying $H_{0m}|r\rangle = \hbar\omega_r|r\rangle$, so that by inserting a closure relation $\mathbb{I} = \sum_r |r\rangle\langle r|$ into Equation (20), we obtain

$$D_{jl}^{ge}(t,t') = \alpha_{jl}^{ge}(t-t') e^{-i\omega_{eg} t'}, \qquad (22)$$

with

$$\alpha_{jl}^{ge}(t-t') = \frac{i}{\hbar} \theta(t-t') \sum_r \left[d_j^{gr} d_l^{re} e^{-i\omega_{rg}(t-t')} - d_l^{gr} d_j^{re} e^{i\omega_{re}(t-t')} \right]. \qquad (23)$$

The instant t_0 plays a role only in an unimportant global phase, which was discarded. In Fourier space, Equation (23) becomes

$$\alpha_{jl}^{ge}(\omega) = \frac{1}{\hbar} \sum_r \left(\frac{d_j^{gr} d_l^{re}}{\omega_{rg} - \omega} + \frac{d_l^{gr} d_j^{re}}{\omega_{re} + \omega} \right). \qquad (24)$$

The desired matrix element given in Equation (19) can be obtained from Equation (14). For emission processes, only positive frequency modes contribute. Using Equation (22), we also obtain

$$\langle g; 1_{k\sigma} 1_{k'\sigma'} | H_M^{\text{eff}}(t) | e; 0 \rangle = \frac{\hbar \sqrt{\omega_k \omega_{k'}}}{4\varepsilon_0 V} e^{i(\omega_k + \omega_{k'} - \omega_{eg})t} \times \left[\epsilon_{k\sigma}^j \epsilon_{k'\sigma'}^l \alpha_{jl}^{ge}(\omega_{eg} - \omega_{k'}) + \epsilon_{k'\sigma'}^j \epsilon_{k\sigma}^l \alpha_{jl}^{ge}(\omega_{eg} - \omega_k) \right], \quad (25)$$

where $\epsilon_{k\sigma}^j$ is the jth Cartesian component of the polarization unit vector for the mode with wavevector k and polarization σ, and V is the volume of the quantization box. In the long-time limit, we are interested in photon pairs satisfying the condition $\omega_k + \omega_{k'} = \omega_{eg}$. With this, we see that $\alpha_{jl}(\omega_{eg} - \omega_{k'}) = \alpha_{jl}(\omega_k) = \alpha_{lj}(\omega_{eg} - \omega_k)$, where we used Equation (24) in the last equality. Therefore, we may simplify Equation (25) to

$$\langle g; 1_{k\sigma} 1_{k'\sigma'} | H_M^{\text{eff}}(t) | e; 0 \rangle = \frac{\hbar \sqrt{\omega_k \omega_{k'}}}{2\varepsilon_0 V} e^{i(\omega_k + \omega_{k'} - \omega_{eg})t} \epsilon_{k'\sigma'}^j \epsilon_{k\sigma}^l \alpha_{jl}^{ge}(\omega_{eg} - \omega_k). \quad (26)$$

The probability rate of emitting one photon in a solid angle $d\Omega$ around $\hat{k}$ and with a frequency in the interval $(\omega, \omega + d\omega)$ and another in a solid angle $d\Omega'$ around $\hat{k}'$, as well as with a frequency in the interval $(\omega', \omega' + d\omega')$, is given by (from now on we denote $\omega \equiv \omega_k$ and $\omega' \equiv \omega_{k'}$)

$$d\Gamma^{\text{TPSE}} = \frac{V^2}{(2\pi)^6} d\Omega d\Omega' d\omega d\omega' \frac{\omega^2 \omega'^2}{c^6} \frac{\left| \int_0^t dt' \langle g; 1_{k\sigma} 1_{k'\sigma'} | H_M^{\text{eff}}(t') | e; 0 \rangle \right|^2}{\hbar^2 t}. \quad (27)$$

Employing Fermi's golden rule, we arrive at

$$\frac{d\Gamma^{\text{TPSE}}}{d\Omega d\Omega' d\omega d\omega'} = \frac{\omega^3 \omega'^3}{c^6 (2\pi)^5 (2\varepsilon_0)^2} \left| \epsilon_{k'\sigma'}^j \epsilon_{k\sigma}^l \alpha_{jl}^{ge}(\omega_{eg} - \omega) \right|^2 \delta(\omega_{eg} - \omega - \omega'). \quad (28)$$

Integration over the solid angles may be readily evaluated from the identity

$$\sum_{\sigma,\sigma'} \int d\Omega d\Omega' \epsilon_{k'\sigma'}^j \epsilon_{k\sigma}^{m*} \epsilon_{k'\sigma'}^{n*} \epsilon_{k\sigma}^l = \frac{(8\pi)^2}{9} \delta_{ml} \delta_{jn}. \quad (29)$$

We also integrate over ω' to find the photon emission rate:

$$\frac{d\Gamma^{\text{TPSE}}}{d\omega} = \frac{\omega^3 (\omega_{eg} - \omega)^3}{18 c^6 \pi^3 \varepsilon_0^2} \alpha_{jl}^{ge}(\omega_{eg} - \omega) \alpha_{jl}^{*ge}(\omega_{eg} - \omega), \quad (30)$$

which is equivalent to the result of Ref. [2].

When performing second-order perturbation theory, the usual notation is to describe the molecular response in terms not of α_{ge}—which is a function of a single frequency variable, but rather a function of two frequency variables, which are obtained from Equation (24) by replacing $\omega_{re} + \omega$ by $\omega_{rg} - \omega'$ in the second term. The calculation from the new effective Hamiltonian (11) not only yields the two-photon spontaneous emission rate with a much shorter first-order calculation but also describes the results in terms of a single frequency variable function that sheds an interesting light on the physical mechanism involved in the phenomenon. In order to unveil the physical significance of α_{jl}^{ge}, let us project the Hamiltonian H_M^{eff} into the field's Hilbert space, thus generalizing Equation (12) for situations where the molecule undergoes an internal transition. This is done by defining the new effective Hamiltonian from Equation (11):

$$H_M^{\text{eff}(ge)}(t) := \langle g | H_M^{\text{eff}}(t) | e \rangle = -\frac{1}{2} \int_{-\infty}^{t} dt' D_{jl}^{ge}(t, t') E_j(\mathbf{0}, t) E_l(\mathbf{0}, t'), \quad (31)$$

which involves only electric field operators. Note that $D_{jl}^{ge}(t,t')$, given by Equation (20), is not a real number, and, therefore, $H_M^{\text{eff}(ge)}$ is non-Hermitian. This is due to the fact that the field degrees of freedom alone constitute an open quantum system, extracting energy from the drive provided by the molecular internal transitions encapsulated in $D_{jl}^{ge}(t,t')$. The non-hermiticity of Hamiltonian (31) also reflects the break of time inversion symmetry imposed by the two-photon decay.

Following the same steps that led us from Equations (12) to (15), we obtain

$$H_M^{\text{eff}(ge)}(t) = -\frac{1}{2} \mathbf{d}^{\text{ind}(ge)}(t) \cdot \mathbf{E}(\mathbf{R},t), \qquad (32)$$

where the induced dipole for the transition $|e\rangle \longrightarrow |g\rangle$ is given by

$$d_j^{\text{ind}(ge)}(t) = \sum_{\mathbf{k}\sigma} \left[\alpha_{jl}^{ge}(\omega_{eg}+\omega_k) E_{\mathbf{k}\sigma}^{l(+)}(\mathbf{R},t) + \alpha_{jl}^{ge}(\omega_{eg}-\omega_k) E_{\mathbf{k}\sigma}^{l(-)}(\mathbf{R},t) \right] e^{-i\omega_{eg}t}. \qquad (33)$$

For this reason, we denominate α_{jl}^{ge} as the *transition polarizability tensor*. This is a useful concept whenever the transition dipole element of a given internal molecular transition vanishes but can be induced by an external electric field. It generalizes the concept of the polarizability tensor, which stands for the dipole induced for a fixed internal molecular state. This induced transition dipole acts as an external source oscillating with frequency ω_{eg} and driving the field appearing in the effective Hamiltonian (31). Here, $\omega_{eg} > 0$ indicates that energy-conserving processes must be accompanied by photon creation, as can be verified in Equation (14). In this case, only the last term in Equation (33) contributes to the process. The other term is relevant for two-photon absorption, and the calculation presented in the section applies with minor modifications to this case.

4. The Molecules Dress the Field

In the previous section, we investigated the convenience of employing effective Hamiltonians in which the electric field dresses the molecules. Now, we shall analyze the opposite case and present an effective Hamiltonian that describes a molecule dressing the electric field operator. Consider two nonpolar molecules A and B. The electric dipole Hamiltonian describing this system in the interaction picture is

$$H^{(2)} = H_A + H_B, \qquad (34)$$

where $H_\zeta = -\mathbf{d}_\zeta(t) \cdot \mathbf{E}(\mathbf{R}_\zeta, t)$, and $\mathbf{d}_\zeta$ is the electric dipole operator of molecule $\zeta = A, B$, whose center of mass is at position $\mathbf{R}_\zeta$. We again represent the system's state with $|\psi(t)\rangle$, satisfying Equation (4), but implicitly including a tensor product of both the molecules' and fields' states. We follow the same reasoning as in the previous section. In this case, however, we want the molecule B to dress the electric field operator. Hence, we choose as the unitary transformation the inverse of the evolution operator for the coupling between molecule B and the field:

$$U_F = \widetilde{\mathcal{T}} e^{\frac{i}{\hbar} \int_{-\infty}^{t} dt' H_B(t')}, \qquad (35)$$

where $\widetilde{\mathcal{T}}$ is the anti-time ordering operator (earlier-time operators on the left). Its presence implies a crucial difference in comparison with Equation (5), and its purpose is to eliminate H_B so that the entire role played by molecule B will be through the field it produces. If only molecule B were present, the unitary transformation U_F would take the interaction picture into the Heisenberg picture. Nonetheless, in the presence of atom A, this unitary transformation yields a new effective Hamiltonian, to which we now turn our attention.

Following steps analogous to those in Section 2, the equivalent Hamiltonian is given by

$$H_F = U_F H^{(2)}(t) U_F^{-1} + i\hbar \frac{\partial U_F}{\partial t} U_F^{-1}. \qquad (36)$$

Due to the anti-time ordering operator, we have $i\hbar\partial_t U_F = -U_F H_B$, and, thus,

$$H_F = -d_A(t) \cdot U_F E(R_A, t) U_F^{-1}, \quad (37)$$

canceling H_B as anticipated. Expanding up to the linear term in d_B, we obtain (see Appendix A)

$$U_F E(R_A, t) U_F^{-1} \approx E(R_A, t) + E_{\text{dip},B}(R_A, t), \quad (38)$$

where

$$E_{\text{dip},B}(R_A, t) = \frac{1}{4\pi\varepsilon_0} \left\{ \frac{3[\hat{r} \cdot d_B(t_r)]\hat{r} - d_B(t_r)}{r^3} + \frac{3[\hat{r} \cdot \dot{d}_B(t_r)]\hat{r} - \dot{d}_B(t_r)}{cr^2} \right.$$
$$\left. + \frac{[\hat{r} \cdot \ddot{d}_B(t_r)]\hat{r} - \ddot{d}_B(t_r)}{c^2 r} \right\}, \quad (39)$$

with $r = R_A - R_B$ and $t_r = t - r/c$ being the retarded time. This expression corresponds to the electric field generated by dipole B at the position of molecule A. This result is readily extended to any number of molecules by exchanging $H_B \longrightarrow H_B + H_C + \cdots$ in Equation (35), thus obtaining

$$H_F^{\text{eff}} = -d_A(t) \cdot \left[E(R_A, t) + \sum_{\zeta = B, C, \ldots} E_{\text{dip},\zeta}(R_A, t) \right]. \quad (40)$$

While Equation (37) is exact and constitutes an equivalent Hamiltonian, Equation (40) is effective and valid only up to linear order in d_B, d_C, etc. It is worth mentioning that Equation (40) can be generalized to other situations. For instance, if atom A is in the presence of a magnetically polarizable atom [30,31], we have to add the electric field produced by the magnetic dipole of atom B in Equation (38). In the next section, we demonstrate the convenience of the new effective Hamiltonian H_F^{eff} by obtaining the RET rate in a first-order calculation.

5. Application to the Resonance Energy Transfer

In a resonance energy transfer (RET) process, an excited molecule decays through nonradiative channels, transferring its energy to a molecule in the ground state [32–41]. This phenomenon is of notable importance to many areas of science due to its broad range of applications across fields such as chemistry [42], medicine [43], and biology [44]. Throughout this section, we discuss the probability that an excited molecule A decays, exciting an identical molecule B that was initially in its ground state, placed at a distance r from A, with both in vacuum.

Up to the second order in perturbation theory, the probability amplitude of interest can be calculated as [15]

$$\mathcal{M}_{fi} \approx \langle \psi_f | H_{\text{int}} | \psi_i \rangle + \lim_{\eta \to 0^+} \sum_r \frac{\langle \psi_f | H_{\text{int}} | \psi_r \rangle \langle \psi_r | H_{\text{int}} | \psi_i \rangle}{E_i - E_r + i\eta}. \quad (41)$$

In this expression, $|\psi_i\rangle = |e_A, g_B, 0_{k\sigma}\rangle$ (with energy E_i) and $|\psi_f\rangle = |g_A, e_B, 0_{k\sigma}\rangle$ describe, respectively, the system's initial and final states, $\hat{H}_{\text{Int}}$ is the interaction Hamiltonian, and $|\psi_r\rangle$ are the intermediate states with energy E_r. In the standard approach, the interaction Hamiltonian is taken as the dipolar Hamiltonian given by Equation (34): $H_{\text{int}} = H^{(2)}$. With this choice, however, the first term in Equation (41) vanishes, and the RET rate is obtained from second-order perturbation theory. Here, we offer an alternative and simpler approach by letting atom B dress the field and taking $H_{\text{int}} = H_{\text{F}}^{\text{eff}}$, as in Equation (40). In this case, it suffices to calculate the first-order matrix element

$$\mathcal{M}_{fi} = -\langle g_A | \boldsymbol{d}_A(t) | e_A \rangle \cdot \langle e_B | \boldsymbol{E}_{\text{dip},B}(\boldsymbol{R}_A, t) | g_B \rangle. \tag{42}$$

Following Equation (21), the first matrix element on the right-hand side of the previous equation becomes (more precisely, we should consider the evolution beginning at time t_0; however, as explained in Section 3, this would only contribute as an irrelevant global phase)

$$\langle g_A | \boldsymbol{d}_A(t) | e_A \rangle = e^{-i\omega_{eg}t} \boldsymbol{d}_A^{ge}, \tag{43}$$

and, by using Equation (39), the terms contained in the second matrix element give the contributions

$$\langle e_B | \boldsymbol{d}_B(t_r) | g_B \rangle = e^{i\omega_{eg}t} e^{-ikr} \boldsymbol{d}_B^{eg},$$

$$\langle e_B | \partial_t \boldsymbol{d}_B(t_r) | g_B \rangle = \frac{\partial}{\partial t} \langle e_B | \boldsymbol{d}_B(t_r) | g_B \rangle = i\omega_{eg} e^{i\omega_{eg}t} e^{-ikr} \boldsymbol{d}_B^{eg},$$

$$\langle e_B | \partial_t^2 \boldsymbol{d}_B(t_r) | g_B \rangle = \frac{\partial^2}{\partial t^2} \langle e_B | \boldsymbol{d}_B(t_r) | g_B \rangle = -\omega_{eg}^2 e^{i\omega_{eg}t} e^{-ikr} \boldsymbol{d}_B^{eg}, \tag{45}$$

where $k = \omega_{eg}/c$. Replacing these results in Equation (42), we arrive at

$$\mathcal{M}_{fi} = \frac{d_{i,A}^{ge} d_{j,B}^{eg} e^{-ikr}}{4\pi\epsilon_0 r^3} \left[(\delta_{ij} - 3\hat{r}_i\hat{r}_j)(1 + ikr) - (\delta_{ij} - \hat{r}_i\hat{r}_j)k^2 r^2 \right]. \tag{46}$$

By applying Fermi's golden rule,

$$\Gamma^{\text{RET}} = \frac{2\pi}{\hbar^2} \rho(\omega_f) \left| \mathcal{M}_{fi} \right|^2, \tag{47}$$

where $\rho(\omega_f)$ is the density of final states with energy $E_f = \hbar\omega_f$, and we directly recover the well-known result for the RET rate [45,46].

6. The Dressed Molecules Dress the Field—And the Reverse

The first-order perturbation in the Hamiltonian (40) vanishes whenever the electric dipole operator of one of the molecules cannot connect the involved molecular states. An example is the force between molecules in their ground state to be analyzed in the next section. Here, instead, we focus on a general discussion without specifying the molecular internal state. The physical mechanism that limits the dipole–dipole correlation depends strongly on the distance R separating the molecules. Indeed, two characteristic time scales are key to understanding the two different regimes: the time it takes light to travel between the molecules, $t_\gamma = r/c$, and the characteristic time for dipole fluctuations, $t_d = 1/\omega_0$, where ω_0 is a typical transition frequency for the molecules. In the asymptotic long-distance regime, $t_\gamma \gg t_d$, it is the electrodynamical retardation that limits the dipole–dipole correlation, and we may neglect dispersion in the atomic response. In the opposite short-distance regime, electrodynamical retardation is negligible, and it is now the delay in the molecular response that limits the dipole–dipole correlation. Now, the molecular dispersion is crucial, but we can take the electrostatic limit for the electric field produced by each electric dipole. To go deeper into the physical particularities of these two complementary regimes, we shall develop a different effective Hamiltonian appropriate to each case.

6.1. The Dressed Molecules Dress the Field: Retarded Long-Distance Regime

In the long-distance regime, we may neglect dispersion in the molecules, which is tantamount to considering an instantaneous molecular response. This means that the timescale variation for the electric field is much slower than the molecular response, enabling us to approximate

$$\frac{i}{\hbar}\int_{-\infty}^{t}dt'[d_j(t),d_l(t')]E_l(\boldsymbol{R},t') \approx \Pi_{jl}E_l(\boldsymbol{R},t), \tag{48}$$

when evaluating the effective Hamiltonian $H_{\mathrm{M}}^{\mathrm{eff}}(t)$ given by Equation (11). We have defined the molecular operator Π_{jl} as

$$\Pi_{jl} = \frac{i}{\hbar}\int_{-\infty}^{\infty}dt'\theta(t-t')[d_j(t),d_l(t')]. \tag{49}$$

From Equation (13), we see that the expectation value $\langle g|\Pi_{jl}|g\rangle$ is the static polarizability of the molecule in its ground state, given by setting $\omega = 0$ in Equation (18). The tensor operator Π_{jl} generalizes this concept by enabling us to capture the static response even for processes involving changes in the molecular internal state. Substituting (48) into (11) leads to

$$H_{\mathrm{M}}^{\mathrm{eff,s}}(t) = -\frac{1}{2}\Pi_{jl}E_j(\boldsymbol{R},t)E_l(\boldsymbol{R},t), \tag{50}$$

which corresponds to the static response limit of $H_{\mathrm{M}}^{\mathrm{eff}}(t)$.

In the case of two molecules, dipole–dipole correlations arise in second-order perturbation theory in the Hamiltonian

$$H_{\mathrm{M}}^{\mathrm{eff,s(2)}}(t) = H_{\mathrm{M,A}}^{\mathrm{eff,s}\,(2)}(t) + H_{\mathrm{M,B}}^{\mathrm{eff,s(2)}}(t) = -\frac{1}{2}\Pi_{A,jl}E_j(\boldsymbol{R}_A,t)E_l(\boldsymbol{R}_A,t) - \frac{1}{2}\Pi_{B,jl}E_j(\boldsymbol{R}_B,t)E_l(\boldsymbol{R}_B,t), \tag{51}$$

where $\Pi_{\zeta,jl}$ is the operator (49) for molecule $\zeta = A, B$.

An equivalent Hamiltonian where the molecules couple directly with each other will be able to capture the dipole–dipole correlation in first-order perturbation theory. This can be done by employing the unitary transformation

$$U_{\mathrm{MF}} = \widetilde{T}e^{\frac{i}{\hbar}\int_{-\infty}^{t}dt'H_{\mathrm{M,B}}^{\mathrm{eff,s}}(t')}, \tag{52}$$

which mimics the one employed in Section 4, with the difference that it is the dressed molecule (through operator Π_{jl}), instead of the naked molecule, that dresses the field. Following the same steps that led us from Equation (34) for molecule B into Equation (37), we get

$$H_{\mathrm{MF}}^{\mathrm{eff,s}}(t) = U_{\mathrm{MF}}H_{\mathrm{M,A}}^{\mathrm{eff,s}}(t)U_{\mathrm{MF}}^{-1}. \tag{53}$$

$H_{\mathrm{MF}}^{\mathrm{eff,s}}$ is very suitable for handling effects related to the interaction between atoms A and B because the expansion of U_{MF} in Equation (53) contains terms combining the product $\Pi_{A,jl}\Pi_{B,mn}$. Such terms also appear through a fourth-order perturbation theory in Hamiltonian (34) but already appear in first order here. To obtain $H_{\mathrm{MF}}^{\mathrm{eff,s}}$, it is enough to implement the transformation rule for the electric field at $\boldsymbol{R}_A$, since $\Pi_{A,jl}$ commutes with $\Pi_{B,mn}$. Substituting Equation (50) into (52), we obtain the following up to linear order in Π_B (see Appendix A):

$$U_{\mathrm{MF}}\boldsymbol{E}(\boldsymbol{R}_A,t)U_{\mathrm{MF}}^{-1} \approx \boldsymbol{E}(\boldsymbol{R}_A,t) + \boldsymbol{E}_{\mathrm{dip},B}^{\mathrm{ind}}(\boldsymbol{R}_A,t), \tag{54}$$

where $\boldsymbol{E}_{\mathrm{dip},B}^{\mathrm{ind}}(\boldsymbol{R}_A,t)$ is given by Equation (39) with the substitution $d_{B,j}(t_r) \longrightarrow \Pi_{B,jk}E_k(\boldsymbol{R}_B,t_r)$. This result is mathematically similar to Equation (38) but with a remarkable physical difference. Here, the field is dressed not by a naked molecule but by a dressed one. This means that the source of the electric dipole field is not the molecular dipole operator but rather a vacuum-induced dipole—as indicated by the superscript "ind" in Equation (54). From Equations (53) and (54), we arrive at

$$H_{\mathrm{MF}}^{\mathrm{eff,s}}(t) = -\frac{1}{2}\Pi_{A,jk}\left(E_k(\boldsymbol{R}_A,t)E_{\mathrm{dip},B,j}^{\mathrm{ind}}(\boldsymbol{R}_A,t) + E_{\mathrm{dip},B,k}^{\mathrm{ind}}(\boldsymbol{R}_A,t)E_j(\boldsymbol{R}_A,t)\right), \tag{55}$$

where we have kept only the terms capturing the dipole–dipole correlation and neglected the higher-order term $\Pi_{A,jk} E^{\text{ind}}_{\text{dip},B,k}(\mathbf{R}_A,t) E^{\text{ind}}_{\text{dip},B,j}(\mathbf{R}_A,t)$. This expression is manifestly symmetric upon the exchange $A \longleftrightarrow B$, as can be verified by substituting Equation (39) into (55).

As an application of the effective Hamiltonian (55), we may consider that both molecules are at their ground states throughout the entire process. In this case, we may take the expectation value of $H^{\text{eff,s}}_{\text{MF}}(t)$ in the ground states of the molecules, which is equivalent to substituting the operators Π_{jl} with the static polarizability tensor of the corresponding molecule in Equation (55). In the particular case of isotropic molecules, this reproduces the asymptotic long-distance limit of the effective Hamiltonian originally employed by P.W. Milonni, which readily yields the known Casimir–Polder result in first order [18]. For comparison, when taking the average of $H^{\text{eff,s}(2)}_{\text{M}}$ (see Equation (51)) over the molecular ground state, the resulting effective Hamiltonian [47] yields the Casimir–Polder energy only to the second order of perturbation theory.

In the opposite short-distance limit, molecular dispersion is essential, and we cannot make the approximation given by Equation (48). In this case, it is more convenient to work with a different effective Hamiltonian, which we shall present in the next subsection.

6.2. The Dressed Field Dresses the Molecules: Non-Retarded Regime

We now turn to the opposite regime, in which the intermolecular distance is so small that we may neglect the electromagnetic retardation in comparison to the molecular response time. Unlike the other examples in this paper, this case does not require quantization of the electromagnetic field. On the other hand, the molecular dispersion is crucial in this regime. The dipole–dipole correlation is usually obtained from a second-order perturbation theory in the dipole–dipole Hamiltonian [48]

$$H_{dd} = \frac{d_{A,j} d_{B,k} \left(\delta_{jk} - 3 \hat{r}_j \hat{r}_k \right)}{4\pi\varepsilon_0 r^3}, \tag{56}$$

where $\mathbf{r} = r\hat{r}$ is the position of molecule B with respect to molecule A. Notice that this Hamiltonian is a particular case of Equation (40) without the vacuum electric field operator and with the dipole electric field taken in the electrostatic approximation. In the non-retarded regime, the field and the molecular operators switch roles with respect to what happens in the retarded asymptotic regime. While, in the latter, we could begin with Hamiltonian (48), which approximates the molecular response with its dressed static response, Equation (56) is precisely the opposite: now it is the field that is dressed by its static response. Indeed, the electrostatic dipole field leading to (56) corresponds to the zero-frequency limit of the Green function of the wave equation, which plays the role of the field susceptibility, as discussed in Appendix B.

Mirroring the procedure of the previous subsection, we let the already-dressed field dress the molecules, thus leading to a new effective Hamiltonian. In the long-distance regime, we employed a unitary transformation extending the formalism of Section 4. Here, on the other hand, we want to extend the formalism developed in Section 2. As a first step, we write Hamiltonian (56) in the interaction picture and then take as the unitary transformation the operator

$$U_{\text{FM}} = \exp\left(\frac{i}{\hbar} \int_{-\infty}^{t} dt' H_{dd}(t') \right), \tag{57}$$

which should be compared to Equation (5). Following steps analogous to the ones leading to Equation (10) yields

$$H^{\text{eff}}_{\text{FM}}(t) = -\frac{i}{2\hbar} \int_{-\infty}^{t} dt' [H_{dd}(t), H_{dd}(t')]. \tag{58}$$

Since operators involving different molecules commute, we have

$$[d_{A,j}(t)d_{B,k}(t), d_{A,m}(t')d_{B,n}(t')] = \frac{[d_{A,j}(t), d_{A,m}(t')]\{d_{B,k}(t), d_{B,n}(t')\}}{2} + \frac{\{d_{A,j}(t), d_{A,m}(t')\}[d_{B,k}(t), d_{B,n}(t')]}{2}. \quad (59)$$

By substituting Equation (56) and the previous identity into Equation (58), we obtain the effective Hamiltonian

$$H_{\text{FM}}^{\text{eff}} = -\frac{i\left(\delta_{jk} - 3\hat{r}_j\hat{r}_k\right)(\delta_{mn} - 3\hat{r}_m\hat{r}_n)}{64\hbar\pi^2\varepsilon_0^2 r^6} \int_{-\infty}^{t} dt' \Big([d_{A,j}(t), d_{A,m}(t')]\{d_{B,k}(t), d_{B,n}(t')\} + \{d_{A,j}(t), d_{A,m}(t')\}[d_{B,k}(t), d_{B,n}(t')]\Big). \quad (60)$$

The new effective Hamiltonian (60) has two terms that capture the physics involved in the dipole–dipole correlation. The product $[d_{A,j}(t), d_{A,m}(t')]\{d_{B,k}(t), d_{B,n}(t')\}$ measures how the dipole fluctuations of molecule B induces a dipole in molecule A, while the other term is its reciprocal. This decomposition is possible because, differently from the standard approach based on second-order perturbation theory with the time-independent dipole–dipole Hamiltonian (56) where the two molecules are considered as an isolated system, here, we take the complementary approach of considering each molecule separately as an open quantum system. This perspective offers two main novelties: (i) $H_{\text{FM}}^{\text{eff}}$ brings to light the dynamical character of the dispersion interaction by making an explicit connection with dipole fluctuations, and (ii) $H_{\text{FM}}^{\text{eff}}$ enables us to assess the contribution from the fluctuations of each molecule separately. In the next section, we analyze an example that illustrates these advantages.

7. Application to the London Interaction Energy

In this section, we consider the dispersion interaction between two ground-state non-polar molecules A and B in vacuum, which interact due to correlations between their fluctuating electric dipoles. As discussed in the previous section, the physical mechanism limiting the dipole–dipole correlation strongly depends on comparing the distance separating the molecules and their internal transition wavelengths. For ground-state molecules, the resulting intermolecular interaction energy exhibits a different power-law dependence with distance in each of the two regimes discussed in Section 6.

As originally demonstrated by London [49], the non-retarded interaction energy can be obtained without quantizing the electromagnetic field and scales with $1/r^6$. The asymptotic long-distance limit was first obtained in the seminal paper by Casimir and Polder [50], where they showed that retardation imposes the necessity of quantizing the electromagnetic field and demonstrated that the interaction energy scales asymptotically with $1/r^7$.

Both regimes have still been at the center of intense investigation in recent years. Casimir–Polder forces have been studied considering excited [51–56] and chiral [57–60] particles. The influence of neighboring surfaces with ever-increasing complexity [61–73] and with dynamical [74–76] and thermal effects [77–83] has also been considered.

The force in the non-retarded regime—sometimes referred to as London or van der Waals force—plays a pivotal role in chemistry [84] and condensed matter physics, where short-range interactions prevail. In van der Waals heterostructures, two-dimensional materials are stacked and held together by London dispersion forces, generating materials with fascinating physical properties that are useful for designing new electronic devices [85–87]. Density functional theory provides a powerful framework capable of obtaining increasingly precise descriptions of molecular polarizabilities and London dispersion forces [88–90]. Modifications of the force due to an intervening electrolyte medium [91–93], with the atomic motion in connection with quantum friction [94–105] or with non-local interferometric phases [106–108], the atomic internal state [109], and coming from boundary conditions imposed by nearby structures [110–114], have disclosed important features of the London–van der Waals interactions.

In the previous section, we discussed how the Casimir–Polder result for the asymptotic long-distance limit can be derived from the effective Hamiltonian $H_{\text{MF}}^{\text{eff}}$ (55). In this section, we obtain the London result for the short-distance non-retarded limit from the new effective Hamiltonian $H_{\text{FM}}^{\text{eff}}$ (60), which provides new physical insights into the dipole-dipole correlation present in the non-retarded regime. By taking the average of Hamiltonian (60) in the ground state of the molecules and employing the result (13) for the molecular polarizability tensors $\alpha_{jl}^A, \alpha_{jl}^B$, we obtain

$$E_{\text{London}} = -\frac{\hbar\left(\delta_{jk} - 3\hat{r}_j\hat{r}_k\right)\left(\delta_{mn} - 3\hat{r}_m\hat{r}_n\right)}{64\pi^2\varepsilon_0^2 r^6} \int_{-\infty}^{\infty} d\tau \left(\alpha_{jm}^A(\tau)\eta_{kn}^B(\tau) + \eta_{jm}^A(\tau)\alpha_{kn}^B(\tau) \right), \quad (61)$$

where we defined the symmetrical dipole correlation function

$$\eta_{jm}^\zeta(\tau) := \frac{1}{\hbar}\langle g|\left\{d_j^\zeta(t'+\tau), d_m^\zeta(t')\right\}|g\rangle, \quad (62)$$

with $\zeta = A, B$. To work in the Fourier space, we can apply Parseval's theorem, so Equation (61) becomes

$$E_{\text{London}} = -\frac{\hbar\left(\delta_{jk} - 3\hat{r}_j\hat{r}_k\right)\left(\delta_{mn} - 3\hat{r}_m\hat{r}_n\right)}{128\pi^3\varepsilon_0^2 r^6} \int_{-\infty}^{\infty} d\omega \left(\alpha_{jm}^A(\omega)\eta_{kn}^B(\omega) + \eta_{jm}^A(\omega)\alpha_{kn}^B(\omega) \right), \quad (63)$$

where we used that $\eta^{A,B}(\omega)$ are real functions, as they are Fourier transforms of real even functions. This result is the analog for two molecules of the decomposition obtained for an atom coupled to the vacuum electric field [115–117]. In the latter, the field susceptibility captures the field radiation reaction. More recently, an analogous decomposition was also obtained for atoms interacting with a scalar quantum field [118]. A decomposition similar to (63) was employed to derive a nonlocal phase for a moving atom interacting with a planar surface [119] and a Sagnac-like atomic phase induced by a rotating nanosphere [120].

In the isotropic case, the polarizability tensors simplify to $\alpha_{rs}^{A(B)}(\tau) = \delta_{rs}\alpha^{A(B)}(\tau)$, and the symmetric correlation functions simplify to $\eta_{rs}^{A(B)}(\tau) = \delta_{rs}\eta^{A(B)}(\tau)$. Then, Equation (63) leads to

$$E_{\text{London}} = -\frac{3\hbar}{64\pi^3\varepsilon_0^2 r^6} \int_{-\infty}^{\infty} d\omega \left(\alpha^A(\omega)\eta^B(\omega) + \eta^A(\omega)\alpha^B(\omega) \right). \quad (64)$$

In Appendix B, we employ the analytical properties of the correlation functions to demonstrate that our results are equivalent to the standard way of expressing the London interaction energy for any molecular model of the polarizabilities. Here, we show the convenience of Equation (64) by considering the simple case of two-level atoms, for which ($\zeta = A, B$) [18]

$$\alpha^\zeta(\omega) = \frac{\alpha_0^\zeta \omega_{0\zeta}^2}{\omega_{0\zeta}^2 - \omega^2}, \quad (65)$$

$$\eta^\zeta(\omega) = \pi\alpha_0^\zeta \omega_{0\zeta} \left[\delta(\omega - \omega_{0\zeta}) + \delta(\omega + \omega_{0\zeta})\right]. \quad (66)$$

Let us analyze each contribution to the London interaction energy in Equation (64) separately. We define

$$E_{\text{London}}^{A\to B} = -\frac{3\hbar}{64\pi^3\varepsilon_0^2 r^6} \int_{-\infty}^{\infty} d\omega\, \eta^A(\omega)\alpha^B(\omega), \quad (67)$$

as the contribution arising from the dipole induced at atom B by the dipole fluctuations of atom A. From Equations (65) and (66), we obtain

$$E_{\text{London}}^{A\to B} = -\frac{3\hbar \alpha_0^A \alpha_0^B \omega_{0A}\omega_{0B}}{32\pi^2\varepsilon_0^2 r^6}\frac{\omega_{0B}}{\omega_{0B}^2 - \omega_{0A}^2}. \tag{68}$$

The interaction energy is $E_{\text{London}} = E_{\text{London}}^{B\to A} + E_{\text{London}}^{A\to B}$, with $E_{\text{London}}^{B\to A}$ being obtained by interchanging the roles of A and B in Equation (68).

Let us consider $\omega_{0A} > \omega_{0B}$. In that case, from Equation (68), we see that $E_{\text{London}}^{A\to B} > 0$, indicating that the dipole induced at a slower atom by a faster one generates a repulsive contribution to the dispersion force. This is due to the fact that the polarizability given by Equation (65) becomes negative for frequencies higher than the atomic transition frequency. Indeed, the induced dipole at the slower atom B cannot follow the fast oscillation of the fluctuating dipole of atom A. The induced dipole at B lags behind the field of atom A and points opposite to its direction at a given time. As a consequence, the induced dipole at B repels the fluctuating dipole at A. However, the opposite holds for the complementary term $E_{\text{London}}^{B\to A}$: the dipole induced in the faster atom A can follow the dipole fluctuations of atom B in phase, leading to an attractive contribution. Attraction overcomes repulsion by a factor ω_{0A}/ω_{0B}, since the slower atom couples less effectively to the field than the faster one. If $\omega_{0A} = \omega_{0B}$, each contribution diverges due to a resonant response. This divergence would be avoided if dissipation were taken into account. Nevertheless, it is remarkable that the divergence cancels once we sum $E_{\text{London}}^{B\to A}$ and $E_{\text{London}}^{A\to B}$, leaving us with the well-behaved total interaction energy

$$E_{\text{London}} = -\frac{3\hbar \alpha_0^A \alpha_0^B}{32\pi^2\varepsilon_0^2 r^6}\frac{\omega_{0A}\omega_{0B}}{\omega_{0A}+\omega_{0B}}, \tag{69}$$

which agrees with the result [2] calculated from second-order perturbation theory based on the dipole–dipole Hamiltonian (56).

Notice that varying ω_{0B} while keeping the other parameters fixed shows that the attraction is maximal when $\omega_{0B} \to \infty$. The previous decomposition clearly illustrates the physical mechanism involved. From Equation (68), we see that in this limit, the repulsive contribution $E_{\text{London}}^{B\to A}$ vanishes, indicating that atom A is effectively transparent, decoupling from the rapid oscillating field produced by B. The attractive term in Equation (68), on the other hand, takes its maximal absolute value in this limit, since the response of atom B is so fast that it perfectly mirrors the fluctuations of the other atom. In this sense, we may conclude that atom B in the limit $\omega_{0B} \to \infty$ is the atomic analog of a perfect conductor.

As was true with the other effective Hamiltonians discussed in this paper, we see that the convenience of employing $H_{\text{FM}}^{\text{eff}}$ is twofold: (i) it lowers the perturbation order required to obtain the London dispersion energy from second to first order, and (ii) it offers physical insights into the mechanisms involved in the phenomenon. The results in this section can be readily extended for multilevel atoms. To this end, it suffices to substitute Equations (65) and (66) with a summation over all internal transition frequencies.

8. Final Remarks and Conclusions

All phenomena in molecular quantum electrodynamics can be obtained from the multipolar Hamiltonian. In this paper, we restricted our attention to phenomena that can be treated perturbatively (which includes the vast majority of cases in this field). In most situations, the dominant effect is obtained from a high-order perturbation theory, requiring intermediate states to connect the initial and the final states. A clear example is the interatomic interaction. While in classical electrodynamics, we may always take the field at each charged particle as the superposition of the field generated by all other particles, in standard quantum electrodynamics, each particle couples only to the free electromagnetic field. Consequently, we must go up to the fourth order to obtain the dominant contribution when considering molecules without permanent electric dipole moments. An alternative is to build effective Hamiltonians. They are customized for each specific application and lose meaning and validity after some point in the perturbative expansion. Not only do they greatly simplify the technical difficulties involved in calculations using the multipolar Hamiltonian, but, equally importantly, the effective Hamiltonians cast the phenomena in a new light, offering insightful physical interpretations.

Several effective Hamiltonians have successfully been employed by many authors in the last decades. In this paper, we have developed a systematic approach to constructing effective Hamiltonians, which allowed us to derive a number of new ones, choosing as a unitary transformation the Hermitian conjugate of the evolution operator for part of the system. This transfers part of the time evolution from the vector state to the operators, dressing them and providing a Hamiltonian that requires a lower order in perturbation theory to account for the process of interest. This method can always be used when the first-order perturbation theory vanishes. Our approach yields time-dependent Hamiltonians that enable us to follow the energy exchange between matter and the field, with each subsystem constituting an open quantum system. We emphasize here that our system of interest is the entire molecule–field system, and it is not interesting to trace over any of the subsystems, as is common in an open quantum system approach [121–123].

As a first application, we have derived the Hamiltonian H_M^{eff}, where the field dresses the molecule and the dipole operators are replaced by their commutator at different times. If we project the commutator in an internal molecular state, it yields the dynamical polarizability of the molecule for the corresponding state. Its nondiagonal elements, on the other hand, allow the dressing to leave the molecule in a final state that is different from the initial one. We have demonstrated that this new time-dependent Hamiltonian provides a simpler treatment of the two-photon spontaneous emission, as the dominant contribution is obtained in first-order perturbation theory. In addition, our formalism introduces the concept of an induced dipole transition, which generalizes the notion of an induced dipole for a given internal state.

Then, we discussed applications involving two molecules A and B. We constructed the new effective Hamiltonian H_F^{eff} through a unitary transformation that transfers all of the effects related to molecule B to the electric field it generates. In this way, molecule A feels an effectively dressed electric field given by the superposition of the free vacuum electric field, and the one generated by the dipole operator of molecule B. H_F^{eff} allows for the description of the resonance energy transfer rate in first-order perturbation theory.

Lastly, we derived two additional Hamiltonians that merge aspects of the previous two, where each one is appropriate for a different intermolecular distance regime. In the asymptotic long-distance regime, molecular dispersion is negligible, enabling us to derive an effective Hamiltonian H_{MF}^{eff} which is formally similar to H_M^{eff}. In this new case, however, the field acting on molecule A is given by the superposition of the free electric field and the one produced by the vacuum-induced dipole generated on molecule B. When we average H_{MF}^{eff} over the molecular ground state, we re-obtain the asymptotic limit of the Hamiltonian employed by P. Milonni [18].

Finally, for the short-distance non-retarded limit, we derived our fourth and last effective Hamiltonian H_{FM}^{eff} based on the fact that, in this limit, we do not need to quantize the electric field. This effective Hamiltonian enables us to clearly identify the different physical mechanisms involved in the correlations responsible for the interaction, separating one term where the dipole fluctuations of molecular A induce a dipole on molecule B and another term where the roles are exchanged.

As a final application, we employed H_{FM}^{eff} to obtain the London dispersion interaction energy in first-order perturbation theory. We showed that, for two-level atoms, the dipole fluctuations of the atom with the higher transition frequency give rise to a repulsive term, since its fast fluctuations cannot be followed by the slower atom. Nonetheless, the force between two isotropic atoms is always attractive, since the fluctuations of the slower atom are strongly correlated and easily followed by the faster atom, overcoming the repulsive contribution. The possibility of quantitatively and separately analyzing the contributions arising from each mechanism correlating fluctuating systems is an advantage of $H_{\text{FM}}^{\text{eff}}$.

The Hamiltonians presented in this paper can be employed in a great variety of situations. For instance, one may treat the effects of boundaries in the two-photon spontaneous emission or resonance energy transfer by simply introducing the appropriate field modes. As in the examples discussed in this paper, these effective Hamiltonians allow for a direct first-order calculation within perturbation theory. More notably, the methodology introduced here can be applied to generate other effective Hamiltonians that may optimize calculations and provide physical intuition.

Author Contributions: Conceptualization, C.F., P.A.M.N. and R.d.M.e.S.; Methodology, A.S.S., P.H.P. and P.P.A.; Validation, P.P.A., C.F. and P.A.M.N.; Formal analysis, A.S.S., P.H.P., C.F. and R.d.M.e.S.; Investigation, A.S.S. and P.H.P.; Writing—original draft, P.P.A., C.F., P.A.M.N. and R.d.M.e.S.; Writing—review & editing, P.P.A., C.F., P.A.M.N. and R.d.M.e.S.; Supervision, R.d.M.e.S.; Project administration, R.d.M.e.S. All authors have read and agreed to the published version of the manuscript.

Funding: This research was funded by the Brazilian agencies CAPES and CNPq. P.A.M.N. was also supported by Instituto Nacional de Ciência e Tecnologia de Fluidos Complexos (INCT-FCx) and the Research Foundations of the States of Rio de Janeiro (FAPERJ) (210.077/2023 and 201.126/2021) and São Paulo (FAPESP).

Data Availability Statement: The original contributions presented in the study are included in the article, further inquiries can be directed to the corresponding author/s.

Acknowledgments: The authors thank François Impens for enlightening discussions.

Conflicts of Interest: The authors declare no conflicts of interest.

Appendix A. Susceptibilities

Appendix A.1. Molecular Polarizability

Let us find the time evolution of the expected value of the dipole operator in the atomic state $|\phi(t)\rangle$, as determined by the electric dipole interaction Hamiltonian $H(t)$ given by Equation (3). We operate in the interaction picture, so $|\phi(t)\rangle \approx \left(\mathbb{I} - \frac{i}{\hbar}\int_{-\infty}^{t} dt' H(t')\right)|\phi(-\infty)\rangle$. This implies that, up to the second order in the dipole operator,

$$\langle \boldsymbol{d}(t)\rangle_t = \left\langle \boldsymbol{d}(t) - \frac{i}{\hbar}\int_{-\infty}^{t} dt' [\boldsymbol{d}(t), H(t')]\right\rangle_{-\infty}, \tag{A1}$$

where $\langle \mathcal{O}\rangle_t = \langle \phi(t)|\mathcal{O}|\phi(t)\rangle$. All of the electric field contributions to the molecular electric dipole are contained in the second term on the right-hand side of Equation (A1). Therefore, we refer to this term as the induced dipole $\langle \boldsymbol{d}^{\text{ind}}(t)\rangle_t$, whose components can be written as

$$\langle d_j^{\text{ind}}(t)\rangle_t = \frac{i}{\hbar}\int_{-\infty}^{t} dt' \langle [d_j(t), d_l(t')]\rangle_{-\infty} E_l(t') =: \int_{-\infty}^{\infty} dt' \alpha_{jl}(t-t') E_l(t'), \tag{A2}$$

where α_{jl} are the elements of the dynamical molecular electric polarizability tensor for the molecular state $|\phi(-\infty)\rangle$. We assume here that $|\phi(-\infty)\rangle$ is an eigenstate of the free molecular Hamiltonian H_0, a situation where time translation symmetry ensures that the average value of $[d(t), d(t)']$ is a function of t, t' only through the difference $t - t'$, as in the last equality.

For many applications, we are interested in the situation where $|\phi(-\infty)\rangle$ is the ground state, in which case the dynamical polarizability reduces to Equation (13). This expression is still valid regardless of whether there is dissipation or not [124]. If there is no dissipation, the polarizability can be directly expressed in terms of the eigenstates $|r\rangle$ of H_0. Inserting a closure relation $\mathbb{I} = \sum_r |r\rangle\langle r|$ into Equation (13), we obtain the familiar expression

$$\alpha_{jl}(t-t') = \frac{2}{\hbar}\theta(t-t')\sum_r d_j^{gr}d_l^{rg}\sin[\omega_{rg}(t-t')]. \tag{A3}$$

In Fourier space, the above expression immediately translates to Equation (18).

Appendix A.2. The Field of a Dipole Is the Field Susceptibility

From the expression of the quantized electric field, it is straightforward to show that (see Section 2.8 of Ref. [18])

$$\frac{i}{\hbar}\theta(t-t')[E_j(\mathbf{r},t), E_l(\mathbf{r}',t')] = \frac{1}{\varepsilon_0}\mathcal{D}_{jl}\mathcal{G}_{\mathrm{r}}(|\mathbf{r}-\mathbf{r}'|, t-t'), \tag{A4}$$

where $\mathcal{G}_\mathrm{r}$ is the retarded Green function of the wave equation, and $\mathcal{D}_{jl}$ is the differential operator

$$\mathcal{D}_{jl} = \partial_j \partial'_l - \frac{\delta_{jl}}{c^2}\partial_t \partial'_t. \tag{A5}$$

On the other hand, from Maxwell's equation, the electric field generated by a charge density ρ and electric current density $\mathbf{J}$ is given by

$$\left(\nabla^2 - \frac{1}{c^2}\frac{\partial^2}{\partial t^2}\right)\mathbf{E} = \frac{\nabla\rho}{\varepsilon_0} + \frac{1}{\varepsilon_0 c^2}\partial_t \mathbf{J}. \tag{A6}$$

As a source, let us consider a point dipole $\mathbf{d}$ existing only at time t' placed at $\mathbf{r}'$, such that [125,126]

$$\rho(\mathbf{r},t) = -\mathbf{d}\cdot\nabla\delta(\mathbf{r}-\mathbf{r}')\delta(t-t'), \tag{A7}$$
$$\mathbf{J}(\mathbf{r},t) = \mathbf{d}\delta(\mathbf{r}-\mathbf{r}')\partial_t\delta(t-t'). \tag{A8}$$

From Equation (A6) and after integrating by parts, the electric field generated by this point dipole is

$$E_{\mathrm{dip},j}(\mathbf{r},t) = -\frac{d_l}{\varepsilon_0}\left(\partial_j\partial_l - \frac{\delta_{jl}}{c^2}\partial_t^2\right)\mathcal{G}_\mathrm{r}(|\mathbf{r}-\mathbf{r}'|, t-t') = \frac{d_l}{\varepsilon_0}\mathcal{D}_{jl}\mathcal{G}_\mathrm{r}(|\mathbf{r}-\mathbf{r}'|, t-t'), \tag{A9}$$

where we used $\partial_l = -\partial'_l$ and $\partial_t = -\partial'_t$ because $\mathcal{G}_\mathrm{r}$ depends only on the differences $\mathbf{r}-\mathbf{r}'$ and $t-t'$. Comparing this result with Equation (A4), one can see that

$$\mathbf{E}_{\mathrm{dip}}(\mathbf{r},t) = \frac{i}{\hbar}\int_{-\infty}^{t}dt' d_l(t')[E(\mathbf{r},t), E_l(\mathbf{r}',t')] \tag{A10}$$

is the electric field at $(\mathbf{r},t)$ generated by the dipole $\mathbf{d}(t')$ at $\mathbf{r}'$, whose explicit expression is given by Equation (39).

Some comments are in order. (i) While Equation (A2) is approximate, Equation (A10) is exact, as a consequence of the linearity of Maxwell's equations. (ii) $\mathbf{E}_{\mathrm{dip}}$ is an operator containing both a molecule operator $\mathbf{d}(t)$ and an identity operator in Fock field space, as the field commutator is a c-number. This last property is also a consequence of the linearity of Maxwell's equations. (iii) This same procedure can be adapted to other sources. For instance, substituting $d_n(t')[E(\mathbf{r},t), E_n(\mathbf{r}',t')]$ into Equation (A10) with (a) $m_n(t')[E(\mathbf{r},t), B_n(\mathbf{r}',t')]$ generates the electric field at $(\mathbf{r},t)$ produced by a magnetic dipole $\mathbf{m}(t')$ at $\mathbf{r}'$, (b) $d_n(t')[B(\mathbf{r},t), E_n(\mathbf{r}',t')]$ generates the magnetic field at $(\mathbf{r},t)$ caused by a electric dipole $\mathbf{d}(t')$ at $\mathbf{r}=\mathbf{r}'$, (c) $Q_{ln}(t')[E(\mathbf{r},t), \partial'_l E_n(\mathbf{r}',t')]$ generates the electric field at $(\mathbf{r},t)$ induced by the quadrupole tensor $Q_{ln}(t')$ at position $\mathbf{r}'$, and so on. (iv) We could have reached these same conclusions from an approach analogous to our approach for molecular susceptibility. Indeed, let us assume that the interaction Hamiltonian $H_{\text{Int}}(t')$ in the interaction picture is linear in the electric and magnetic fields. Now, we are not restricted to point sources. For example, it can be the field generated by prescribed classical charge and current fluctuations in a macroscopic body. Analogously to Equation (A1), we have

$$E(t) \approx E_0(t) - \frac{i}{\hbar} \int_{-\infty}^{t} dt' [E_0(t), H_{\text{Int}}(t')]. \quad \text{(A11)}$$

The second term in Equation (A11) can be recognized as the field produced by the source present in H_{Int}. For the electric dipole case, $H_{\text{Int}}(t) = -d(t) \cdot E(r,t)$, and we immediately recover Equation (A10). However, notice that while Equation (A10) is exact, Equation (A11) is an approximation. Indeed, $[H_{\text{Int}}(t_1), [H_{\text{Int}}(t_2), E_0(t)]] \neq 0$, since the dipole operators in different instants of time do not commute. This reflects that, although Maxwell's equations are linear, the dipole induced in matter depends nonlinearly on the field, which, in turn, produces a nonlinearity in the time evolution of the electric field. Still, if H_{Int} depends only on material classical and prescribed variables, as in the aforementioned example of a macroscopic body, then Equation (A11) becomes an exact equation.

Appendix B. London Interaction in the Imaginary Frequency Domain

Here, we demonstrate the equivalence between Equation (64) and the expression that is usually employed in the literature [2]

$$E_{\text{London}} = -\frac{3\hbar}{32\pi^3 \varepsilon_0^2 r^6} \int_{-\infty}^{\infty} d\omega\, \alpha^A(i\omega) \alpha^B(i\omega). \quad \text{(A12)}$$

From the fluctuation–dissipation theorem at zero temperature, we have

$$\eta^\zeta(\omega) = 2\,\text{sgn}(\omega)\,\text{Im}[\alpha^\zeta(\omega)], \quad \text{(A13)}$$

where the sign function is defined as $\text{sgn}(\omega) = \omega/|\omega|$. Substituting Equation (A13) into (64),

$$E_{\text{London}} = -\frac{3\hbar}{32\pi^3 \varepsilon_0^2 r^6} \int_{-\infty}^{\infty} d\omega \Big(\text{Re}[\alpha^A(\omega)]\,\text{sgn}(\omega)\,\text{Im}[\alpha^B(\omega)] + \text{sgn}(\omega)\,\text{Im}[\alpha^A(\omega)]\,\text{Re}[\alpha^B(\omega)] \Big). \quad \text{(A14)}$$

Recalling that $\text{Re}[\alpha(\omega)]$ ($\text{Im}[\alpha(\omega)]$) is an even (odd) function, since $\alpha(\tau)$ is a real number, we obtain

$$E_{\text{London}} = -\frac{3\hbar}{16\pi^3 \varepsilon_0^2 r^6} \,\text{Im} \int_0^{\infty} d\omega\, \alpha^A(\omega) \alpha^B(\omega). \quad \text{(A15)}$$

Causality implies that the polarizabilities are analytical in the superior half-plane [127], allowing us to perform a Wick rotation, leading to Equation (A12).

References

1. Dirac, P.A.M. The Quantum Theory of the Emission and Absorption of Radiation. *Proc. Royal Soc. Lond. A* **1927**, *114*, 243.
2. Craig, D.P.; Thirunamachandran, T. *Molecular Quantum Electrodynamics*; Dover Publications: New York, NY, USA, 1984.
3. Compagno, G.; Passante, R.; Persico, F. *Atom-Field Interactions and Dressed Atoms*; Cambridge University Press: Cambridge, UK, 1995.
4. Buhmann, S.Y. *Dispersion Forces I: Macroscopic Quantum Electrodynamics and Ground-State Casimir, Casimir–Polder and van der Waals Forces*; Springer: Cham, Switzerland, 2012.
5. Buhmann, S.Y. *Dispersion Forces II: Many-Body Effects, Excited Atoms, Finite Temperature and Quantum Friction*; Springer: Cham, Switzerland, 2012.
6. Milonni, P.W. *An Introduction to Quantum Optics and Quantum Fluctuations*; Oxford University Press: Oxford, UK, 2019.
7. Passante, R.; Power, E.A.; Thirunamachandran, T. Radiation-molecule coupling using dynamic polarizabilities: Application to many-body forces. *Phys. Lett. A* **1998**, *249*, 77. [CrossRef]
8. Passante, R.; Rizzuto, L. Effective Hamiltonians in nonrelativistic quantum electrodynamics. *Symmetry* **2021**, *13*, 2375. [CrossRef]
9. Alonso, L.; Matos, G.C.; Impens, F.; Neto, P.A.M.; de Melo e Souza, R. Multipole Approach to the Dynamical Casimir Effect with Finite-Size Scatterers. *Entropy* **2024**, *26*, 251. [CrossRef]

10. de Melo e Souza, R.; Impens, F.; Neto, P.A.M. Microscopic dynamical Casimir effect. *Phys. Rev. A* **2018**, *97*, 032514. [CrossRef]
11. Lo, L.; Law, C.K. Quantum radiation from a shaken two-level atom in vacuum. *Phys. Rev. A* **2018**, *98*, 063807. [CrossRef]
12. Belén Farías, M.; Fosco, C.D.; Lombardo, F.C.; Mazzitelli, F.D. Motion induced radiation and quantum friction for a moving atom. *Phys. Rev. D* **2019**, *100*, 036013. [CrossRef]
13. Fosco, C.D.; Lombardo, F.C.; Mazzitelli, F.D. Motion-induced radiation due to an atom in the presence of a graphene plane. *Universe* **2021**, *7*, 158. [CrossRef]
14. Dalvit, D.A.R.; Kort-Kamp, W.J.M. Shaping dynamical Casimir photons. *Universe* **2021**, *7*, 189. [CrossRef]
15. Cohen-Tannoudji, C.; Dupont-Roc, J.; Grynberg, G. *Photons and Atoms: Introduction to Quantum Electrodynamics*; Wiley: New York, NY, USA, 1997.
16. Cresser, J.D. Electric field commutation relation in the presence of a dipole atom. *Phys. Rev. A* **1984**, *29*, 1984. [CrossRef]
17. Cresser, J.D. Unequal Time EM Field Commutators in Quantum Optics. *Phys. Scr.* **1988**, *T21*, 52. [CrossRef]
18. Milonni, P.W. *The Quantum Vacuum: An Introduction to Quantum Electrodynamics*; Academic Press: New York, NY, USA, 1994.
19. Brune, M.; Raimond, J.M.; Goy, P.; Davidovich, L.; Haroche, S. Realization of a two-photon maser oscillator. *Phys. Rev. Lett.* **1987**, *59*, 1899. [CrossRef]
20. Davidovich, L.; Raimond, J.M.; Brune, M.; Haroche, S. Quantum theory of a two-photon micromaser. *Phys. Rev. A* **1987**, *36*, 3771. [CrossRef]
21. Neto, P.A.M.; Davidovich, L.; Raimond, J.-M. Theory of the nondegenerate two-photon micromaser. *Phys. Rev. A* **1991**, *43*, 5073. [CrossRef]
22. Hayat, A.; Ginzburg, P.; Orenstein, M. Observation of two-photon emission from semiconductors. *Nat. Photonics* **2008**, *2*, 238. [CrossRef]
23. Wang, H.; Hu, H.; Chung, T.H.; Qin, J.; Yang, X.; Li, J.P.; Liu, R.Z.; Zhong, H.S.; He, Y.M.; Ding, X.; et al. On-demand semiconductor source of entangled photons which simultaneously has high fidelity, efficiency, and indistinguishability. *Phys. Rev. Lett.* **2019**, *122*, 113602. [CrossRef]
24. Zhang, J.; Ma, J.; Parry, M.; Cai, M.; Camacho-Morales, R.; Xu, L.; Neshev, D.N.; Sukhorukov, A.A. Spatially entangled photon pairs from lithium niobate nonlocal metasurfaces. *Sci. Adv.* **2022**, *8*, eabq4240. [CrossRef]
25. Poddubny, A.N.; Ginzburg, P.; Belov, P.A.; Zayats, A.V.; Kivshar, Y.S. Tailoring and enhancing spontaneous two-photon emission using resonant plasmonic nanostructures. *Phys. Rev. A* **2012**, *86*, 033826. [CrossRef]
26. Muniz, Y.; Manjavacas, A.; Farina, C.; Dalvit, D.A.R.; Kort-Kamp, W.J.M. Two-photon spontaneous emission in atomically thin plasmonic nanostructures. *Phys. Rev. Lett.* **2020**, *125*, 033601. [CrossRef]
27. Hu, F.; Li, L.; Liu, Y.; Meng, Y.; Gonga, M.; Yang, Y. Two-plasmon spontaneous emission from a nonlocal epsilon-near-zero material. *Commun. Phys.* **2021**, *4*, 84. [CrossRef]
28. Muniz, Y.; Abrantes, P.P.; Martín-Moreno, L.; Pinheiro, F.A.; Farina, C.; Kort-Kamp, W.J.M. Entangled two-plasmon generation in carbon nanotubes and graphene-coated wires. *Phys. Rev. B* **2022**, *105*, 165412. [CrossRef]
29. Smeets, S.; Maes, B.; Rosolen, G. General framework for two-photon spontaneous emission near plasmonic nanostructures. *Phys. Rev. A* **2023**, *107*, 063516. [CrossRef]
30. Feinberg, G.; Sucher, J. General Theory of the van der Waals Interaction: A model-independent approach. *Phys. Rev. A* **1970**, *2*, 2395. [CrossRef]
31. Farina, C.; Santos, F.C.; Tort, A.C. On the force between an electrically polarizable atom and a magnetically polarizable one. *J. Phys. A* **2002**, *35*, 2477. [CrossRef]
32. Förster, T. Energiewanderung und fluoreszenz. *Naturwissenschaften* **1946**, *33*, 166. [CrossRef]
33. Martínez, P.L.H.; Govorov, A.; Demir, H.V. *Understanding and Modeling Förster-Type Resonance Energy Transfer (FRET)*; Springer: Singapore, 2017; Volume 1.
34. Milonni, P.W.; Rafsanjani, S.M.H. Distance dependence of two-atom dipole interactions with one atom in an excited state. *Phys. Rev. A* **2015**, *92*, 062711. [CrossRef]
35. Biehs, S.-A.; Menon, V.M.; Agarwal, G.S. Long-range dipole-dipole interaction and anomalous Förster energy transfer across a hyperbolic metamaterial. *Phys. Rev. B* **2016**, *93*, 245439. [CrossRef]
36. Weeraddana, D.; Premaratne, M.; Gunapala, S.D.; Andrews, D.L. Controlling resonance energy transfer in nanostructure emitters by positioning near a mirror. *J. Chem. Phys.* **2017**, *147*, 074117. [CrossRef]
37. Li, Y.; Nemilentsau, A.; Argyropoulos, C. Resonance energy transfer and quantum entanglement mediated by epsilon-near-zero and other plasmonic waveguide systems. *Nanoscale* **2019**, *11*, 14635. [CrossRef]
38. Abrantes, P.P.; Szilard, D.; Rosa, F.S.S.; Farina, C. Resonance energy transfer at percolation transition. *Mod. Phys. Lett. A* **2020**, *35*, 2040022. [CrossRef]
39. Abrantes, P.P.; Bastos, G.; Szilard, D.; Farina, C.; Rosa, F.S.S. Tuning resonance energy transfer with magneto-optical properties of graphene. *Phys. Rev. B* **2021**, *103*, 174421. [CrossRef]
40. Pini, F.; Francés-Soriano, L.; Andrigo, V.; Natile, M.M.; Hildebrandt, N. Optimizing upconversion nanoparticles for FRET biosensing. *ACS Nano* **2023**, *17*, 4971. [CrossRef]
41. Nayem, S.H.; Sikder, B.; Uddin, S.Z. Anisotropic energy transfer near multi-layer black phosphorus. *2D Mater.* **2023**, *10*, 045022. [CrossRef]

42. Song, Q.; Yan, X.; Cui, H.; Ma, M. Efficient cascade resonance energy transfer in dynamic nanoassembly for intensive and long-lasting multicolor chemiluminescence. *ACS Nano* **2020**, *14*, 3696. [CrossRef]
43. Rusanen, J.; Kareinen, L.; Levanov, L.; Mero, S.; Pakkanen, S.H.; Kantele, A.; Amanat, F.; Krammer, F.; Hedman, K.; Vapalahti, O.; et al. A 10-Minute "Mix and Read" Antibody Assay for SARS-CoV-2. *Viruses* **2021**, *13*, 143. [CrossRef]
44. Bednarz, A.; Sønderskov, S.M.; Dong, M.; Birkedal, V. Ion-mediated control of structural integrity and reconfigurability of DNA nanostructures. *Nanoscale* **2023**, *15*, 1317. [CrossRef]
45. Andrews, D.L.; Sherborne, B.S. Resonant excitation transfer: A quantum electrodynamical study. *J. Chem. Phys.* **1987**, *86*, 4011. [CrossRef]
46. Franz, J.C.; Buhmann, S.Y.; Salam, A. Macroscopic quantum electrodynamics theory of resonance energy transfer involving chiral molecules. *Phys. Rev. A* **2023**, *107*, 032809. [CrossRef]
47. Craig, D.P.; Power, E.A. The asymptotic Casimir–Polder potential from second-order perturbation theory and its generalization for anisotropic polarizabilities. *Int. J. Quantum Chem.* **1969**, *3*, 903. [CrossRef]
48. Cohen-Tannoudji, C.; Diu, B.; Laloë, F. *Quantum Mechanics*; Wiley-VCH: Weinheim, Germany, 2019; Volume II.
49. London, F. Zur Theorie und Systematik der Molekularkräfte. *Z. Phys.* **1930**, *63*, 245. [CrossRef]
50. Casimir, H.B.G.; Polder, D. The influence of retardation on the London-van der Waals forces. *Phys. Rev.* **1948**, *73*, 360. [CrossRef]
51. Power, E.A.; Thirunamachandran, T. Dispersion forces between molecules with one or both molecules excited. *Phys. Rev. A* **1995**, *51*, 3660. [CrossRef]
52. Power, E.A.; Thirunamachandran, T. Two- and three-body dispersion forces with one excited molecule. *Chem. Phys.* **1995**, *198*, 5. [CrossRef]
53. Rizzuto, L.; Passante, R.; Persico, F. Dynamical Casimir–Polder energy between an excited- and a ground-state atom. *Phys. Rev. A* **2004**, *70*, 012107. [CrossRef]
54. Barcellona, P.; Passante, R.; Rizzuto, L.; Buhmann, S.Y. van der Waals interactions between excited atoms in generic environments. *Phys. Rev. A* **2016**, *94*, 012705. [CrossRef]
55. Kien, F.L.; Kornovan, D.F.; Chormaic, S.N.; Busch, T. Repulsive Casimir-Polder potentials of low-lying excited states of a multilevel alkali-metal atom near an optical nanofiber. *Phys. Rev. A* **2022**, *105*, 042817. [CrossRef]
56. Lu, B.-S.; Arifa, K.Z.; Ducloy, M. An excited atom interacting with a Chern insulator: Toward a far-field resonant Casimir–Polder repulsion. *Eur. Phys. J. D* **2022**, *76*, 210. [CrossRef]
57. Jenkins, J.K.; Salam, A.; Thirunamachandran, T. Retarded dispersion interaction energies between chiral molecules. *Phys. Rev. A* **1994**, *50*, 4767. [CrossRef]
58. Salam, A. On the effect of a radiation field in modifying the intermolecular interaction between two chiral molecules. *J. Chem. Phys.* **2006**, *124*, 014302. [CrossRef]
59. Butcher, D.T.; Buhmann, S.Y.; Scheel, S. Casimir-Polder forces between chiral objects. *New J. Phys.* **2012**, *14*, 11301. [CrossRef]
60. Barcellona, P.; Safari, H.; Salam, A.; Buhmann, S.Y. Enhanced chiral discriminatory van der Waals interactions mediated by chiral surfaces. *Phys. Rev. Lett.* **2017**, *118*, 193401. [CrossRef]
61. Wylie, J.M.; Sipe, J.E. Quantum electrodynamics near an interface. II. *Phys. Rev. A* **1985**, *32*, 2030. [CrossRef]
62. Buhmann, S.Y.; Welsch, D.-G.; Kampf, T. Ground-state van der Waals forces in planar multilayer magnetodielectrics. *Phys. Rev. A* **2005**, *72*, 032112. [CrossRef]
63. Dalvit, D.A.R.; Neto, P.A.M.; Lambrecht, A.; Reynaud, S. Probing quantum-vacuum geometrical effects with cold atoms. *Phys. Rev. Lett.* **2008**, *100*, 040405. [CrossRef]
64. Messina, R.; Dalvit, D.A.R.; Neto, P.A.M.; Lambrecht, A.; Reynaud, S. Dispersive interactions between atoms and nonplanar surfaces. *Phys. Rev. A* **2009**, *80*, 022119. [CrossRef]
65. Contreras-Reyes, A.M.; Guérout, R.; Neto, P.A.M.; Dalvit, D.A.R.; Lambrecht, A.; Reynaud, S. Casimir-Polder interaction between an atom and a dielectric grating. *Phys. Rev. A* **2010**, *82*, 052517. [CrossRef]
66. Cysne, T.; Kort-Kamp, W.J.M.; Oliver, D.; Pinheiro, F.A.; Rosa, F.S.S.; Farina, C. Tuning the Casimir-Polder interaction via magneto-optical effects in graphene. *Phys. Rev. A* **2014**, *90*, 052511. [CrossRef]
67. Bimonte, G.; Emig, T.; Kardar, M. Casimir-Polder interaction for gently curved surfaces. *Phys. Rev. D* **2014**, *90*, 081702(R). [CrossRef]
68. Bimonte, G.; Emig, T.; Kardar, M. Casimir-Polder force between anisotropic nanoparticles and gently curved surfaces. *Phys. Rev. D* **2015**, *92*, 025028. [CrossRef]
69. Garcion, C.; Fabre, N.; Bricha, H.; Perales, F.; Scheel, S.; Ducloy, M.; Dutier, G. Intermediate-range Casimir-Polder interaction probed by high-order slow atom diffraction. *Phys. Rev. Lett.* **2021**, *127*, 170402. [CrossRef]
70. Abrantes, P.P.; Pessanha, V.; de Melo e Souza, R.; Farina, C. Controlling the atom-sphere interaction with an external electric field. *Phys. Rev. A* **2021**, *104*, 022820. [CrossRef]
71. Marachevsky, V.N.; Sidelnikov, A.A. Casimir-Polder interaction with Chern-Simons boundary layers. *Phys. Rev. D* **2023**, *107*, 105019. [CrossRef]
72. Alves, D.T.; Queiroz, L.; Nogueira, E.C.M.; Peres, N.M.R. Curvature-induced repulsive effect on the lateral Casimir-Polder–van der Waals force. *Phys. Rev. A* **2023**, *107*, 062821. [CrossRef]
73. Fosco, C.D.; Lombardo, F.C.; Mazzitelli, F.D. Casimir physics beyond the proximity force approximation: The derivative expansion. *Physics* **2024**, *6*, 290. [CrossRef]

74. Messina, R.; Vasile, R.; Passante, R. Dynamical Casimir-Polder force on a partially dressed atom near a conducting wall. *Phys. Rev. A* **2010**, *82*, 062501. [CrossRef]
75. Behunin, R.O.; Hu, B.-L. Nonequilibrium forces between atoms and dielectrics mediated by a quantum field. *Phys. Rev. A* **2011**, *84*, 012902. [CrossRef]
76. Barcellona, P.; Passante, R.; Rizzuto, L.; Buhmann, S.Y. Dynamical Casimir–Polder interaction between a chiral molecule and a surface. *Phys. Rev. A* **2016**, *93*, 032508. [CrossRef]
77. Goedecke, G.H.; Wood, R.C. Casimir–Polder interaction at finite temperature. *Phys. Rev. A* **1999**, *11*, 2577. [CrossRef]
78. Barton, G. Long-range Casimir–Polder-Feinberg-Sucher intermolecular potential at nonzero temperature. *Phys. Rev. A* **2001**, *64*, 032102. [CrossRef]
79. Obrecht, J.M.; Wild, R.J.; Antezza, M.; Pitaevskii, L.P.; Stringari, S.; Cornell, E.A. Measurement of the Temperature Dependence of the Casimir-Polder Force. *Phys. Rev. Lett.* **2007**, *98*, 063201. [CrossRef]
80. Haakh, H.; Intravaia, F.; Henkel, C.; Spagnolo, S.; Passante, R.; Power, B.; Sols, F. Temperature dependence of the magnetic Casimir-Polder interaction. *Phys. Rev. A* **2009**, *80*, 062905. [CrossRef]
81. Chaichian, M.; Klimchitskaya, G.L.; Mostepanenko, V.M.; Tureanu, A. Thermal Casimir-Polder interaction of different atoms with graphene. *Phys. Rev. A* **2012**, *86*, 012515. [CrossRef]
82. Laliotis, A.; de Silans, T.P.; Maurin, I.; Ducloy, M.; Bloch1, D. Casimir–Polder interactions in the presence of thermally excited surface modes. *Nat. Commun.* **2014**, *5*, 4364. [CrossRef]
83. Khusnutdinov, N.; Kashapov, R.; Woods, L.M. Thermal Casimir and Casimir–Polder interactions in *N* parallel 2D Dirac materials. *2D Mater.* **2018**, *5*, 035032. [CrossRef]
84. Israelachvili, J.N. *Intermolecular and Surface Forces*; Academic Press: Waltham, MA, USA, 2011.
85. Geim, A.K.; Grigorieva, I.V. Van der Waals heterostructures. *Nature* **2013**, *499*, 419. [CrossRef]
86. Liu, Y.; Weiss, N.O.; Duan, X.; Cheng, H.-C.; Huang, Y.; Duan, X. Van der Waals heterostructures and devices. *Nat. Rev. Mater.* **2016**, *1*, 16042. [CrossRef]
87. Castellanos-Gomez, A.; Duan, X.; Fei, Z.; Gutierrez, H.R.; Huang, Y.; Huang, X.; Quereda, J.; Qian, Q.; Sutter, E.; Sutter, P. Van der Waals heterostructures. *Nat. Rev. Methods Primers* **2022**, *2*, 58. [CrossRef]
88. Caldeweyher, E.; Ehlert, S.; Hansen, A.; Neugebauer, H.; Spicher, S.; Bannwarth, C.; Grimme, S. A generally applicable atomic-charge dependent London dispersion correction. *J. Chem. Phys.* **2019**, *150*, 154122. [CrossRef]
89. Caldeweyher, E.; Mewes, J.-M.; Ehlert, S.; Grimme, S. Extension and evaluation of the D4 London-dispersion model for periodic systems. *Phys. Chem. Chem. Phys.* **2020**, *22*, 8499. [CrossRef]
90. Chowdhury, S.T.u.R.; Tang, H.; Perdew, J.P. van der Waals corrected density functionals for cylindrical surfaces: Ammonia and nitrogen dioxide adsorbed on a single-walled carbon nanotube. *Phys. Rev. B* **2021**, *103*, 195410. [CrossRef]
91. Dryden, D.M.; Hopkins, J.C.; Denoyer, L.K.; Poudel, L.; Steinmetz, N.F.; Ching, W.-Y.; Podgornik, R.; Parsegian, A.; French, R.H. van der Waals Interactions on the Mesoscale: Open-Science Implementation, Anisotropy, Retardation, and Solvent Effects. *Langmuir* **2015**, *31*, 10145. [CrossRef]
92. Spreng, B.; Neto, P.A.M.; Ingold, G.-L. Plane-wave approach to the exact van der Waals interaction between colloid particles. *J. Chem. Phys.* **2020**, *153*, 024115. [CrossRef]
93. Nunes, R.O.; Spreng, B.; de Melo e Souza, R.; Ingold, G.-L.; Neto, P.A.M.; Rosa, F.S.S. The Casimir Interaction between Spheres Immersed in Electrolytes. *Universe* **2021**, *7*, 156. [CrossRef]
94. Scheel, S.; Buhmann, S.Y. Casimir-Polder forces on moving atoms. *Phys. Rev. A* **2009**, *80*, 042902. [CrossRef]
95. Barton, G. On van der Waals friction: I. Between two atoms. *New J. Phys.* **2010**, *12*, 113044. [CrossRef]
96. Pieplow, G.; Henkel, C. Fully covariant radiation force on a polarizable particle. *New J. Phys.* **2013**, *15*, 023027. [CrossRef]
97. Intravaia, F.; Behunin, R.O.; Dalvit, D.A.R. Quantum friction and fluctuation theorems. *Phys. Rev. A* **2014**, *89*, 050101(R). [CrossRef]
98. Intravaia, F.; Behunin, R.O.; Henkel, C.; Busch, K.; Dalvit, D.A.R. Failure of Local Thermal Equilibrium in Quantum Friction. *Phys. Rev. Lett.* **2016**, *117*, 100402. [CrossRef]
99. Donaire, M.; Lambrecht, A. Velocity-dependent dipole forces on an excited atom. *Phys. Rev. A* **2016**, *93*, 022701. [CrossRef]
100. Reiche, D.; Intravaia, F.; Hsiang, J.-T.; Busch, K.; Hu, B.L. Nonequilibrium thermodynamics of quantum friction. *Phys. Rev. A* **2020**, *102*, 050203(R). [CrossRef]
101. Reiche, D.; Busch, K.; Intravaia, F. Nonadditive Enhancement of Nonequilibrium Atom-Surface Interactions. *Phys. Rev. Lett.* **2020**, *124*, 193603. [CrossRef]
102. Farías, M.B.; Lombardo, F.C.; Soba, A.; Villar, P.I.; Decca, R.S. Towards detecting traces of non-contact quantum friction in the corrections of the accumulated geometric phase. *NPJ Quantum Inf.* **2020**, *6*, 25. [CrossRef]
103. Lombardo, F.C.; Decca, R.S.; Viotti, L.; Villar, P.I. Detectable Signature of Quantum Friction on a Sliding Particle in Vacuum. *Adv. Quantum Technol.* **2021**, *4*, 2000155. [CrossRef]
104. Dedkov, G.V.; Kyasov, A.A. Nonlocal friction forces in the particle-plate and plate-plate configurations: Nonretarded approximation. *Surf. Sci.* **2020**, *700*, 121681. [CrossRef]
105. Dedkov, G.V. Van der Waals Interactions of Moving Particles with Surfaces of Cylindrical Geometry. *Universe* **2021**, *7*, 106. [CrossRef]

106. Impens, F.; Behunin, R.O.; Ttira, C.C.; Neto, P.A.M. Non-local double-path Casimir phase in atom interferometers. *EPL* **2013**, *101*, 60006. [CrossRef]
107. Impens, F.; Ttira, C.C.; Neto, P.A.M. Non-additive dynamical Casimir atomic phases. *J. Phys. B At. Mol. Opt. Phys.* **2013**, *46*, 245503. [CrossRef]
108. Impens, F.; de Melo e Souza, R.; Matos, G.C.; Neto, P.A.M. Dynamical Casimir effects with atoms: From the emission of photon pairs to geometric phases. *EPL* **2022**, *138*, 30001. [CrossRef]
109. Salam, A. van der Waals Dispersion Potential between Excited Chiral Molecules via the Coupling of Induced Dipoles. *Physics* **2023**, *5*, 247. [CrossRef]
110. Dung, H.T. Interatomic dispersion potential in a cylindrical system: Atoms being off axis. *J. Phys. B* **2016**, *49*, 165502. [CrossRef]
111. Zuki, F.M.; Edyvean, R.G.J.; Pourzolfaghar, H.; Kasim, N. Modeling of the Van Der Waals Forces during the Adhesion of Capsule-Shaped Bacteria to Flat Surfaces. *Biomimetics* **2021**, *6*, 5. [CrossRef]
112. Laliotis, A.; Lu, B.-S.; Ducloy, M.; Wilkowski, D. Atom-surface physics: A review. *AVS Quantum Sci.* **2021**, *3*, 043501. [CrossRef]
113. Nogueira, E.C.M.; Queiroz, L.; Alves, D.T. Peak, valley, and intermediate regimes in the lateral van der Waals force. *Phys. Rev. A* **2021**, *104*, 012816. [CrossRef]
114. Nogueira, E.C.M.; Queiroz, L.; Alves, D.T. Sign inversion in the lateral van der Waals force. *Phys. Rev. A* **2022**, *105*, 062816. [CrossRef]
115. Milonni, P.W.; Ackerhalt, J.R.; Smith, W.A. Interpretation of Radiative Corrections in Spontaneous Emission. *Phys. Rev. Lett.* **1973**, *31*, 958. [CrossRef]
116. Dalibard, J.; Dupont-Roc, J.; Cohen-Tannoudji, C. Vacuum fluctuations and radiation reaction: Identification of their respective contributions. *J. Phys.* **1982**, *43*, 1617. [CrossRef]
117. Cohen-Tannoudji, C. Fluctuations in Radiative Processes. *Phys. Scr.* **1986**, *12*, 19. [CrossRef]
118. Zhou, W.; Cheng, S.; Yu, H. Interatomic interaction of two ground-state atoms in vacuum: Contributions of vacuum fluctuations and radiation reaction. *Phys. Rev. A* **2021**, *103*, 012227. [CrossRef]
119. Impens, F.; Ttira, C.C.; Behunin, R.O.; Neto, P.A.M. Dynamical local and nonlocal Casimir atomic phases. *Phys. Rev. A* **2014**, *89*, 022516. [CrossRef]
120. Matos, C.G.; de Melo e Souza, R.; Neto, P.A.M.; Impens, F. Quantum Vacuum Sagnac Effect. *Phys. Rev. Lett.* **2021**, *127*, 270401.
121. Breuer, H.P.; Petruccione, F. *The Theory of Open Quantum Systems*; Oxford University Press: Oxford, UK, 2002.
122. Calzetta, E.A.; Hu, B.-L.B. *Nonequilibrium Quantum Field Theory*; Cambridge University Press: Cambridge, UK, 2022.
123. Weiss, U. *Quantum Dissipative Systems*; World Scientific: Singapore, 2007.
124. Kubo, R. The fluctuation-dissipation theorem. *Rep. Prog. Phys.* **1966**, *29*, 255. [CrossRef]
125. Dubovik, V.M.; Tugushev, V.V. Toroid moments in electrodynamics and solid-state physics. *Phys. Rep.* **1990**, *187*, 142. [CrossRef]
126. Pitombo, R.S.; Vasconcellos, M.; Farina, C.; de Melo e Souza, R. Source method for the evaluation of multipole fields. *Eur. J. Phys.* **2021**, *42*, 025202. [CrossRef]
127. Nussenzveig, H.M. *Causality and Dispersion Relations*; Academic Press: New York, NY, USA, 1972.

Disclaimer/Publisher's Note: The statements, opinions and data contained in all publications are solely those of the individual author(s) and contributor(s) and not of MDPI and/or the editor(s). MDPI and/or the editor(s) disclaim responsibility for any injury to people or property resulting from any ideas, methods, instructions or products referred to in the content.

Article

Dephasing Dynamics in a Non-Equilibrium Fluctuating Environment

Xiangjia Meng [1,2], Yaxin Sun [3], Qinglong Wang [3], Jing Ren [3], Xiangji Cai [3,*] and Artur Czerwinski [4,*]

1. School of information Engineering, Shandong Youth University of Political Science, Jinan 250103, China; mxj@sdyu.edu.cn
2. New Technology Research and Development Center of Intelligent Information Controlling in Universities of Shandong, Shandong Youth University of Political Science, Jinan 250103, China
3. School of Science, Shandong Jianzhu University, Jinan 250101, China; 201912102001@stu.sdjzu.edu.cn (Y.S.); 201912102020@stu.sdjzu.edu.cn (Q.W.); renjing19@sdjzu.edu.cn (J.R.)
4. Institute of Physics, Faculty of Physics, Astronomy and Intypeatics, Nicolaus Copernicus University in Torun, ul. Grudziadzka 5, 87-100 Torun, Poland
* Correspondence: xiangjicai@foxmail.com (X.C.); aczerwin@umk.pl (A.C.)

Abstract: We performed a theoretical study of the dephasing dynamics of a quantum two-state system under the influences of a non-equilibrium fluctuating environment. The effect of the environmental non-equilibrium fluctuations on the quantum system is described by a generalized random telegraph noise (RTN) process, of which the statistical properties are both non-stationary and non-Markovian. Due to the time-homogeneous property in the master equations for the multi-time probability distribution, the decoherence factor induced by the generalized RTN with a modulatable-type memory kernel can be exactly derived by means of a closed fourth-order differential equation with respect to time. In some special limit cases, the decoherence factor recovers to the expression of the previous ones. We analyzed in detail the environmental effect of memory modulation in the dynamical dephasing in four types of dynamics regimes. The results showed that the dynamical dephasing of the quantum system and the conversion between the Markovian and non-Markovian characters in the dephasing dynamics under the influence of the generalized RTN can be effectively modulated via the environmental memory kernel.

Keywords: open quantum systems; decoherence; non-equilibrium environmental fluctuations

Citation: Meng, X.; Sun, Y.; Wang, Q.; Ren, J.; Cai, X.; Czerwinski, A. Dephasing Dynamics in a Non-Equilibrium Fluctuating Environment. *Entropy* **2023**, *25*, 634. https://doi.org/10.3390/e25040634

Academic Editors: Paula I. Villar and Fernando C. Lombardo

Received: 18 March 2023
Revised: 5 April 2023
Accepted: 7 April 2023
Published: 8 April 2023

Copyright: © 2023 by the authors. Licensee MDPI, Basel, Switzerland. This article is an open access article distributed under the terms and conditions of the Creative Commons Attribution (CC BY) license (https:// creativecommons.org/licenses/by/ 4.0/).

1. Introduction

Quantum coherence is an important phenomenon in the microcosmic world, which has been attracting continuous attention with the advance of experimental technologies. In a wide variety of applications related to quantum physics, the destruction of coherence is inevitable owing to the reason that any quantum system keeps interacting with the surrounding environments. The unavoidable interactions of an open quantum system with its surroundings bring about its correlations with environmental states and make the system lose coherence in dynamical evolution [1–6]. The loss of the quantum coherence of open systems induced by the environments is usually called decoherence, which is widely used to describe the quantum–classical transition and is regarded as a great obstacle to the design and realization of experimental devices for quantum information processing. Recently, the investigations of the decoherence process of open quantum systems have received more and more considerable attention, which plays a significant role in a series of essential issues in quantum information science, such as quantum computation, quantum measurement, quantum control, and so on [7–21].

Over the past several decades, the quantum decoherence dynamics of open systems has been investigated by making the assumption that system–environment coupling is weak and by ignoring the memory effect of the actual dynamical evolution. These treatments are usually called Markovian approximations, and the quantum dynamics of open systems is generally described in the Lindblad-type master equations. However, the couplings with the environment are not weak, and the quantum evolution of the open system displays a memory effect in the vast majority of realistic cases. In these situations, the Markovian approximations are no longer valid, and the non-Markovian character exhibited in the decoherence dynamics plays a non-negligible role [22–25]. Under the influence of environments exhibiting equilibrium fluctuations, the study of non-Markovian quantum dynamics has drawn increasing attention by treating the environmental noise with a stationary statistical property [13,20,26–46]. Recently, it was shown that the non-equilibrium environmental fluctuations become dominant in some transient and ultra-fast physical or biological processes. The instantaneous environmental state influenced by the initial couplings to the system cannot return to equilibrium rapidly, corresponding to the statistics of the environmental noise no longer being stationary [47,48]. Thus, to study quantum dynamics in these situations, the effects of non-equilibrium environmental fluctuations should be taken into full consideration.

Random telegraph noise (RTN) as the widely used classical noise with non-Gaussianity has been the subject of the theoretical simulation of the influences of environmental fluctuations on open quantum systems [49–59]. In some previous research, the environmental fluctuations governed by the RTN were usually assumed to have stationary and Markovian statistical properties. Actually, this assumption is just an idealization of the environmental fluctuations in statistics. In some realistic situations, the statistical properties of the fluctuating environments may be non-stationary and non-Markovian. On the basis of this fact, the non-Markovian RTN governed by an exponential-type memory kernel with stationary and non-stationary statistics was proposed and discussed in succession. The generalized RTN with non-stationary and non-Markovian statistics has been employed extensively to investigate the related questions concerning the quantum decoherence dynamics of open systems in the presence of non-equilibrium environmental fluctuations [60–67]. In recent research, the stationary RTN with non-Markovian statistics governed by a memory kernel of a modulatable-type has also been put forward. It has been demonstrated that the dynamical dephasing of the quantum two-state system can be modulated by the environmental memory kernel in an equilibrium environment [68]. The exact expression for the decoherence factor for open quantum systems in the presence of generalized RTN with non-stationary and non-Markovian statistics is rather difficult to obtain. It is shown that the decoherence factor satisfies a time differential equation of third-order under the influence of the generalized RTN with an exponential-type memory kernel [61]. However, in a non-equilibrium environment governed by the generalized RTN with a modulatable-type memory kernel, the decoherence factor of a quantum two-state system has not been derived. The environmental effect of memory modulation in the dynamical dephasing in a non-equilibrium environment has not been investigated yet. Therefore, there are some important physical issues arising naturally and that we should further address. Under the influence of the generalized RTN with a modulatable memory kernel, is it possible to derive the decoherence factor exactly by establishing a closed differential equation with respect to time? How do the memory effects of the generalized RTN modulate the quantum dynamical dephasing of the system in a non-equilibrium fluctuating environment? Can we convert the Markovian and non-Markovian characters in the dephasing dynamics by changing the modulation frequency in the memory kernel of the generalized RTN?

In the present paper, we theoretically investigated the dephasing dynamics of a quantum two-state system under the influence of a fluctuating environment displaying non-equilibrium fluctuations described by the generalized RTN with non-stationary and non-Markovian statistics. The decoherence factor satisfies a closed fourth-order time differential equation under the generalized RTN with a modulatable-type memory kernel. The expression of the decoherence factor can be exactly simplified as the previous ones in some special limit cases of the environmental memory kernel. We analyzed the environmental effect of the memory modulation in the dynamical dephasing in four types of dynamics regimes: weak coupling weak memory regime, weak coupling strong memory regime, strong coupling weak memory regime, and strong coupling strong memory regime, respectively. The results display that the quantum dephasing dynamics of the system and the conversion between the Markovian and non-Markovian characters in the dynamical dephasing can be effectively modulated via the environmental memory kernel. In addition, the boundary in the dephasing dynamics between the Markovian and non-Markovian characters is determined by the combined effects of the system–environment coupling, the environmental memory, and the environmental modulation.

The organization of the paper is as follows. We first present the theoretical framework, in Section 2, of the quantum dephasing dynamics under the influence of non-equilibrium environmental fluctuations. We derived the decoherence factor of the quantum system exactly under the generalized RTN with a modulatable-type memory kernel by establishing a closed differential equation with respect to time. In Section 3, we give the results of the quantum dynamical dephasing in four types of dynamics regimes and the dynamical conversion between the Markovian and non-Markovian characters. Finally, we give the concluding remarks in Section 4.

2. Quantum Dephasing under the Influence of Non-Equilibrium Environmental Fluctuations

The physical model we considered here is a quantum two-state system in interaction with a classical fluctuating environment, which displays non-equilibrium fluctuations. We assumed the environmental effects do not lead to population transfer and the quantum system undergoes pure dephasing during its dynamical evolution. The influences of the environment on the system cause the energy gap between the two states in the type $E_1(t) - E_2(t) = \hbar\omega(t)$, where $E_k(t)$ ($k = 1, 2$) denotes the instantaneous energy of the state k and $\omega(t)$ is the transition frequency between the two states $|1\rangle$ and $|2\rangle$, which fluctuates stochastically due to the coupling between the system and environment [47,48,69,70].

In terms of the spectral diffusion framework of Kubo–Anderson, the instantaneous frequency difference of the quantum system can be rewritten as $\omega(t) = \omega_0 + \zeta(t)$, with ω_0 denoting the standard frequency difference and $\zeta(t)$ the fluctuation part arising from the environmental effects generally governed by a classical stochastic process. Stochastic processes with a stationary statistical property have been widely used to describe the equilibrium environmental fluctuations [71]. Under the influence of the environments exhibiting non-equilibrium fluctuations, the fluctuation part $\zeta(t)$ in the instantaneous frequency difference is generally governed by a stochastic process with non-stationary statistics, which corresponds, in the physical description, to environmentally excited phonons with sharply defined phases initially [47,48].

For the quantum system prepared in an initial coherent state with the superposition of $|2\rangle$ and $|1\rangle$, the non-diagonal element in the density matrix quantifies the time-dependent coherence of the system:

$$\rho_{21}(t) = D(t)e^{i\omega_0 t}\rho_{21}(0), \tag{1}$$

where $D(t)$ represents the decoherence factor, which can be written in terms of the moments of the fluctuation part $\zeta(t)$ in the Dyson series expansion:

$$D(t) = \left\langle \exp\left[i\int_0^t dt' \zeta(t')\right] \right\rangle = 1 + \sum_{n=1}^{\infty} i^n \int_0^t dt_1 \cdots \int_0^{t_{n-1}} dt_n \langle \zeta(t_1) \cdots \zeta(t_n) \rangle, \tag{2}$$

where $\langle \cdots \rangle$ represents a statistical average taken over $\zeta(t)$. The decoherence factor $D(t)$ closely depends on the statistical properties of the stochastic fluctuations induced by the environment. Under the influence of non-equilibrium fluctuating environments, the decoherence factor $D(t)$ is no longer real, but complex in time, resulting from the non-stationary statistics of the fluctuation part $\zeta(t)$.

For the dynamical dephasing process of the system in a non-equilibrium fluctuating environment, there are two important physical qualities, namely the frequency shift $s(t)$ and the dephasing rate $\gamma(t)$, linked to the decoherence factor $D(t)$, with the definitions as

$$s(t) = -\text{Im}\left[\frac{dD(t)/dt}{D(t)}\right], \quad \gamma(t) = -\text{Re}\left[\frac{dD(t)/dt}{D(t)}\right]. \tag{3}$$

The frequency shift $s(t)$ expressed in Equation (3) can be used to distinguish the stationary and non-stationary statistics of the environmental noise between equilibrium and non-equilibrium fluctuating environments. In general, there will not appear a frequency shift for the environments exhibiting equilibrium fluctuations, whereas under the influence of non-equilibrium environmental fluctuations, the frequency shift is time-dependent. The decoherence rate $\gamma(t)$ of the dephasing dynamics in Equation (3) is linked to the information exchange that takes place between the system and the environment. There is a one-way continuous information flow to the environment out of the system without environmental coherence back-action for the case that the decoherence rate $\gamma(t)$ is positive at all times. For the case that $\gamma(t)$ sometimes takes negative values, the information flows back into the system from the environment with the emergence of the environmental coherence back-action. According to the definition of Breuer–Laine–Piilo, the non-Markovianity, namely the total of the maximum flow of the environmental information backward to the quantum system, is written as [72]:

$$\mathcal{N} = -\int_{\gamma(t)<0} \gamma(t)|D(t)|dt = \sum_{j=1}^{\infty} |D(t_{2j})| - |D(t_{1j})|, \tag{4}$$

where $[t_{1j}, t_{2j}]$ are the jth time intervals in which $|D(t)|$ increases.

Combined with the expansion in the Dyson series on the basis of the moments of Equation (2), it is also possible to expand the decoherence factor $D(t)$ by means of the cumulants of the fluctuation part $\zeta(t)$ [71]. Because both expansions involve environmental correlations of order tending to infinity, therefore, it is difficult to obtain the exact expression for the decoherence factor based on them. For the general case, we need to truncate the environmental correlations to some finite order to derive the decoherence factor approximately. Some approaches have been developed to derive the decoherence factor of a quantum two-state system under the influence of environmental noise exactly. The exact expression of the decoherence factor governed by environmental fluctuations with stationary and Markovian statistical properties can be obtained, for example, by means of the stochastic Liouville equation [73]. There are, however, very few physical models for which the decoherence factor can be exactly achieved under the influence of non-equilibrium environmental fluctuations with non-stationary and non-Markovian statistics. In the following, we derive the exact expression of the decoherence factor of the quantum two-state system under the influence of the generalized RTN by means of establishing a closed time differential equation of the decoherence factor.

2.1. Non-Equilibrium Environmental Fluctuations Described by Generalized RTN

It should be noted that the standard RTN is a classical stochastic process with time-homogeneity and non-Gaussianity. The standard RTN transits stochastically between the values ± 1 with a mean transition rate λ and the amplitude ν in stationary and Markovian statistics [74–76]. The ratio of the amplitude ν to the rate λ of the transition is used to identify the weak-coupling ($\nu/\lambda < 1$) and strong-coupling ($\nu/\lambda > 1$) regimes, respectively [75,76].

It is possible to extract the characteristics of the generalized RTN with non-Markovian and non-stationary statistics from that of the standard RTN according to the classical theory of probability. The non-Markovian statistics of the generalized RTN is characterized by the master equations for the multi-time probability distributions [60]:

$$\frac{\partial}{\partial t}\mathbb{P}(\zeta,t;\zeta_1,t_1;\cdots;\zeta_n,t_n) = \int_{t_1}^{t} K(t-\tau)\lambda\mathbb{T}\mathbb{P}(\zeta,t;\zeta_1,t_1;\cdots;\zeta_n,t_n)d\tau, \qquad (5)$$

with $K(t-\tau)$ being the memory kernel of the generalized RTN and the multi-time probability $\mathbb{P}(\zeta,t;\zeta_1,t_1;\cdots;\zeta_n,t_n)$ and the matrix $\mathbb{T}$ for transition respectively written as

$$\mathbb{P}(\zeta,t;\zeta_1,t_1;\cdots;\zeta_n,t_n) = \begin{pmatrix} P(+\nu,t;\zeta_1,t_1;\cdots;\zeta_n,t_n) \\ P(-\nu,t;\zeta_1,t_1;\cdots;\zeta_n,t_n) \end{pmatrix}, \quad \mathbb{T} = \begin{pmatrix} -1 & 1 \\ 1 & -1 \end{pmatrix}. \qquad (6)$$

The statistical property of the environmental noise depends on its prior history because of the fact that the memory effect has been taken into consideration. The non-stationary environmental statistical property of the generalized RTN arises from the single-point probability distribution [77]:

$$P(\zeta,t) = \frac{1}{2}[1 + aP(t)]\delta_{\zeta,\nu} + \frac{1}{2}[1 - aP(t)]\delta_{\zeta,-\nu}. \qquad (7)$$

where a is the non-stationary parameter with $|a| \leq 1$ and $P(t-t') = \mathscr{L}^{-1}[e^{-zt'}\mathcal{P}(z)]$ denotes the auxiliary function with $\mathcal{P}(z) = 1/[z + 2\lambda\mathcal{K}(z)]$ and $\mathscr{L}^{-1}$ representing the inverse Laplace transform. For the memoryless case, namely $K(t-\tau) = \delta(t-\tau)$, then the generalized RTN returns to the Markovian one. For the special case $a = 0$, the generalized RTN recovers to the stationary one, which corresponds to the environmental fluctuations displaying the equilibrium feature [61,62].

Based on the statistical properties given above and on the basis of Bayes' rule in classical probability theory, the statistical features of the generalized RTN are represented in terms of the moments of first- and second-orders:

$$\begin{aligned} M_1(t) &= \langle \zeta(t) \rangle = a\nu P(t), \\ M_2(t,t') &= \langle \zeta(t)\zeta(t') \rangle = \nu^2 P(t-t'), \end{aligned} \qquad (8)$$

and the factorization for the higher-order moments [61,62]:

$$M_n(t_1,t_2,\cdots,t_n) = P(t_1-t_2)M_{n-2}(t_3,t_4\cdots,t_n), \qquad (9)$$

for the ordered time instants $t_1 > t_2 > \cdots > t_n$ ($n \geq 3$). Obviously, the statistical features of the generalized RTN are closely linked to the auxiliary probability function $P(t-t')$. Thus, we can gain all the information of the generalized RTN once we obtain the expression of the auxiliary probability function in theory.

2.2. Closed Dynamical Equation for the Decoherence Factor under Generalized RTN with a Modulatable Memory Kernel

In general, the type of environmental memory kernel in Equation (5) can be arbitrary. There are many types of environmental memory kernels, the exponential type, the modulatable type, the power law type, and so on [78–82]. The generalized RTN governed by the non-Markovian non-stationary statistical properties with an exponential memory kernel has been proposed [60,62]. It has been shown that the decoherence factor obeys a closed time differential equation of third-order in a non-equilibrium environment under the influence of the generalized RTN governed by an exponential-type memory kernel by means of the differential relations of the moments with respect to time [61,62].

We considered here the case that the type of the memory kernel in Equation (5) of the generalized RTN is a modulatable one:

$$K(t-\tau) = \kappa \cos[\Omega(t-\tau)]e^{-\kappa(t-\tau)}, \qquad (10)$$

where κ is the environmental memory decay rate and Ω denotes the memory modulation frequency [80,81]. Physically, this corresponds to a model with the environmental modulation of the memory effect. In the case with the modulation frequency $\Omega = 0$, the type of environmental memory kernel becomes an exponential one. The smaller κ is, the stronger the memory effect of the generalized RTN is. In the case with the decay rate $\kappa \to +\infty$, the generalized RTN becomes memoryless, namely $K(t - \tau) = \delta(t - \tau)$, and it only displays Markovian statistics.

According to the previous work in [61,62], the dynamical equation for the decoherence factor is closely linked to the time differential relationships of the moments of the generalized RTN. Because of the fact that the ancillary probability function is related to the statistical features of the generalized RTN as in Equation (9), a closed time differential equation for the decoherence factor of the quantum system can be derived in terms of the differential relation of the auxiliary probability functional $P(t)$. The type of memory kernel implies that the auxiliary probability function $P(t)$ of environmental noise satisfies a closed time differential equation of third-order as follows:

$$\frac{d^3}{dt^3}P(t) + c_2 \frac{d^2}{dt^2}P(t) + c_1 \frac{d}{dt}P(t) + c_0 P(t) = 0, \qquad (11)$$

with the coefficients $c_2 = 2\kappa$, $c_1 = \kappa^2 + \Omega^2 + 2\kappa\lambda$, and $c_0 = 2\kappa^2\lambda$ and the initial conditions $P(0) = 1$, $(d/dt)P(0) = 0$, and $(d^2/dt^2)P(0) = -2\kappa\lambda$. As a consequence, a fourth-order closed differential equation with respect to time for the decoherence factor can be obtained:

$$\frac{d^4}{dt^4}D(t) + C_3 \frac{d^3}{dt^3}D(t) + C_2 \frac{d^2}{dt^2}D(t) + C_1 \frac{d}{dt}D(t) + C_0 D(t) = 0, \qquad (12)$$

where the coefficients can be written as

$$C_3 = 2\kappa, \quad C_2 = \kappa^2 + \Omega^2 + 2\kappa\lambda + \nu^2, \quad C_1 = 2\kappa^2\lambda + 2\kappa\nu^2, \quad C_0 = \nu^2(\kappa^2 + \Omega^2), \qquad (13)$$

and the initial conditions satisfy

$$D(0) = 1, \quad \frac{d}{dt}D(0) = -ia\nu, \quad \frac{d^2}{dt^2}D(0) = -\nu^2, \quad \frac{d^3}{dt^3}D(0) = -\nu^2 - ia\nu. \qquad (14)$$

With the help of Laplace transformation taken over Equation (12), the decoherence factor $D(t)$ can be analytically solved, in terms of the initial conditions in Equation (14), as

$$D(t) = \mathscr{L}^{-1}[\mathcal{D}(z)],$$
$$\mathcal{D}(z) = \frac{z^3 + 2\kappa z^2 + (\kappa^2 + \Omega^2 + 2\kappa\lambda)z + 2\kappa^2\lambda + ia\nu(z^2 + 2\kappa z + \kappa^2 + \Omega^2)}{z^4 + 2\kappa z^3 + (\kappa^2 + \Omega^2 + 2\kappa\lambda + \nu^2)z^2 + 2\kappa(\kappa\lambda + \nu^2)z + (\kappa^2 + \Omega^2)\nu^2}. \qquad (15)$$

By means of the approach established in [68], the decoherence factor of the quantum system in time domain can be written as

$$D(t) = \sum_{j=1}^{n_r} \left[\frac{r_{j1} t^{\epsilon_j - 1}}{(\epsilon_j - 1)!} + \cdots + r_{j\epsilon_1} \right] e^{a_j t}$$
$$+ \sum_{j=1}^{n_c} \left\{ \left[\frac{c_{j1} t^{\epsilon_j - 1}}{(\epsilon_j - 1)!} + \cdots + c_{j\epsilon_1} \right] e^{b_j t} + \left[\frac{c_{j1}^* t^{\epsilon_j - 1}}{(\epsilon_j - 1)!} + \cdots + c_{j\epsilon_1}^* \right] e^{b_j^* t} \right\}, \qquad (16)$$

where r_{jk} and c_{jk} are the real and complex coefficients, which are respectively expressed as

$$r_{jk} = \frac{1}{(k-1)!}\left\{\frac{d^{k-1}}{dz^{k-1}}[\mathcal{D}(z)(z-a_j)^{e_j}]\right\}_{z=a_j},$$
$$c_{jk} = \frac{1}{(k-1)!}\left\{\frac{d^{k-1}}{dz^{k-1}}[\mathcal{D}(z)(z-b_j)^{\epsilon_j}]\right\}_{z=b_j}. \quad (17)$$

with a_j and b_j denoting the real and non-real roots of the denominator of $\mathcal{D}(z)$ in Equation (15) and the relation $\sum_j^{n_r} e_j + 2\sum_j^{n_c} \epsilon_j = 4$.

2.3. Comparisons with Previous Work

To compare this study in the present paper with that in previous work, we derived the expression of the decoherence factor in some special cases of the generalized RTN in the following.

We first considered the limit case that $\kappa \to +\infty$, namely the memoryless generalized RTN. Then, the expression of the decoherence factor under the influence of the generalized RTN in Equation (15) can be simplified as

$$D(t) = \mathscr{L}^{-1}[\mathcal{D}(z)], \quad \mathcal{D}(z) = \frac{z + 2\lambda + ia\nu}{z^2 + 2\lambda z + \nu^2}. \quad (18)$$

Consequently, the time domain decoherence factor $D(t)$ can be expressed as

$$D(t) = e^{-\lambda t}\begin{cases}[\cosh(\chi t) + \frac{\lambda}{\chi}\sinh(\chi t)] + i\frac{a\nu}{\chi}\sinh(\chi t), & \nu < \lambda, \\ (1+\lambda t) + ia\lambda t, & \nu = \lambda, \\ [\cos(\chi t) + \frac{\lambda}{\chi}\sin(\chi t)] + i\frac{a\nu}{\chi}\sin(\chi t), & \nu > \lambda,\end{cases} \quad (19)$$

with $\chi = \sqrt{|\lambda^2 - \nu^2|}$. This expression of the decoherence factor of the quantum system in Equation (19) recovers to that in [62]. Under the influence of the RTN only exhibiting the Markovian statistical property, two important regimes of dynamics have been distinguished: the weak-coupling ($\nu < \lambda$) and the strong-coupling ($\nu > \lambda$) regimes, and the dephasing dynamics displays the Markovian and non-Markovian characters in the two coupling regimes, respectively.

We now consider the case in which there is no environmental modulation of the memory effect with $\Omega = 0$, corresponding to an exponential-type memory kernel of the generalized RTN, namely $K(t-\tau) = \kappa e^{-\kappa(t-\tau)}$. In this case, the expression of the decoherence factor of the system in Equation (15) can be simplified as

$$D(t) = \mathscr{L}^{-1}[\mathcal{D}(z)], \quad \mathcal{D}(z) = \frac{z^2 + \kappa z + 2\kappa\lambda + ia\nu(z+\kappa)}{z^3 + \kappa z^2 + (2\kappa\lambda + \nu^2)z + \kappa\nu^2}. \quad (20)$$

This expression of the decoherence factor under the influence of the generalized RTN with an exponential-type memory kernel in Equation (20) recovers to that in [61]. In this case of the RTN exhibiting the non-Markovian statistical property, the dephasing dynamics can also display a non-Markovian character even though the system–environment coupling is weak, and the boundary of the Markovian and non-Markovian dynamics regimes is determined by both the system–environment coupling and the memory effect of the generalized RTN [61,62].

3. Results and Discussion

In the following, we display the results of the dephasing dynamics of the quantum two-state system induced by nonequilibrium fluctuations in the environment exhibiting the generalized RTN statistical properties with a memory kernel of the modulatable-type. Our main focus is on the environmental effect of memory modulation on the quantum dynamical dephasing of the system under the influence of the generalized RTN in four types of regimes of the dephasing dynamics relying on the coupling ν of the environment and the decay rate of the environmental memory κ. In addition, we discuss the environmental effect of memory modulation on the conversion between Markovian dynamics and non-Markovian dynamics.

3.1. Dynamical Dephasing in Weak-Coupling Weak-Memory Regime

We first show the results of the dynamical dephasing in the weak-coupling weak-memory regime with the transition amplitude $\nu = 0.2\lambda$ and memory decay rate $\kappa = 3\lambda$. As shown in Figure 1a, the dephasing dynamics displays a Markovian character when there is no environmental effect of memory modulation, namely $\Omega = 0$. As the modulation frequency Ω increases, the dephasing dynamics is first enhanced and then suppressed, and the dynamical dephasing undergoes a conversion from a Markovian to a non-Markovian character related to a critical value Ω_{th}. When $\Omega > \Omega_{th}$, the non-Markovian character begins to appear in the quantum dephasing dynamics of the system, and it becomes obvious with the increase of the modulation frequency. As depicted in Figure 1b, the decoherence rate $\gamma(t)$ displays a monotonic increase to a constant value in a long time limit for small values of the modulation frequency, whereas it displays periodic oscillations for the modulation frequency greater than the critical value Ω_{th}. The decoherence rate $\gamma(t)$ first increases with positive values and then begins to be negative in some time intervals as the modulation frequency Ω increases. When $\Omega > \Omega_{th}$, the time intervals in which the decoherence rate is negative increase with the increase of the modulation frequency. The changes in the decoherence rate are in accordance with the character in the dynamical dephasing. Figure 1c displays the environmental effect of memory modulation on the energy re-normalization of the quantum system. Obviously, the frequency shift $s(t)$ also shows a conversion from monotonic decay to disappearance in a long time limit to non-monotonically periodic oscillations with the increase of the modulation frequency.

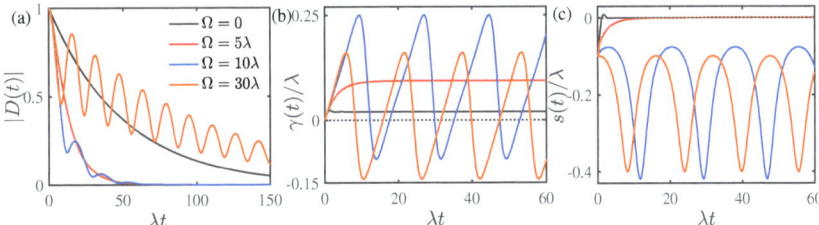

Figure 1. (Color online) The (**a**) decoherence factor $|D(t)|$, (**b**) decoherence rate $\gamma(t)$, and (**c**) frequency shift $s(t)$ as functions of time for different modulation frequencies Ω in the memory kernel in the weak-coupling weak-memory regime with the transition amplitude $\nu = 0.2\lambda$ and memory decay rate $\kappa = 3\lambda$. The initial non-stationary parameter of the environmental noise was set as $a = 0.5$.

3.2. Dynamical Dephasing in Weak-Coupling Strong-Memory Regime

We now discuss the case of the dynamical dephasing in the weak-coupling strong-memory regime with transition amplitude $\nu = 0.2\lambda$ and memory decay rate $\kappa = 0.1\lambda$. As displayed in Figure 2a, the dephasing dynamics always displays a non-Markovian character even though the system–environment coupling is weak, which is mainly a result of the strong memory effect of the generalized RTN. As the modulation frequency Ω increases, the dephasing dynamics of the system is first increased and then reduced. Meanwhile, the non-Markovian character in the dephasing dynamics becomes prominent. As depicted in Figure 2b, the decoherence rate $\gamma(t)$ decays monotonically for small modulation frequencies, whereas it displays non-monotonic periodic oscillations for large modulation frequencies. With the increase of the modulation frequency, the time intervals for which the decoherence rate takes positive values first increase and then decrease, whereas the time intervals in which the decoherence rate is negative increase. The character in the decoherence rate is consistent with that in the dephasing dynamics of the quantum system. As depicted in Figure 2c, the frequency shift $s(t)$ shows a non-monotonic decay and vanishes in a long time limit for small modulation frequencies, whereas it shows non-monotonically periodic oscillations when the modulation frequency is greater than some values.

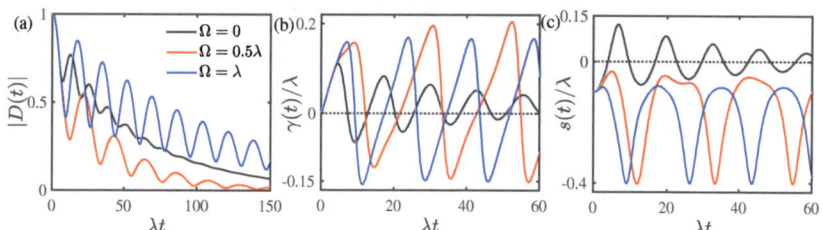

Figure 2. (Color online) The time-dependent (**a**) decoherence factor $|D(t)|$, (**b**) decoherence rate $\gamma(t)$, and (**c**) frequency shift $s(t)$ for different environmental modulation frequencies Ω in the memory kernel in the weak-coupling strong-memory regime with the transition amplitude $\nu = 0.2\lambda$ and memory decay rate $\kappa = 0.1\lambda$. The initial non-stationary parameter of the environmental noise was chosen as $a = 0.5$.

3.3. Dynamical Dephasing in Strong-Coupling Weak-Memory Regime

In this subsection, we discuss the case of the dynamical dephasing in the strong-coupling weak-memory regime with transition amplitude $\nu = 3\lambda$ and memory decay rate $\kappa = 4\lambda$. As depicted in Figure 3a, the dephasing dynamics of the system always shows a non-Markovian character arising from the strong coupling with the environment. As the modulation frequency Ω increases, the dynamical dephasing is suppressed and the non-Markovian character in the dephasing dynamics of the quantum system becomes obvious. As depicted in Figure 3b, the decoherence rate $\gamma(t)$ always shows periodic oscillations with discrete zeros. The time intervals in which the decoherence rate is negative increase as the modulation frequency increases. As displayed in Figure 3c, the frequency shift $s(t)$ displays non-monotonic periodic oscillations.

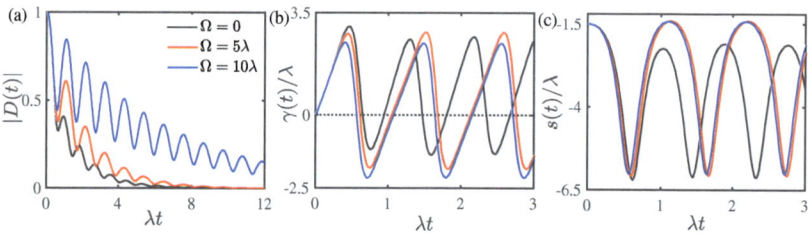

Figure 3. (Color online) The (**a**) decoherence factor $|D(t)|$, (**b**) decoherence rate $\gamma(t)$, and (**c**) frequency shift $s(t)$ as functions of time for different modulation frequencies Ω in the memory kernel in the strong-coupling weak-memory regime with the transition amplitude $\nu = 3\lambda$ and memory decay rate $\kappa = 4\lambda$. The initial non-stationary parameter of the environmental noise was set as $a = 0.5$.

3.4. Dynamical Dephasing in Strong-Coupling Strong-Memory Regime

Finally, we show the results of the dynamical dephasing in the strong-coupling strong-memory regime with $\nu = 3\lambda$ and $\kappa = \lambda$. As displayed in Figure 4a, the dephasing dynamics always show a non-Markovian character owing to both the strong interaction with the environment and the strong memory effect of the generalized RTN. With the increase of the modulation frequency Ω, the dynamical dephasing of the quantum system and the non-Markovian character in the dephasing dynamics is first suppressed and then enhanced.

As depicted in Figure 4b, the decoherence rate $\gamma(t)$ always shows periodic oscillations with discrete zeros. The time intervals that the decoherence rate is negative first decrease and then increase as the modulation frequency increases. As shown in Figure 4c, the frequency shift $s(t)$ displays non-monotonic periodic oscillations, which is similar to the case in the strong-coupling weak-memory regime.

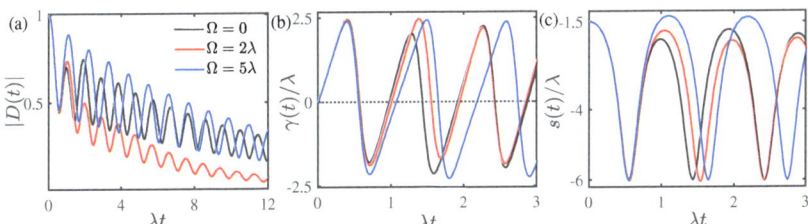

Figure 4. (Color online) The time-dependent (**a**) decoherence factor $|D(t)|$, (**b**) decoherence rate $\gamma(t)$, and (**c**) frequency shift $s(t)$ for different modulation frequencies Ω in the memory kernel in the strong-coupling strong-memory regime with the transition amplitude $\nu = 3\lambda$ and memory decay rate $\kappa = 1\lambda$. The initial non-stationary parameter of the environmental noise was chosen as $a = 0.5$.

3.5. Conversion between Markovian and Non-Markovian Characters in Dephasing Dynamics

According to the above results discussed in four types of dynamics regimes, we can see that the dynamical dephasing of the quantum system and the non-Markovian character exhibited in the dephasing dynamics under the influence of the generalized RTN can be effectively modulated via the environmental memory kernel. It is worth noting that we can encounter a non-Markovian character in the dephasing dynamics by controlling the modulation frequency of the environmental memory kernel in the weak-coupling weak-memory regime. Under the influence of the environmental effect of memory modulation, the boundary of the Markovian and non-Markovian characters in the dynamical dephasing closely depends on the modulation frequency of the generalized RTN. In the following, we show the conversion from the Markovian to the non-Markovian character in the quantum dephasing dynamics of the system in the parameter space of $\nu \sim \kappa$ for different environmental modulation frequencies Ω.

Figure 5 shows the phase diagram of Markovian and non-Markovian dynamical conversion in the $\nu \sim \kappa$ space in terms of the non-Markovianity defined in Equation (4) in the presence of different environmental modulation effects. In the strong-coupling regime ($\nu > \lambda$), the dephasing dynamics of the quantum system always displays a non-Markovian character ($\mathcal{N} > 0$), whereas it undergoes a conversion from a Markovian ($\mathcal{N} = 0$) to a non-Markovian ($\mathcal{N} > 0$) character with the increase of the transition amplitude ν in the weak-coupling regime ($\nu < \lambda$). Furthermore, for a given coupling strength ν, the larger the modulation frequency Ω is, the larger the critical value of the memory decay rate κ_{th} of the conversion for the dynamical boundary is. For example, for $\nu = 0.8\lambda$, the critical values are $\kappa_{th} = 1.227\lambda$ for $\Omega = 0$, $\kappa_{th} = 3.263\lambda$ for $\Omega = 2\lambda$, and $\kappa_{th} = 5.106\lambda$ for $\Omega = 3\lambda$, respectively. That is, the non-Markovian region of dynamical dephasing increases as the modulation frequency Ω increases. It is worth mentioning that we can realize the conversion of the Markovian and non-Markovian characters in the dephasing dynamics by changing the environmental modulation frequency in the weak-coupling weak-memory regime. However, in the other three dynamics regimes, we cannot realize the conversion from the non-Markovian character ($\mathcal{N} > 0$) in the dephasing dynamics with no environmental modulation, namely $\Omega = 0$, to the Markovian character ($\mathcal{N} = 0$) in the dynamical dephasing by changing the modulation frequency in the environmental memory kernel.

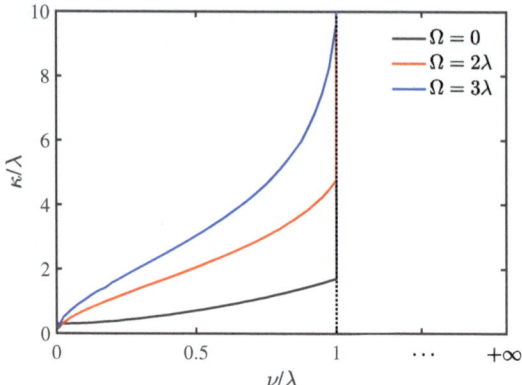

Figure 5. (Color online) Phase diagram of the conversion from the Markovian to the non-Markovian character in the dephasing dynamics of the quantum system for different environmental modulation frequencies Ω. The upper-left and lower-right regions of the curves are the Markovian and non-Markovian dynamical regions, respectively. The black dotted line stands for the dynamical boundary of the conversion induced by the standard RTN, namely the boundary between weak and strong couplings.

4. Conclusions

We performed a theoretical study of the quantum dynamical dephasing of a two-state system that interacts with a classical environment, which displays non-equilibrium fluctuations. Under the influence of the environmental fluctuations governed by a generalized RTN process with a modulatable-type memory kernel, we derived a closed time differential equation of fourth-order for the decoherence factor of the system and obtained the analytical solution of the decoherence factor exactly. For some special limit cases of the environmental memory kernel, the expression of the decoherence factor of the system can be simplified as the ones that have been derived in previous work. We analyzed the environmental effect of memory modulation in the dephasing dynamics in four types of regimes, respectively. The results showed that the dynamical dephasing of the system and the non-Markovian character exhibited in the dephasing dynamics can be effectively modulated via the environmental memory kernel. It is worth mentioning that we can encounter non-Markovian characters by changing the modulation frequency of the environmental memory kernel in the weak-coupling weak-memory regime, which have rarely been reported in previous studies. We also plotted the phase diagram to investigate the environmental influence of the memory modulation on the Markovian and non-Markovian dynamical transition in the parameter space in terms of the system–environment coupling and the memory effect of the generalized RTN. The results showed that, in the strong-coupling regime, the dynamical dephasing of the quantum system always displays a non-Markovian character, whereas in the weak-coupling regime, it suffers from a conversion from a Markovian to a non-Markovian character, for which the boundary is determined by the combined effects of the system–environment coupling, the decay rate in the environmental memory kernel, and the environmental modulation frequency of the memory kernel.

Author Contributions: Conceptualization, X.C. and A.C.; formal analysis, X.M., Y.S., Q.W. and J.R.; writing—original draft preparation, X.M. and X.C.; writing—review and editing, Y.S., X.C. and A.C. All authors have read and agreed to the published version of the manuscript.

Funding: X.C. is supported by the National Natural Science Foundation of China under Grant No. 12005121. X.M. acknowledges support from the Youth Innovation Science and Technology Support Program of Universities in Shandong Province under Grant No. 2021KJ082. J.R. is supported by the Doctoral Research Fund of Shandong Jianzhu University under Grant No. X19040Z.

Institutional Review Board Statement: Not applicable.

Informed Consent Statement: Not applicable.

Data Availability Statement: Not applicable.

Conflicts of Interest: The authors declare no conflict of interest.

Abbreviations

The following abbreviation is used in this manuscript:
RTN   Random telegraph noise

References

1. Breuer, H.P.; Petruccione, F. *The Theory of Open Quantum Systems*; Oxford University Press: New York, NY, USA, 2002.
2. Zurek, W.H. Decoherence, einselection, and the quantum origins of the classical. *Rev. Mod. Phys.* **2003**, *75*, 715. [CrossRef]
3. Schlosshauer, M. Decoherence, the measurement problem, and interpretations of quantum mechanics. *Rev. Mod. Phys.* **2005**, *76*, 1267. [CrossRef]
4. Streltsov, A.; Adesso, G.; Plenio, M.B. *Colloquium*: Quantum coherence as a resource. *Rev. Mod. Phys.* **2017**, *89*, 041003. [CrossRef]
5. Hu, M.L.; Hu, X.; Wang, J.; Peng, Y.; Zhang, Y.R.; Fan, H. Quantum coherence and geometric quantum discord. *Phys. Rep.* **2018**, *762*, 1. [CrossRef]
6. Schlosshauer, M. Quantum decoherence. *Phys. Rep.* **2019**, *831*, 1. [CrossRef]
7. Weiss, U. *Quantum Dissipative Systems*; World Scientific: Singapore, 1999.
8. Nielsen, M.A.; Chuang, I.L. *Quantum Computation and Quantum Information*; Cambridge University Press: Cambridge, UK, 2000.
9. Schlosshauer, M. *Decoherence and the Quantum-to-Classical Transition*; Springer: Berlin/Heidelberg, Germany, 2007.
10. Leggett, A.J.; Chakravarty, S.; Dorsey, A.T.; Fisher, M.P.A.; Garg, A.; Zwerger, W. Dynamics of the dissipative two-state system. *Rev. Mod. Phys.* **1987**, *59*, 1. [CrossRef]
11. Carollo, A.; Valenti, D.; Spagnolo, B. Geometry of quantum phase transitions. *Phys. Rep.* **2020**, *838*, 1. [CrossRef]
12. Gurvitz, S.A.; Fedichkin, L.; Mozyrsky, D.; Berman, G.P. Relaxation and the Zeno Effect in Qubit Measurements. *Phys. Rev. Lett.* **2003**, *91*, 066801. [CrossRef]
13. Lombardo, F.C.; Villar, P.I. Decoherence induced by a composite environment. *Phys. Rev. A* **2005**, *72*, 034103. [CrossRef]
14. Kang, L.; Zhang, Y.; Xu, X.; Tang, X. Quantum measurement of a double quantum dot coupled to two kinds of environment. *Phys. Rev. B* **2017**, *96*, 235417. [CrossRef]
15. Lan, K.; Du, Q.; Kang, L.; Tang, X.; Jiang, L.; Zhang, Y.; Cai, X. Dynamics of an open double quantum dot system via quantum measurement. *Phys. Rev. B* **2020**, *101*, 174302. [CrossRef]
16. Viotti, L.; Lombardo, F.C.; Villar, P.I. Boundary-induced effect encoded in the corrections to the geometric phase acquired by a bipartite two-level system. *Phys. Rev. A* **2020**, *101*, 032337. [CrossRef]
17. Villar, P.I.; Soba, A. Geometric phase accumulated in a driven quantum system coupled to a structured environment. *Phys. Rev. A* **2020**, *101*, 052112. [CrossRef]
18. Sedziak-Kacprowicz, K.; Czerwinski, A.; Kolenderski, P. Tomography of time-bin quantum states using time-resolved detection. *Phys. Rev. A* **2020**, *102*, 052420. [CrossRef]
19. Czerwinski, A.; Sedziak-Kacprowicz, K.; Kolenderski, P. Phase estimation of time-bin qudits by time-resolved single-photon counting. *Phys. Rev. A* **2021**, *103*, 042402. [CrossRef]
20. Viotti, L.; Lombardo, F.C.; Villar, P.I. Geometric phase in a dissipative Jaynes-Cummings model: Theoretical explanation for resonance robustness. *Phys. Rev. A* **2022**, *105*, 022218. [CrossRef]
21. Czerwinski, A.; Szlachetka, J. Efficiency of photonic state tomography affected by fiber attenuation. *Phys. Rev. A* **2022**, *105*, 062437. [CrossRef]
22. Rivas, A.; Huelga, S.F.; Plenio, M.B. Quantum non-Markovianity: Characterization, quantification and detection. *Rep. Prog. Phys.* **2014**, *77*, 094001. [CrossRef] [PubMed]
23. Breuer, H.; Laine, E.; Piilo, J.; Vacchini, B. *Colloquium*: Non-Markovian dynamics in open quantum systems. *Rev. Mod. Phys.* **2016**, *88*, 021002. [CrossRef]
24. de Vega, I.; Alonso, D. Dynamics of non-Markovian open quantum systems. *Rev. Mod. Phys.* **2017**, *89*, 015001. [CrossRef]
25. Li, L.; Hall, M.J.W.; Wiseman, H.M. Concepts of quantum non-Markovianity: A hierarchy. *Phys. Rep.* **2018**, *759*, 1. [CrossRef]
26. Diósi, L.; Gisin, N.; Strunz, W.T. Non-Markovian quantum state diffusion. *Phys. Rev. A* **1998**, *58*, 1699. [CrossRef]
27. Yu, T.; Diósi, L.; Gisin, N.; Strunz, W.T. Non-Markovian quantum-state diffusion: Perturbation approach. *Phys. Rev. A* **1999**, *60*, 91. [CrossRef]
28. Strunz, W.T.; Diósi, L.; Gisin, N. Open System Dynamics with Non-Markovian Quantum Trajectories. *Phys. Rev. Lett.* **1999**, *82*, 1801. [CrossRef]
29. Breuer, H.P.; Kappler, B.; Petruccione, F. Stochastic wave-function method for non-Markovian quantum master equations. *Phys. Rev. A* **1999**, *59*, 1633. [CrossRef]

30. Breuer, H.P.; Burgarth, D.; Petruccione, F. Non-Markovian dynamics in a spin star system: Exact solution and approximation techniques. *Phys. Rev. B* **2004**, *70*, 045323. [CrossRef]
31. Budini, A.A. Random Lindblad equations from complex environments. *Phys. Rev. E* **2005**, *72*, 056106. [CrossRef]
32. Breuer, H.P.; Gemmer, J.; Michel, M. Non-Markovian quantum dynamics: Correlated projection superoperators and Hilbert space averaging. *Phys. Rev. E* **2006**, *73*, 016139. [CrossRef] [PubMed]
33. Cai, X. Quantum Dynamics in a Fluctuating Environment. *Entropy* **2019**, *21*, 1040. [CrossRef]
34. Piilo, J.; Härkönen, K.; Maniscalco, S.; Suominen, K.A. Open system dynamics with non-Markovian quantum jumps. *Phys. Rev. A* **2009**, *79*, 062112. [CrossRef]
35. Tu, M.W.Y.; Zhang, W.M. Non-Markovian decoherence theory for a double-dot charge qubit. *Phys. Rev. B* **2008**, *78*, 235311. [CrossRef]
36. Zhang, W.M.; Lo, P.Y.; Xiong, H.N.; Tu, M.W.Y.; Nori, F. General Non-Markovian Dynamics of Open Quantum Systems. *Phys. Rev. Lett.* **2012**, *109*, 170402. [CrossRef]
37. Shabani, A.; Lidar, D.A. Completely positive post-Markovian master equation via a measurement approach. *Phys. Rev. A* **2005**, *71*, 020101. [CrossRef]
38. Vacchini, B.; Breuer, H.P. Exact master equations for the non-Markovian decay of a qubit. *Phys. Rev. A* **2010**, *81*, 042103. [CrossRef]
39. Vacchini, B. Non-Markovian master equations from piecewise dynamics. *Phys. Rev. A* **2013**, *87*, 030101. [CrossRef]
40. Lombardo, F.C.; Villar, P.I. Corrections to the Berry phase in a solid-state qubit due to low-frequency noise. *Phys. Rev. A* **2014**, *89*, 012110. [CrossRef]
41. Man, Z.; Xia, Y.; Lo Franco, R. Harnessing non-Markovian quantum memory by environmental coupling. *Phys. Rev. A* **2015**, *92*, 012315. [CrossRef]
42. Vacchini, B. Generalized Master Equations Leading to Completely Positive Dynamics. *Phys. Rev. Lett.* **2016**, *117*, 230401. [CrossRef]
43. Yan, Y.A.; Shao, J. Equivalence of stochastic formulations and master equations for open systems. *Phys. Rev. A* **2018**, *97*, 042126. [CrossRef]
44. Man, Z.X.; Xia, Y.J.; Lo Franco, R. Temperature effects on quantum non-Markovianity via collision models. *Phys. Rev. A* **2018**, *97*, 062104. [CrossRef]
45. Zhang, Q.; Man, Z.X.; Xia, Y.J. Non-Markovianity and the Landauer principle in composite thermal environments. *Phys. Rev. A* **2021**, *103*, 032201. [CrossRef]
46. Czerwinski, A. Open quantum systems integrable by partial commutativity. *Phys. Rev. A* **2020**, *102*, 062423. [CrossRef]
47. Martens, C.C. Communication: Decoherence in a nonequilibrium environment: An analytically solvable model. *J. Chem. Phys.* **2010**, *133*, 241101. [CrossRef] [PubMed]
48. Lombardo, F.C.; Villar, P.I. Nonunitary geometric phases: A qubit coupled to an environment with random noise. *Phys. Rev. A* **2013**, *87*, 032338. [CrossRef]
49. Zheng, Y.; Brown, F.L.H. Single-Molecule Photon Counting Statistics via Generalized Optical Bloch Equations. *Phys. Rev. Lett.* **2003**, *90*, 238305. [CrossRef]
50. Brokmann, X.; Hermier, J.P.; Messin, G.; Desbiolles, P.; Bouchaud, J.P.; Dahan, M. Statistical Aging and Nonergodicity in the Fluorescence of Single Nanocrystals. *Phys. Rev. Lett.* **2003**, *90*, 120601. [CrossRef]
51. Burkard, G. Non-Markovian qubit dynamics in the presence of $1/f$ noise. *Phys. Rev. B* **2009**, *79*, 125317. [CrossRef]
52. Rossi, M.A.C.; Paris, M.G.A. Non-Markovian dynamics of single- and two-qubit systems interacting with Gaussian and non-Gaussian fluctuating transverse environments. *J. Chem. Phys.* **2016**, *144*, 024113. [CrossRef]
53. Benedetti, C.; Buscemi, F.; Bordone, P.; Paris, M.G.A. Dynamics of quantum correlations in colored-noise environments. *Phys. Rev. A* **2013**, *87*, 052328. [CrossRef]
54. Benedetti, C.; Paris, M.G.A.; Maniscalco, S. Non-Markovianity of colored noisy channels. *Phys. Rev. A* **2014**, *89*, 012114. [CrossRef]
55. Benedetti, C.; Buscemi, F.; Bordone, P.; Paris, M.G.A. Effects of classical environmental noise on entanglement and quantum discord dynamics. *Int. J. Quantum Inf.* **2012**, *8*, 1241005. [CrossRef]
56. Lo Franco, R.; D'Arrigo, A.; Falci, G.; Compagno, G.; Paladino, E. Entanglement dynamics in superconducting qubits affected by local bistable impurities. *Phys. Scr.* **2012**, *T147*, 014019. [CrossRef]
57. Silveri, M.P.; Tuorila, J.A.; Thuneberg, E.V.; Paraoanu, G.S. Quantum systems under frequency modulation. *Rep. Prog. Phys.* **2017**, *80*, 056002. [CrossRef]
58. Cialdi, S.; Rossi, M.A.C.; Benedetti, C.; Vacchini, B.; Tamascelli, D.; Olivares, S.; Paris, M.G.A. All-optical quantum simulator of qubit noisy channels. *Appl. Phys. Lett.* **2017**, *110*, 081107. [CrossRef]
59. Cialdi, S.; Benedetti, C.; Tamascelli, D.; Olivares, S.; Paris, M.G.A.; Vacchini, B. Experimental investigation of the effect of classical noise on quantum non-Markovian dynamics. *Phys. Rev. A* **2019**, *100*, 052104. [CrossRef]
60. Fuliński, A. Non-Markovian noise. *Phys. Rev. E* **1994**, *50*, 2668. [CrossRef] [PubMed]
61. Cai, X.; Zheng, Y. Decoherence induced by non-Markovian noise in a nonequilibrium environment. *Phys. Rev. A* **2016**, *94*, 042110. [CrossRef]
62. Cai, X.; Zheng, Y. Non-Markovian decoherence dynamics in nonequilibrium environments. *J. Chem. Phys.* **2018**, *149*, 094107. [CrossRef] [PubMed]
63. Cai, X.; Zheng, Y. Quantum dynamical speedup in a nonequilibrium environment. *Phys. Rev. A* **2017**, *95*, 052104. [CrossRef]

64. Lin, D.; Zou, H.M.; Yang, J. Based-nonequilibrium-environment non-Markovianity, quantum Fisher information and quantum coherence. *Phys. Scr.* **2019**, *95*, 015103. [CrossRef]
65. Cai, X.; Meng, R.; Zhang, Y.; Wang, L. Geometry of quantum evolution in a nonequilibrium environment. *Europhys. Lett.* **2019**, *125*, 30007. [CrossRef]
66. Basit, A.; Ali, H.; Badshah, F.; Yang, X.F.; Ge, G.Q. Controlling sudden transition from classical to quantum decoherence via non-equilibrium environments. *New J. Phys.* **2020**, *22*, 033039. [CrossRef]
67. Basit, A.; Ali, H.; Badshah, F.; Yang, X.F.; Ge, G. Nonequilibrium effects on one-norm geometric correlations and the emergence of a pointer-state basis in the weak- and strong-coupling regimes. *Phys. Rev. A* **2021**, *104*, 042417. [CrossRef]
68. Cai, X. Quantum dephasing induced by non-Markovian random telegraph noise. *Sci. Rep.* **2020**, *10*, 88. [CrossRef]
69. Martens, C.C. Quantum dephasing of a two-state system by a nonequilibrium harmonic oscillator. *J. Chem. Phys.* **2013**, *139*, 024109. [CrossRef] [PubMed]
70. Lombardo, F.C.; Villar, P.I. Correction to the geometric phase by structured environments: The onset of non-Markovian effects. *Phys. Rev. A* **2015**, *91*, 042111. [CrossRef]
71. van Kampen, N.G. *Stochastic Process in Physics and Chemistry*; North-Holland: Amsterdam, The Netherland, 1992.
72. Breuer, H.; Laine, E.; Piilo, J. Measure for the Degree of Non-Markovian Behavior of Quantum Processes in Open Systems. *Phys. Rev. Lett.* **2009**, *103*, 210401. [CrossRef]
73. Zheng, L.; Peng, Y. Quantum decoherence of a two-level system in colored environments. *Phys. Rev. A* **2022**, *105*, 052443. [CrossRef]
74. Möttönen, M.; de Sousa, R.; Zhang, J.; Whaley, K.B. High-fidelity one-qubit operations under random telegraph noise. *Phys. Rev. A* **2006**, *73*, 022332. [CrossRef]
75. Bergli, J.; Faoro, L. Exact solution for the dynamical decoupling of a qubit with telegraph noise. *Phys. Rev. B* **2007**, *75*, 054515. [CrossRef]
76. Cywiński, L.; Lutchyn, R.M.; Nave, C.P.; Das Sarma, S. How to enhance dephasing time in superconducting qubits. *Phys. Rev. B* **2008**, *77*, 174509. [CrossRef]
77. Chen, M.; Chen, H.; Han, T.; Cai, X. Disentanglement Dynamics in Nonequilibrium Environments. *Entropy* **2022**, *24*, 1330. [CrossRef]
78. Maniscalco, S.; Petruccione, F. Non-Markovian dynamics of a qubit. *Phys. Rev. A* **2006**, *73*, 012111. [CrossRef]
79. Mazzola, L.; Laine, E.M.; Breuer, H.P.; Maniscalco, S.; Piilo, J. Phenomenological memory-kernel master equations and time-dependent Markovian processes. *Phys. Rev. A* **2010**, *81*, 062120. [CrossRef]
80. Lindenberg, K.; West, B.J. Statistical properties of quantum systems: The linear oscillator. *Phys. Rev. A* **1984**, *30*, 568. [CrossRef]
81. Kalandarov, S.A.; Kanokov, Z.; Adamian, G.G.; Antonenko, N.V.; Scheid, W. Non-Markovian dynamics of an open quantum system with nonstationary coupling. *Phys. Rev. E* **2011**, *83*, 041104. [CrossRef] [PubMed]
82. Min, W.; Luo, G.; Cherayil, B.J.; Kou, S.C.; Xie, X.S. Observation of a Power-Law Memory Kernel for Fluctuations within a Single Protein Molecule. *Phys. Rev. Lett.* **2005**, *94*, 198302. [CrossRef]

Disclaimer/Publisher's Note: The statements, opinions and data contained in all publications are solely those of the individual author(s) and contributor(s) and not of MDPI and/or the editor(s). MDPI and/or the editor(s) disclaim responsibility for any injury to people or property resulting from any ideas, methods, instructions or products referred to in the content.

Article

Entanglement Degradation in Two Interacting Qubits Coupled to Dephasing Environments

Rahma Abdelmagid [1], Khadija Alshehhi [1] and Gehad Sadiek [1,2,*]

1 Department of Applied Physics and Astronomy, University of Sharjah, Sharjah 27272, United Arab Emirates; u22104927@sharjah.ac.ae (R.A.); u17103635@sharjah.ac.ae (K.A.)
2 Department of Physics, Ain Shams University, Cairo 11566, Egypt
* Correspondence: gsadiek@sharjah.ac.ae

Abstract: One of the main obstacles toward building efficient quantum computing systems is decoherence, where the inevitable interaction between the qubits and the surrounding environment leads to a vanishing entanglement. We consider a system of two interacting asymmetric two-level atoms (qubits) in the presence of pure and correlated dephasing environments. We study the dynamics of entanglement while varying the interaction strength between the two qubits, their relative frequencies, and their coupling strength to the environment starting from different initial states of practical interest. The impact of the asymmetry of the two qubits, reflected in their different frequencies and coupling strengths to the environment, varies significantly depending on the initial state of the system and its degree of anisotropy. For an initial disentangled, or a Werner, state, as the difference between the frequencies increases, the entanglement decay rate increases, with more persistence at the higher degrees of anisotropy in the former state. However, for an initial anti-correlated Bell state, the entanglement decays more rapidly in the symmetric case compared with the asymmetric one. The difference in the coupling strengths of the two qubits to the pure (uncorrelated) dephasing environment leads to higher entanglement decay in the different initial state cases, though the rate varies depending on the degree of anisotropy and the initial state. Interestingly, the correlated dephasing environment, within a certain range, was found to enhance the entanglement dynamics starting from certain initial states, such as the disentangled, anti-correlated Bell, and Werner, whereas it exhibits a decaying effect in other cases such as the initial correlated Bell state.

Keywords: quantum decoherence; open quantum systems; quantum information

Citation: Abdelmagid, R.; Alshehhi, K.; Sadiek, G. Entanglement Degradation in Two Interacting Qubits Coupled to Dephasing Environments. *Entropy* **2023**, *25*, 1458. https://doi.org/10.3390/e25101458

Academic Editors: Paula I. Villar and Fernando C. Lombardo

Received: 12 September 2023
Revised: 4 October 2023
Accepted: 12 October 2023
Published: 17 October 2023

Copyright: © 2023 by the authors. Licensee MDPI, Basel, Switzerland. This article is an open access article distributed under the terms and conditions of the Creative Commons Attribution (CC BY) license (https://creativecommons.org/licenses/by/4.0/).

1. Introduction

Quantum information science aims to harness the unique properties of quantum systems for advanced computation, communication, and simulation [1]. However, quantum systems, such as qubits, are highly sensitive to the inevitable interactions with their environment, leading to decoherence and loss of quantum entanglement [2]. In particular, dephasing, which refers to the loss of relative phase information between quantum states, presents a significant challenge for realizing robust quantum technologies. Understanding the behavior of interacting qubits in the presence of dephasing is crucial for mitigating these effects and advancing the quantum information processing (QIP) field. The controllable coherent coupling between qubits is mandatory for enabling powerful quantum computations. However, dephasing can disrupt this coupling and cause the loss of entanglement, hindering the performance of quantum gates and introducing errors in quantum computations [3]. The impact of dephasing is influenced by factors such as the qubit–qubit interaction strength, the specific form of the interaction, and the dephasing mechanism itself. The investigation of the interplay between dephasing and interacting qubits to understand how dephasing affects entanglement dynamics, gate operations, and the overall performance of quantum systems has been in the focus of research in the

QIP field. Mitigating the effects of dephasing is a key objective in quantum information science. Researchers have explored various strategies to suppress dephasing-induced decoherence in interacting qubit systems. Dynamical decoupling techniques, such as spin echo sequences and dynamical decoupling pulses, aim to manipulate the qubits' interactions with the environment to reduce the impact of dephasing [4–6]. Another approach is the use of quantum error correction codes [7,8], which redundantly encode quantum information to protect it against errors and decoherence. Additionally, the concept of decoherence-free subspaces allows for encoding information in a subspace that is insensitive to specific forms of dephasing [9,10]. These mitigation strategies seek to enhance coherence, extend qubit lifetimes, and improve the overall reliability of quantum computations in the presence of dephasing.

Several theoretical and experimental works were devoted to studying and exploring the behavior of interacting qubits coupled to dissipative and dephasing environments, especially in systems that are promising candidates for implementing quantum computation and simulation such as superconducting circuits, trapped ions, and semiconductor-based qubits [11–19]. In particular, there has been a special interest in studying systems of two qubits coupled to dissipative and dephasing environments, where in most of these works, the two qubits are considered to be identical, while coupled to a single or two separate independent environments (such as the optical, thermal, dephasing, or dissipative) [20–31]. In a pioneering work by Yu and Eberly, it was shown that the bipartite entanglement between two originally entangled qubits, which are isolated from each other while coupled to quantum or classical noise, may vanish within a finite time, which they called entanglement sudden death (ESD) [32,33]. In a very relevant work, the entanglement between two interacting identical qubits coupled to separate dephasing environments, starting in a mixed entangled state, was found to exhibit periods of sudden death and rebirth (dark and bright periods) before vanishing completely. The time it takes to entirely vanish was found to be longer than the time needed for the entanglement sudden death in a system of two non-interacting qubits [34]. The entanglement dynamics in a system of two qubits initiated in an extended Werner-like state under the effect of a dephasing channel was studied [35,36]. It was shown that the purity of the initial state significantly affects the entanglement robustness in the noisy channel. The time evolution of a system of two qubits coupled to a classical dephasing environment starting from different initial states and driven by a Gaussian stochastic process was investigated, where it was demonstrated that the engineering of the environment has a very small effect on the sudden death of the entanglement, though it may significantly preserve the entanglement for a long time [37]. The quantum correlation between two independent qubits coupled to classical dephasing environments (singly or collectively) was studied using the local quantum uncertainty (LQU) as a measure [38]. The dynamics of LQU versus that of the entanglement, represented in terms of the concurrence, were considered. It was shown that, while the entanglement exhibits a sudden death, the LQU decays asymptotically. Very recently, it was shown how the uncorrelated pure dephasing of one component of a hybrid system can impact the dephasing rate of the transition in light–matter systems [19].

In this paper, we study a system of two interacting two-level non-identical atoms (qubits) in the presence of pure (uncorrelated) and correlated dephasing environments. We investigate how the asymmetry of the two-qubit system, attributed to their different frequencies and coupling strengths to the environment, affects the entanglement dynamics and asymptotic behavior. We show that the impact of this asymmetry varies significantly depending on the initial state of the system and the degree of anisotropy of their mutual interaction. For certain initial states, the difference in the qubits' energy gaps (frequencies) may cause higher decay rates and entanglement sudden death, while for others, it could provide an enhancing effect. Furthermore, we demonstrate how the difference in the coupling strength of the uncorrelated dephasing environment generally harms the entanglement and causes rapid decay with a rate that varies depending on the degree of anisotropy and the initial state type. Finally, we present the effect of the coupling to the correlated dephasing environment and show how it may enhance, for a short period of time, or damage the system entanglement depending on its value, the initial state, and the anisotropy of the system.

In fact, several platforms are relevant to our study, for instance, spin qubits, such as electron or nuclear spins in quantum dots, are susceptible to dephasing due to interactions with nearby nuclei, electrons, and other environmental factors. This dephasing can lead to the loss of quantum information stored in the spin states [39–42]. Furthermore, flux qubits, which are superconducting qubits, are sensitive to magnetic flux changes. They can experience dephasing due to fluctuations in the magnetic environment, leading to the loss of coherence in the qubit states [43–46]. Moreover, qubits coupled to resonant microwave cavities can experience dephasing due to fluctuations in the cavity modes. This dephasing can impact the fidelity of two-qubit gates and overall quantum circuit performance [47–50]. Besides, nuclear magnetic resonance (NMR) qubits, which are based on the manipulation of nuclear spins, may dephase due to interactions with other nuclear spins, leading to transverse relaxation, which reduces the coherence time of the qubits [51,52].

This paper is organized as follows: In the next section, we present our model and the solution. In Section 3, we discuss the entanglement dynamics and asymptotic behavior, starting from different initial states, based on our model. We conclude in Section 4.

2. The Model and Solution

We considered a system of two interacting asymmetric (non-identical) atoms (qubits), each one characterized by two levels: a ground state and an excited state labeled as $|g_i\rangle$ and $|e_i\rangle$, where $i = 1, 2$ refers to the first and second atoms, respectively. The Hamiltonian of the system is given by

$$H = \omega_1 S_1^z + \omega_2 S_2^z + J\left(\frac{(1+\gamma)}{2} S_1^x S_2^x + \frac{(1-\gamma)}{2} S_1^y S_2^y + \delta S_1^z S_2^z\right), \tag{1}$$

The first two terms in the Hamiltonian represent the asymmetry of the two non-interacting atoms with ω_1 and ω_2 accounting for the transition frequency of each atom, while the final three terms describe the atom–atom interactions. The spin operator S is defined by $S_i^z = 1/2(|g_i\rangle\langle g_i| - |e_i\rangle\langle e_i|)$ and $S_i^+ = |e_i\rangle\langle g_i| = S_i^x + iS_i^y = S_i^{-\dagger}$. Clearly, these operators are monomorphic to the spin 1/2 operators; therefore, we can describe all our system characteristics using the spin system terminology. The parameter J represents the atom–atom interaction strength, while the anisotropy parameters γ and δ specify the different types of systems that we may consider: Ising ($\gamma = 1$ and $\delta = 0$), XYZ ($\gamma = 0.5$ and $\delta = 1$), and XXX ($\gamma = 1$ and $\delta = 0.5$). Throughout this paper, we set the parameters $\hbar = J = 1$ for convenience.

We studied the time evolution of the system starting from either an initial pure or a mixed state. Starting with the entangled atoms in a pure state, the wave function of the composite system can be defined, at $t = 0$, as

$$|\psi(0)\rangle = a\,|e_1, e_2\rangle + b\,|e_1, g_2\rangle + c\,|g_1, e_2\rangle + d\,|g_1, g_2\rangle, \tag{2}$$

which is a linear combination of all possible product states of the two atoms and a, b, c, and d are arbitrary complex quantities that satisfy the normalization condition:

$$|a|^2 + |b|^2 + |c|^2 + |d|^2 = 1. \tag{3}$$

In this basis, the Hamiltonian has the form:

$$H = \frac{1}{4}\begin{pmatrix} \delta + 2(\omega_1 + \omega_2) & 0 & 0 & \gamma \\ 0 & -\delta + 2(\omega_1 - \omega_2) & 1 & 0 \\ 0 & 1 & -\delta - 2(\omega_1 - \omega_2) & 0 \\ \gamma & 0 & 0 & \delta - 2(\omega_1 + \omega_2) \end{pmatrix}, \tag{4}$$

On the other hand, the density matrix corresponding to an initial Werner mixed state is given by

$$\rho(0) = \frac{1}{3}(a\,|e_1,e_2\rangle\langle e_1,e_2| + d\,|g_1,g_2\rangle\langle g_1,g_2| + (b+c)|\psi\rangle\langle\psi|), \tag{5}$$

where the wavefunction takes the following form:

$$|\psi\rangle = \frac{1}{\sqrt{b+c}}(\sqrt{b}|e_1,g_2\rangle + e^{i\chi}\sqrt{c}|g_1,e_2\rangle) \tag{6}$$

where a, b, c, and d are the independent parameters governing the nature of the initial state of the two entangled atoms. They satisfy the relation $(a + b + c + d)/3 = 1$, and χ is the initial phase. In fact, the initial states considered in this work are of practical interest and have been already constructed before experimentally in different types of quantum systems. For instance, the Bell state has been created in different systems, such as the trapped ions in the pioneering work of Blatt and Wineland [53] and other works, in particular, that studied the Bell inequality testing in spins in nitrogen-vacancy centers, optical photons, neutral atoms, and superconducting qubits [54–58]. The Werner state was also prepared experimentally via spontaneous parametric conversion and controllable depolarization and decoherence of photons [59,60].

For an open quantum system coupled to a Markovian environment, the system dynamics is represented by the Lindblad master equation [2,61,62]:

$$\dot\rho(t) = -i[H,\rho] + \mathcal{D}_\rho. \tag{7}$$

where $\mathcal{D}_\rho$ describes the non-unitary dynamics of the system:

$$\mathcal{D}_\rho = -\frac{1}{2}\sum_{j=1}^{M}\sum_{k=1}^{N}\left\{[L_k^{(j)}\rho, L_k^{(j)\dagger}] + [L_k^{(j)}, \rho L_k^{(j)\dagger}]\right\}, \tag{8}$$

where the jth Lindblad operator $L_k^{(j)}$ represents the effect of the considered environment on the system site k. Recasting the density operator as a vector in the Liouville space [62], Equation (7) can be rewritten in a matrix form as

$$\vec{\dot\rho}(t) = (\hat{\mathcal{L}}^H + \hat{\mathcal{L}}^D)\vec\rho = \hat{\mathcal{L}}\vec\rho, \tag{9}$$

where $\hat{\mathcal{L}}^H$ and $\hat{\mathcal{L}}^D$ are superoperators acting on the vector ρ in the Liouville space, representing the unitary and dephasing (dissipative) processes, respectively. The solution of Equation (9) yields the density matrix at any time t as

$$\vec\rho(t) = \sum_i A_i\,\vec{\eta}_i\,e^{\lambda_i t}, \tag{10}$$

The coefficient A_i is determined by the system's initial conditions. $\{\lambda_i\}$ is the set of eigenvalues, and $\{\vec\eta_i\}$ is the set of eigenvectors of the tetrahedral matrix $\mathcal{L}$, which are obtained by exact numerical diagonalization. In our case, the Lindblad operator representing the dephasing environment and acting on the atom j is given by S_j^z, where in the pure dephasing case, each atom (qubit) is exposed to an independent dephasing environment, whereas in the second case, the two qubits are exposed to a common correlated dephasing environment.

As a result, $\mathcal{L}^D$ takes the form

$$\begin{aligned}\mathcal{L}^D &= -\sum_{j=1,2}\Gamma_j(S_j^z S_j^z\rho + \rho S_j^z S_j^z - 2S_j^z\rho S_j^z) \\ &\quad -2\Gamma_0(S_1^z S_2^z\rho + \rho S_1^z S_2^z - S_1^z\rho S_2^z - S_2^z\rho S_1^z).\end{aligned} \tag{11}$$

where Γ_j is the dephasing rate of the jth atom and $2\Gamma_0$ is the correlated dephasing rate. After some calculations, the Liouville operator takes the matrix form:

$$\mathcal{L} = \frac{1}{4}\begin{pmatrix} 0 & 0 & 0 & i\gamma & 0 & 0 & 0 & 0 & 0 & 0 & 0 & 0 & -i\gamma & 0 & 0 & 0 \\ 0 & \epsilon_2^- & i & 0 & 0 & 0 & 0 & 0 & 0 & 0 & 0 & 0 & 0 & -i\gamma & 0 & 0 \\ 0 & i & \epsilon_1^- & 0 & 0 & 0 & 0 & 0 & 0 & 0 & 0 & 0 & 0 & 0 & -i\gamma & 0 \\ i\gamma & 0 & 0 & \beta_1^- & 0 & 0 & 0 & 0 & 0 & 0 & 0 & 0 & 0 & 0 & 0 & -i\gamma \\ 0 & 0 & 0 & 0 & \epsilon_2^+ & 0 & 0 & i\gamma & -i & 0 & 0 & 0 & 0 & 0 & 0 & 0 \\ 0 & 0 & 0 & 0 & 0 & 0 & i & 0 & 0 & -i & 0 & 0 & 0 & 0 & 0 & 0 \\ 0 & 0 & 0 & 0 & 0 & i & \beta_1^+ & 0 & 0 & 0 & -i & 0 & 0 & 0 & 0 & 0 \\ 0 & 0 & 0 & 0 & i\gamma & 0 & 0 & \alpha_1^+ & 0 & 0 & 0 & -i & 0 & 0 & 0 & 0 \\ 0 & 0 & 0 & 0 & -i & 0 & 0 & 0 & \epsilon_1^+ & 0 & 0 & i\gamma & 0 & 0 & 0 & 0 \\ 0 & 0 & 0 & 0 & 0 & -i & 0 & 0 & 0 & \beta_2^+ & i & 0 & 0 & 0 & 0 & 0 \\ 0 & 0 & 0 & 0 & 0 & 0 & -i & 0 & 0 & i & 0 & 0 & 0 & 0 & 0 & 0 \\ 0 & 0 & 0 & 0 & 0 & 0 & 0 & -i & i\gamma & 0 & 0 & \alpha_2^+ & 0 & 0 & 0 & 0 \\ -i\gamma & 0 & 0 & 0 & 0 & 0 & 0 & 0 & 0 & 0 & 0 & 0 & \beta_2^- & 0 & 0 & i\gamma \\ 0 & -i\gamma & 0 & 0 & 0 & 0 & 0 & 0 & 0 & 0 & 0 & 0 & 0 & \alpha_1^- & i & 0 \\ 0 & 0 & -i\gamma & 0 & 0 & 0 & 0 & 0 & 0 & 0 & 0 & 0 & 0 & i & \alpha_2^- & 0 \\ 0 & 0 & 0 & -i\gamma & 0 & 0 & 0 & 0 & 0 & 0 & 0 & 0 & i\gamma & 0 & 0 & 0 \end{pmatrix}, \quad (12)$$

where

$$\begin{aligned} \epsilon_k^\pm &= -4\Gamma_k \pm 2i(\delta + 2\omega_k) \\ \alpha_k^\pm &= -4\Gamma_k \pm 2i(\delta - 2\omega_k) \\ \beta_1^\mp &= 4(\mp 2\Gamma_0 - \Gamma_1 - \Gamma_2 - i(\omega_1 \pm \omega_2)) \\ \beta_2^\pm &= 4(\pm 2\Gamma_0 - \Gamma_1 - \Gamma_2 + i(\omega_1 \mp \omega_2)) \end{aligned} \quad (13)$$

and $k = 1, 2$.

3. Dynamics of Entanglement

A comprehensive view of the system can be gained by investigating the dynamics of the bipartite entanglement that arises naturally between the two atoms and the atomic population inversion starting from different initial states of particular interest. In this section, we implemented our solution to study the dynamics of the entanglement of the system.

The entanglement can be quantified via the aid of the concurrence $C(\rho)$ as proposed by Wootters [63]. It can be calculated from

$$C(\rho) = \max\left[0, \varepsilon_1 - \varepsilon_2 - \varepsilon_3 - \varepsilon_4\right], \quad (14)$$

where ε_i in decreasing order are the square roots of the four eigenvalues of the non-Hermitian matrix:

$$R \equiv \rho\tilde{\rho}, \quad (15)$$

where $\tilde{\rho}$ is the spin flipped state defined as

$$\tilde{\rho} = (\hat{\sigma}_y \otimes \hat{\sigma}_y)\rho^*(\hat{\sigma}_y \otimes \hat{\sigma}_y), \quad (16)$$

Here, ρ^* is the complex conjugate of ρ and $\hat{\sigma}_y$ is the Pauli spin matrix in the y direction. In general, it is known that $C(\rho)$ goes from 0 for a separable disentangled state to 1 for a maximally entangled state. Since the main goal of our work is to investigate the impact of the asymmetry of the two qubits on the system dynamics and entanglement properties, the asymmetry is reflected in two aspects, the difference in the qubit transition frequencies (energy gaps) and their coupling strengths to the environment. Therefore, in our model, which is generic, we assumed one of the two frequencies, ω_1, is equal to one, while the other one is weighted in terms of the first. Consequently, we can study the different scenarios at different ratios of ω_1 and ω_2. Furthermore, since we are working in a unit system where $\hbar = 1$, all the frequencies and coupling strengths are expressed in units of ω_1 and the time t in units of ω_1^{-1}. When our model is applied to one of the relevant physical systems, we can assign a numerical value to the frequencies. For instance, in a system such as the trapped ions, the energy gap, and therefore $\hbar\omega_1$, is around 10^{14} Hz, whereas in the superconducting qubits, it is a few GHz [64].

3.1. Disentangled State

We start by considering the time evolution of the system, at different degrees of anisotropy, starting from the disentangled initial state $|\psi(0)\rangle_D = (|e_1, e_2\rangle + |e_1, g_2\rangle + |g_1, e_2\rangle + |g_1, g_2\rangle)/2$. The system is coupled to a pure dephasing environment characterized by dephasing rates $\Gamma_1 = 0.1$ and $\Gamma_2 = 0.01$. For the rest of this paper, we use these values of the dephasing rates unless otherwise stated explicitly. The dynamics of the entanglement between the two qubits for different values of ω_2, while keeping $\omega_1 = 1$, are depicted in Figure 1a,b. Figure 1a shows the dynamics for a closed system in absence of a dephasing environment. Each ω_2 value exhibits a distinct, continuous, and irregular oscillatory behavior. Notably, maximum entanglement is achieved when $\omega_2 \neq \omega_1$, with varying amplitudes among the peaks. Further examination of the entanglement evolution reveals a quasi-periodic pattern of oscillations. In cases where $\omega_2 \neq \omega_1$, both the entanglement amplitude and the shape of the oscillations are disturbed. Specifically, when $\omega_2 = 0.1$, the entanglement is more pronounced compared to the case when $\omega_2 = 5$. It is worth noting that, while the entanglement begins from zero, once it emerges, it persists and never vanishes. Meanwhile, as can be noticed from Figure 1b, in the presence of a pure dephasing environment, the entanglement ends up vanishing after a period of time that varies depending on the difference between the two frequencies. The highest peak of entanglement is again observed when $\omega_2 = \omega_1$, but here with smaller entanglement content. For $\omega_2 = 0.1$, three peaks that are intermediated by ESD appear, and finite disentanglement occurs just after the third peak. Though when $\omega_2 = 2$, the entanglement persists for a longer duration compared to the dynamics of the other frequencies, initially, the ESB is delayed and the peaks have small amplitudes that are comparable to those of $\omega_2 = 0.1$. Moreover, ESD occurs twice, with the amplitude of the final revived peak being notably smaller. Figure 1c displays the time evolution of entanglement versus the frequency of the second qubit in the completely anisotropic (Ising) system. In this state, the two atoms initially possess zero entanglements, then become entangled for a finite time before becoming disentangled again. We observe that, when the frequencies of the two qubits are close to each other, with a difference of less than one atomic transition level, the entanglement reaches a maximum value of approximately 0.55, and the entanglement persists for a longer duration. As the second atom's frequency ω_2 deviates further away from the first atom's frequency ω_1, the entanglement oscillations become shorter and experience more frequent occurrences of entanglement sudden death (ESD) and revivals, ultimately leading to faster disentanglement.

The dynamics of an Isotropic (XXX) system shows a distinct behavior from the Ising system, as shown in Figure 1d. A preliminary overview of the 3D plot shows that, as ω_2 increases, the period of the initial disentanglement between the two atoms decreases, while the maxima of entanglement become smaller. Furthermore, the plot shows that, when the frequencies of the two atoms are close to each other, the atoms maintain their disentanglement status. However, when the atoms exhibit asymmetry, particularly when the frequency of the second atom is roughly double that of the first atom, the system successfully establishes entanglement between the atoms, reaching a maximum entanglement value of 0.6, which surpasses the value attained in the Ising model. Furthermore, the entanglement in the XXX system appears to persist for a longer duration compared to the Ising system, with disentanglement occurring around t = 20, whereas in the Ising system, the longest period of entanglement lasts until t = 15.

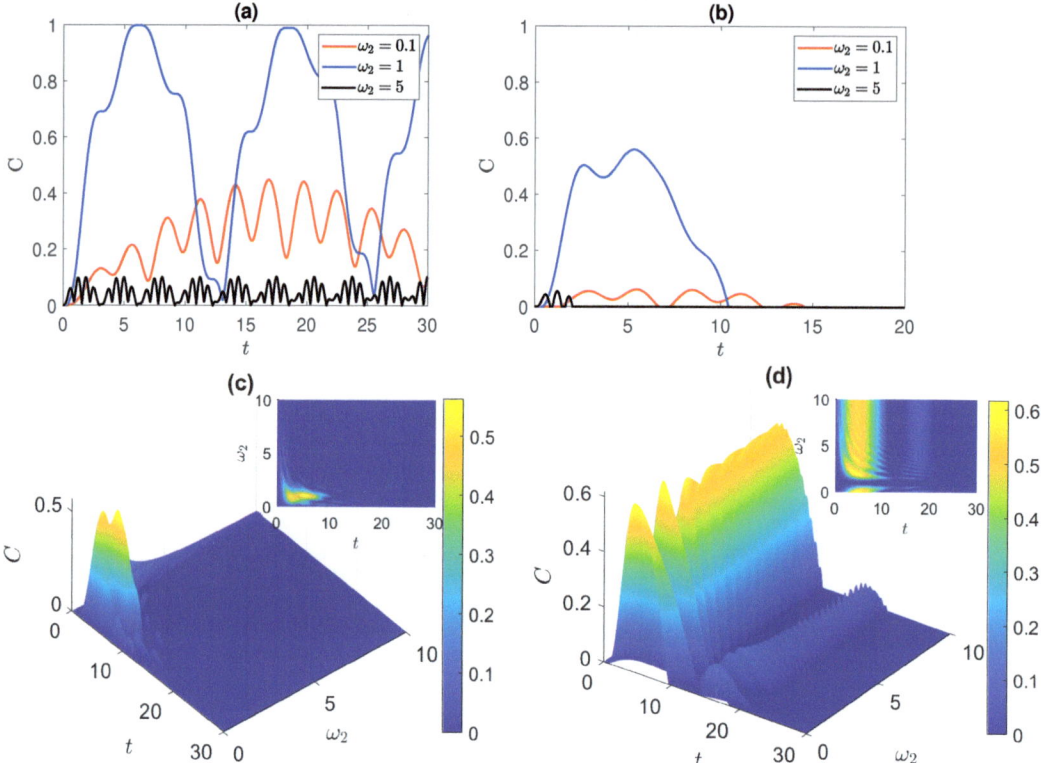

Figure 1. Dynamics of entanglement starting from the disentangled state $|\psi(0)\rangle_D = (|e_1, e_2\rangle + |e_1, g_2\rangle + |g_1, e_2\rangle + |g_1, g_2\rangle)/2$, where $\omega_1 = 1$ and varying ω_2 in the: (**a**) Ising system in the absence of environments; (**b**,**c**) Ising system in the presence of uncorrelated dephasing environment; (**d**) XXX system in the presence of uncorrelated dephasing environment. The dephasing parameters are set to $\Gamma_1 = 10\, \Gamma_2 = 0.1$.

In Figure 2, we continue our investigation into the dynamics of the system starting from the initially disentangled state. Here, we considered a system with a partial degree of anisotropy (XYZ system). In Figure 2a, which examines the entanglement evolution for $0 < \omega_2 < 10$, we observe that the initially disentangled atoms gain entanglement regardless of whether the atoms are symmetric or asymmetric. We note that, when the atoms are symmetric or close to symmetry, the entanglement of the atoms occurs once with a peak that reaches a certain height before decaying, as shown in the inset. However, when the atoms are asymmetric with the value of ω_2 above 2, the atoms experience ESD at least three times; the ESD period becomes longer after each revival, and the amplitude of the peaks decreases. It is noteworthy that, for this XYZ system, the entanglement reaches a maximum amplitude of approximately 0.7 when $\omega_2 = 3$. This maximum value is higher than what was observed in both the Ising and XXX models. Next, we study in Figure 2b the effect of varying the independent dephasing rates Γ_1 and Γ_2. Since the highest peak of entanglement was observed for $\omega_2 = 3$ at $t = 5/2$, we investigated the state of the system at that instance. When the dephasing rates are very low, the entanglement reaches a value of 0.96, indicating that the atoms are almost completely entangled. As the dephasing rates increase, the entanglement decreases. In fact, a similar effect was observed in the XXX and Ising models. It was also found that the variation of the pure dephasing rates depend on the anisotropy of the system, where in the XYZ system, increasing Γ_1 increases the decay rate slower than increasing Γ_2, while in the Ising system, it occurs the other way around. On the other hand, in the isotropic XXX system, the variation of Γ_1 and Γ_2 exhibits a symmetric dephasing rate, such that increasing either one of them increases the dephasing rate the same amount. In Figure 2c, we examine the time evolution of entanglement as a function of the coupled dephasing rate Γ_0 when $\omega_2 = 3$. At a given Γ_0, the entanglement exhibits an oscillatory behavior with a collapse revival pattern. Interestingly, the entanglement appears to be enhanced as we increase Γ_0, in particular when $0.05 < \Gamma_0 < 0.07$, where the enhancement reaches its maximum. New collapse and revival peaks are created in this range, leading to a delay in the disentanglement of the two atoms. However, increasing Γ_0 further has a detrimental effect on the entanglement. The XXX and Ising models exhibit a comparable effect as the one depicted in Figure 2a,b. However, for the variation of entanglement with the independent dephasing rates, the rate at which Γ_i accelerates the dephasing was found to be faster in the other two models compared with the XYZ one.

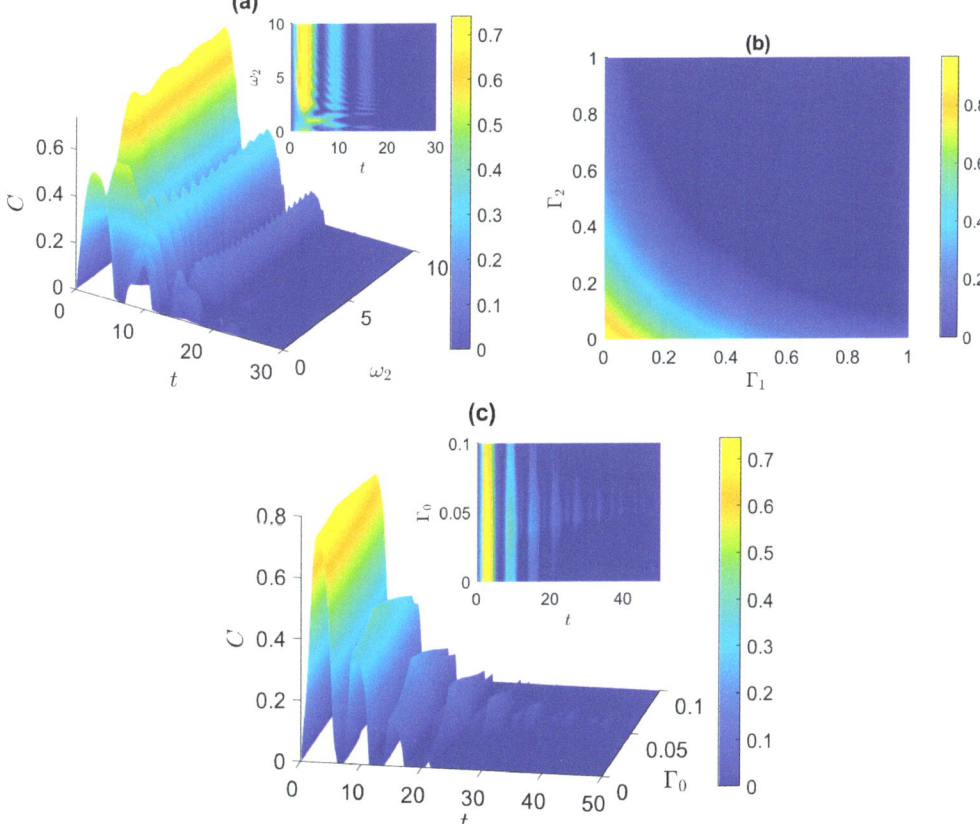

Figure 2. Starting from the disentangled state $|\psi(0)\rangle_D = (|e_1,e_2\rangle + |e_1,g_2\rangle + |g_1,e_2\rangle + |g_1,g_2\rangle)/2$, in the XYZ system: (**a**) dynamics of entanglement vs. ω_2, at $\omega_1 = 1$; (**b**) entanglement vs. Γ_1 and Γ_2, at $t = 5/2$ and $\omega_2 = 3$; (**c**) dynamics of entanglement vs. Γ_0, at $\omega_2 = 3$.

3.2. Correlated Bell State

Figure 3 shows the dynamics of entanglement starting from the maximally entangled correlated Bell state $\psi_{Bc} = (|e_1\rangle|e_2\rangle + |g_1\rangle|g_2\rangle)/\sqrt{2}$.

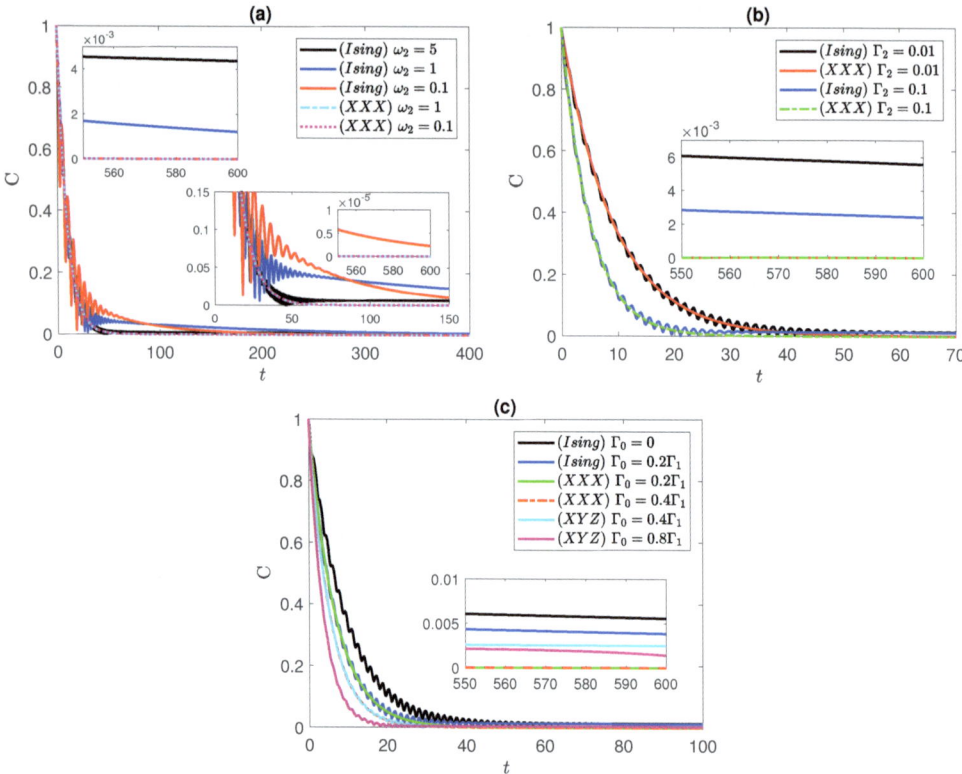

Figure 3. Dynamics of entanglement starting from the correlated Bell state, $\psi_{Bc} = (|e_1\rangle|e_2\rangle + |g_1\rangle|g_2\rangle)/\sqrt{2}$, in the Ising, XXX, and XYZ systems at different values of: (**a**) ω_2 at $\Gamma_1 = 0.1$, and $\Gamma_2 = 0.01$; (**b**) Γ_2 at $\Gamma_1 = 0.1$ and $\Gamma_0 = 0$; (**c**) Γ_0 at $\Gamma_1 = 0.1$, and $\Gamma_2 = 0.01$. $\omega_1 = 1$ in all panels.

In the Ising system case, depicted in Figure 3a, the entanglement starts at a maximum value and gradually decays until completely vanishing, where at $\omega_2 = 0.1$ and 1, we note an oscillatory behavior. A very similar effect was observed in the XYZ system, which we do not show here to avoid redundancy. In contrast, we found that the entanglement in the XXX system exhibits no oscillations and is independent of the frequency of the second atom, as depicted in the figure for $\omega_2 = 0.1$ and 1, where the curves coincide with each other. Interestingly, the impact of varying the pure dephasing rate on the entanglement dynamics was found to be similar across the Ising, XXX, and XYZ systems, irrespective of the anisotropy variation; again, we do not plot the XYZ system dynamics due to the close similarity to the Ising case. It is remarkable that the Ising system sustains its entanglement for a long period of time before vanishing, and as can be noticed in the inner insets in Figure 3a, that period of time increase with the frequency ω_2, while the entanglement decays very slowly with time. This can be an advantage for quantum information processing in such systems where the entanglement persists for a long period of time despite the dephasing effect. We show the entanglement behavior, in Figure 3b, in the Ising and XXX systems, where Γ_1 is fixed to 0.1, while Γ_2 is varying. When $\Gamma_2 = 0.01$, the dynamics of both systems follow the same trace. However, since $\omega_2 = 3$, the Ising system exhibits oscillatory behavior in its dynamics for which, at this value of ω_2, the oscillations of the Ising system remains clearly observable. Increasing Γ_2 to 0.1 results in a damping effect on the entanglement and an accelerated rate of dephasing, leading to faster disentanglement of the atoms. Nevertheless, the dynamics of both systems ultimately decay at the same rate. It is worth mentioning that the oscillations in the Ising system slightly delay the disentanglement process as the entanglement approaches low values, whereas for the XXX system with $\Gamma_2 = 0.1$, the entanglement vanishes faster compared to the Ising system with the same Γ_2 value. Again, one can notice in the insets of Figure 3b how the entanglement of the Ising system persists for a long period of time before vanishing, where the period increases as the uncorrelated dephasing strength decreases, whereas that of the XXX system vanishes much earlier at the same values of the dephasing strengths. The impact of varying the correlated dephasing rate is investigated in Figure 3c. Unlike the disentangled state, this parameter induces only a damping effect on the entanglement. The change in the dephasing rate across all systems demonstrates a nearly identical behavior as Γ_0 increases. This is illustrated by first inspecting the change in the dephasing rate for the Ising system by varying Γ_0 from 0 to $0.2\Gamma_1$. Subsequently, observing the effect on the XXX system at $\Gamma_0 = 0.2\Gamma_1$, we observe alignment between the Ising and XXX system lines. It is important to note that the XXX system does not exhibit oscillations in the entanglement dynamics as discussed in Figure 3a. Therefore, at low entanglement levels, oscillations arising from the asymmetric XY interaction lead to a delay in disentanglement. Additionally, we further explore the change by increasing Γ_0 from 0.4 to 0.8. As expected, the dynamics of the XXX and XYZ systems align with each other at $\Gamma_0 = 0.4\Gamma_1$. The behavior of the entanglement in the insets of Figure 3c shows that the entanglement in both of Ising system and XYZ system, with partial and complete anisotropy, persists for a long period of time, which is higher in the Ising system and decreases as the correlated dephasing strength decreases. On the other hand, the entanglement in the isotropic (XXX) system vanishes very early, which indicates that a stronger spin–spin coupling in one direction resists the dephasing impact efficiently.

3.3. Anti-Correlated Bell State

Another maximally entangled initial state is examined in Figure 4, namely the anti-correlated Bell state (ACBS) $|\psi(0)\rangle_{Ba} = (|e_1, g_2\rangle + |g_1, e_2\rangle)/\sqrt{2}$. Surprisingly, all three systems exhibit the same dynamics shown in Figure 4. As illustrated in Figure 4a, it is evident that in a closed system, when $\omega_2 = \omega_1$, the entanglement remains constant at a value of 1. However, when $\omega_2 \neq \omega_1$, the dynamics experience an oscillatory behavior with varying periods determined by ω_2. For example, when $\omega_2 = 0.1$, the period of the oscillations is longer compared to the case when $\omega_2 = 5$, where smaller oscillations are observed, maintaining the entanglement close to 1 over time. When introducing the pure dephasing environment in Figure 4b while varying ω_2, we observe that the entanglement decays faster for symmetric atoms (with $\omega_2 = \omega_1 = 1$) compared to the asymmetric atoms case. Remarkably, when $\omega_2 = 2$, the entanglement between the atoms is preserved for a longer duration, while increasing ω_2 further leads to accelerated decay of entanglement. Investigating the effect of the independent dephasing environments on the ACBS case shows that increasing either Γ_1 or Γ_2 results in a symmetric impact on the entanglement dynamics, similar to the findings for the disentangled and correlated Bell states. Varying Γ_0 in Figure 4c reveals an intriguing observation. Although we retrieved an evolution that is similar to the initially disentangled state case, where an increase in Γ_0 enhances the entanglement and delays the disentanglement, particularly when Γ_0 is within the range of $0.5\Gamma_1$ to $0.7\Gamma_1$, for this case, the enhancement is more pronounced compared to the previous case, with the entanglement being maintained at maximum values for an extended period of oscillations. To further explore the dynamics, we examine the combined effect of varying both Γ_2 and Γ_0 in Figure 4d.

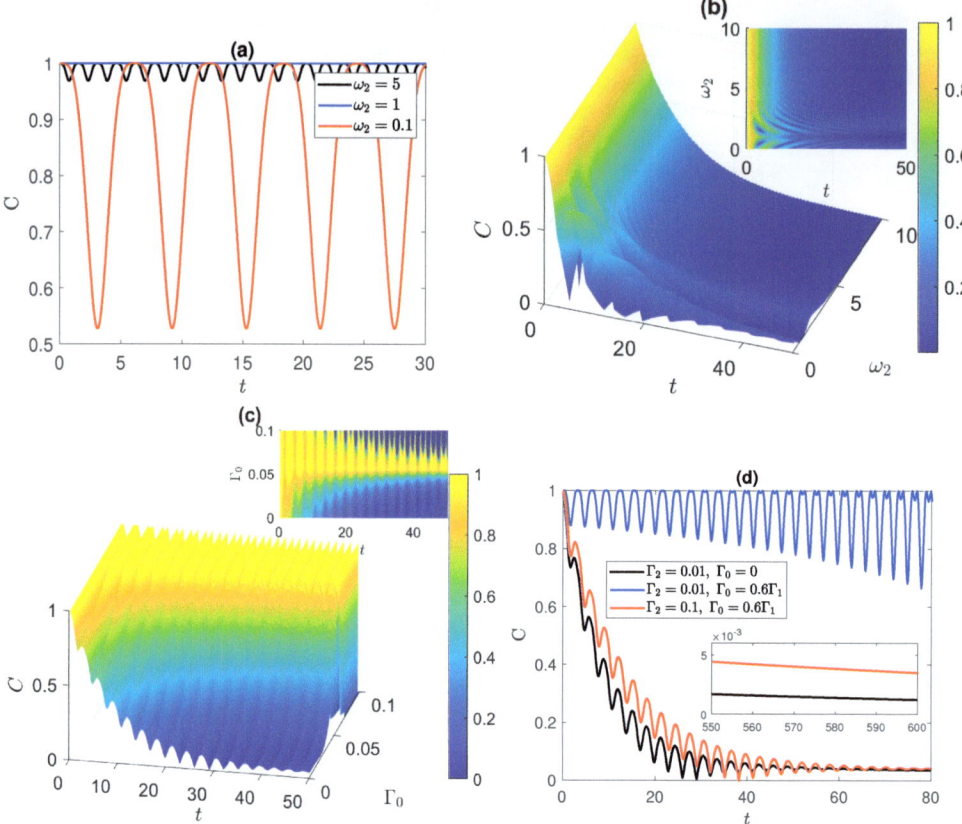

Figure 4. Dynamics of entanglement starting from the anti-correlated Bell state, $|\psi(0)\rangle_{Ba} = (|e_1, g_2\rangle + |g_1, e_2\rangle)/\sqrt{2}$, in the XXX and XYZ systems: (**a**) at different values of ω_2 in the absence of environments; (**b**) vs. ω_2 at $\Gamma_1 = 0.1$, $\Gamma_2 = 0.01$, and $\Gamma_0 = 0$; (**c**) vs. Γ_0 at $\Gamma_1 = 0.1$ and $\Gamma_2 = 0.01$; (**d**) at different values of Γ_0 and Γ_2, at $\Gamma_1 = 0.1$ and $\omega_2 = 3$. $\omega_1 = 1$ in all panels.

Increasing Γ_0 from 0 to $0.6\Gamma_1$ leads to a significant enhancement in entanglement, resulting in a prolonged period of maximum entanglement. Nevertheless, the minima of the oscillations decrease with time, evoking a less-stable entanglement that leads eventually to disentanglement. On the other hand, increasing Γ_2 from 0.01 to 0.1 portrays the dephasing effect of the independent environment. The behavior of the entanglement in the insets of Figure 4d demonstrates that the entanglement in both of the XXX and XYZ system, starting form the state $|\psi(0)\rangle_{Ba}$, persists for a long period of time against the dephasing effects, and surprisingly, the higher dephasing values $\Gamma_2 = 0.1$ and $\Gamma_0 = 0.6\Gamma_1$ lead to longer periods of time compared with $\Gamma_2 = 0.01$ and $\Gamma_0 = 0$.

3.4. W and Werner States

In Figure 5, we examine the initial partial entangled state and mixed state. Figure 5a presents a system initialized in the W-state $|\psi(0)\rangle_W = (|e_1, g_2\rangle + |g_1, e_2\rangle + |g_1, g_2\rangle)/\sqrt{3}$. We note that the behavior of the XXX system, as Γ_0 varies, resembles the behavior of the ACBS in which the entanglement oscillations exhibit multiple peaks with a maximum amplitude when $0.5\,\Gamma_1 < \Gamma_0 < 0.7\,\Gamma_1$. In the Ising and XYZ models, the peaks gradually decrease with each oscillation, and the decay rate is faster in the Ising system compared to the XYZ one. On the other hand, the variation of Γ_1 and Γ_2 exhibits a symmetric dephasing rate in the isotropic XXX system starting from the W-state, whereas in the Ising and XYZ systems, increasing Γ_1 increases the decay rate slower than in the case of increasing Γ_2. When investigating the effect of varying ω_2 versus t in the W-state, we observe that the entanglement dynamics of the XXX system share a similar two-dimensional projection with the ACBS, albeit with a faster decay due to the W-state being a partially entangled state with less initial entanglement content. The dynamics of the Ising and XYZ systems display more oscillations that experience ESD several times, leading to a faster decay, with the decay rate being higher in the Ising model than in the XYZ model.

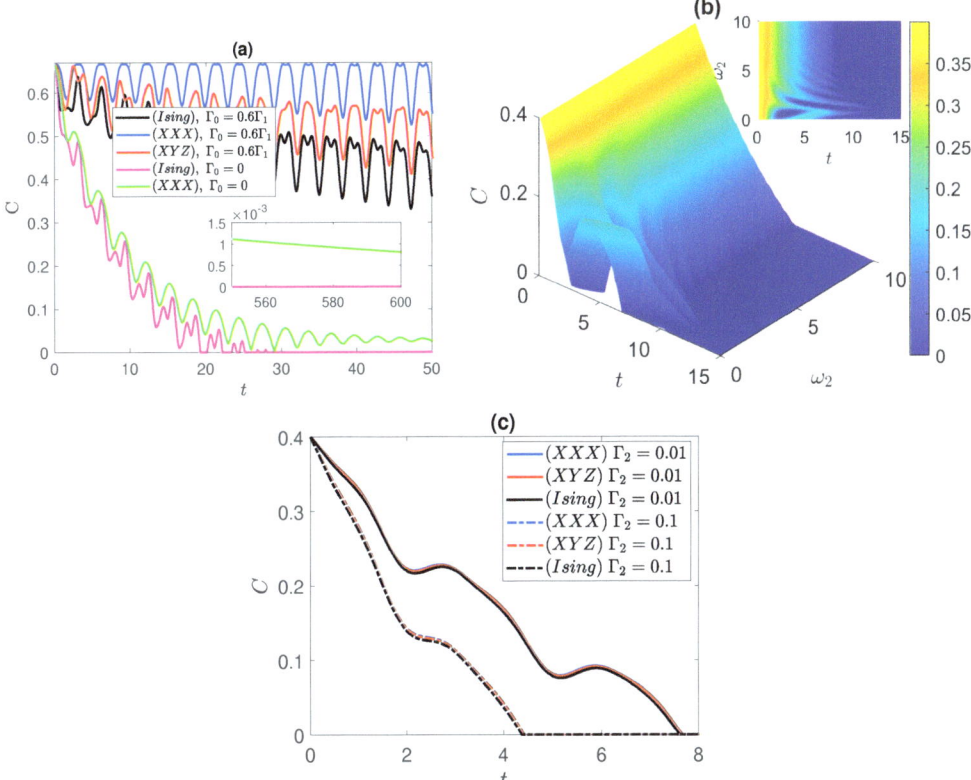

Figure 5. Dynamics of entanglement starting from (**a**) the W-state, $|\psi(0)\rangle_W = (|e_1, g_2\rangle + |g_1, e_2\rangle + |g_1, g_2\rangle)/\sqrt{3}$, in the Ising, XXX, and XYZ systems at $\omega_1 = 1, \omega_2 = 3, \Gamma_1 = 0.1, \Gamma_2 = 0.01$, and different values of Γ_0; (**b**) Werner state in the XYZ system, vs. ω_2, at $\omega_1 = 1, \Gamma_1 = 0.1, \Gamma_2 = 0.01$, and $\Gamma_0 = 0$; (**c**) Werner state in the three systems at $\omega_1 = 1, \omega_2 = 3, \Gamma_1 = 0.1, \Gamma_0 = 0$, and different values of Γ_2.

The inset of Figure 5a illustrates, starting from the state $|\psi(0)\rangle_W$, in contrast to what we have observed before, that the entanglement in the Ising system vanishes very early, while that of the XXX system persists for a long period of time before vanishing in the absence of correlated dephasing. The final state under consideration is the Werner state, with the initial parameters taken as $a = 0.2, b = 1, c = 1, d = 1 - a$, and $\chi = \pi/4$. Varying ω_2 in Figure 5b, one can notice that, when the value of ω_2 is close to ω_1, the entanglement decreases and experiences ESD, which is followed by a revival peak. This peak delays the disentanglement of the atoms as shown in the figure, while for higher values of ω_2, the disentanglement occurs earlier. For the effect of Γ_0, we obtained dynamics that follow the same pattern as the ACBS, except that the maximum of the entanglement is 0.4, not 1. This pattern applies to all the distinct systems, with fine variations in the entanglement evolution among them. In Figure 5c, we demonstrate the matching effect of the independent dephasing environment over the different systems. First, we display the dynamics at $\Gamma_2 = 0.01$ in the three systems, which yields distinct, yet highly similar curves for each system. Then, by increasing Γ_2 to 0.1, we obtain three additional curves that undergo a similar behavior of the decaying rate.

4. Conclusions

We considered a system of two interacting two-level non-identical atoms (qubits) coupled to pure (uncorrelated) and correlated dephasing environments. We studied the system dynamics starting from different initial states that vary in the degree of purity and entanglement content. We tested the impact of the asymmetry of the two-qubit system on the entanglement dynamics and asymptotic behavior. It was found that the differences in the two qubits' frequencies and coupling strengths to the uncorrelated dephasing environment vary considerably depending on the initial state and degree of anisotropy of interaction between the two qubits. Starting from certain initial states, such as the disentangled and Werner states, increasing the difference between the frequencies of the two qubits leads to higher entanglement decay rates, which are reduced as the degree of anisotropy increases in the initial disentangled state case. In contrast, starting from an anti-correlated Bell state, equal frequencies would lead to higher entanglement decay rates at different degrees of anisotropy. In general, the deviation between the coupling strengths of the two qubits to the uncorrelated dephasing environment yields a higher decay of entanglement, though its rate varies with the degree of anisotropy and the initial state type. The coupling of the two qubits to the correlated dephasing environment was found to be useful, enhancing the entanglement, within a certain range of values of the coupling strength, for specific initial states, such as the disentangled, Werner, and anti-correlated Bell states, whereas it is devastating in the case of other initial states such as the correlated Bell state.

Author Contributions: Conceptualization, R.A. and G.S; Methodology, R.A. and G.S.; Formal analysis, R.A., K.A. and G.S.; Investigation, R.A., K.A. and G.S.; Data curation, R.A. and K.A.; Writing—original draft, R.A., K.A. and G.S.; Writing—review & editing, R.A. and G.S.; Supervision, G.S. All authors have read and agreed to the published version of the manuscript.

Funding: This work was supported by the University of Sharjah, Office of Vice Chancellor of Research, Grant No. 2202440128. Furthermore, we would like to acknowledge the financial support through the award of the 4th Forum for Women in Research (QUWA): Sustaining Women's Empowerment in Research and Innovation at the University of Sharjah sponsored by the Abu Dhabi National Oil Company (ADNOC), Emirates NBD, the Sharjah Electricity Water and Gas Authority (SEWA), the Technology Innovation Institute (TII), and GSK.

Data Availability Statement: The data presented in this study is contained within the article.

Conflicts of Interest: The authors declare no conflict of interest.

References

1. Nielsen, M.; Chuang, I. *Quantum Computation and Quantum Information*; Cambridge University Press: Cambridge, UK, 2000.
2. Breuer, H.P.; Petruccione, F. *The Theory of Open Quantum Systems*; Oxford University Press: Oxford, UK, 2002.
3. Knill, E.; Laflamme, R.; Zurek, W. Resilient quantum computation: Error models and thresholds. *R. Soc. Lond. A* **1998**, *454*, 365–384. [CrossRef]
4. Paz-Silva, G.; Lidar, D. Optimally combining dynamical decoupling and quantum error correction. *Sci. Rep.* **2013**, *3*, 1530. [CrossRef]
5. Paz-Silva, G.; Lee, S.W.; Green, T.; Viola, L. Dynamical decoupling sequences for multi-qubit dephasing suppression and long-time quantum memory. *New J. Phys.* **2016**, *18*, 073020. [CrossRef]
6. Qiu, J.; Zhou, Y.; Hu, C.K.; Yuan, J.; Zhang, L.; Chu, J.; Huang, W.; Liu, W.; Luo, K.; Ni, Z.; et al. Suppressing Coherent Two-Qubit Errors via Dynamical Decoupling. *Phys. Rev. Appl.* **2021**, *16*, 054047. [CrossRef]
7. Shor, P. Scheme for reducing decoherence in quantum computer memory. *Phys. Rev. A* **1995**, *52*, R2493(R). [CrossRef]
8. Calderbank, A.; Shor, P. Good quantum error-correcting codes exist. *Phys. Rev. A* **1996**, *54*, 1098. [CrossRef]

9. Lidar, D.; Chuang, I.; Whaley, K. Decoherence-Free Subspaces for Quantum Computation. *Phys. Rev. Lett.* **1998**, *81*, 2594. [CrossRef]
10. Bacon, D.; Lidar, D.; Whaley, K. Robustness of decoherence-free subspaces for quantum computation. *Phys. Rev. A* **1999**, *60*, 1944. [CrossRef]
11. Harrow, A.; Nielsen, M. Robustness of quantum gates in the presence of noise. *Phys. Rev. A* **2003**, *68*, 012308. [CrossRef]
12. Das, S.; Agarwal, G. Bright and dark periods in the entanglement dynamics of interacting qubits in contact with the environment. *J. Phys. B At. Mol. Opt. Phys.* **2009**, *42*, 141003. [CrossRef]
13. Liu, Z.D.; Lyyra, H.; Sun, Y.N.; Liu, B.H.; Li, C.F.; Guo, G.C.; Maniscalco, S.; Piilo, J. Experimental implementation of fully controlled dephasing dynamics and synthetic spectral densities. *Nat. Commun.* **2018**, *9*, 3453. [CrossRef] [PubMed]
14. Guo, Q.; Zheng, S.B.; Wang, J.; Song, C.; Zhang, P.; Li, K.; Liu, W.; Deng, H.; Huang, K.; Zheng, D.; et al. Dephasing-Insensitive Quantum Information Storage and Processing with Superconducting Qubits. *Phys. Rev. Lett.* **2018**, *121*, 130501. [CrossRef] [PubMed]
15. Chen, Y.; Bae, Y.; Heinrich, A. Harnessing the Quantum Behavior of Spins on Surfaces. *Adv. Mater.* **2022**, *35*, 2107534. [CrossRef] [PubMed]
16. Kaplan, H.; Guo, L.; Tan, W.; De, A.; Marquardt, F.; Pagano, G.; Monroe, C. Many-Body Dephasing in a Trapped-Ion Quantum Simulator. *Phys. Rev. Lett.* **2020**, *125*, 120605. [CrossRef] [PubMed]
17. Zheng, R.H.; Ning, W.; Yang, Z.B.; Xia, Y.; Zheng, S.B. Demonstration of dynamical control of three-level open systems with a superconducting qutrit. *New J. Phys.* **2022**, *24*, 063031. [CrossRef]
18. Moskalenko, I.; Simakov, I.; Abramov, N.; Grigorev, A.; Moskalev, D.; Pishchimova, A.; Smirnov, N.; Zikiy, E.; Rodionov, I.; Besedin, I. High fidelity two-qubit gates on fluxoniums using a tunable coupler. *npj Quantum Inf.* **2022**, *8*, 130. [CrossRef]
19. Mercurio, A.; Abo, S.; Mauceri, F.; Russo, E.; Macrì, V.; Miranowicz, A.; Savasta, S.; Stefano, O. Pure Dephasing of Light-Matter Systems in the Ultrastrong and Deep-Strong Coupling Regimes. *Phys. Rev. Lett.* **2023**, *130*, 123601. [CrossRef]
20. Tahira, R.; Ikram, M.; Azim, T.; Zubairy, S. Entanglement dynamics of a pure bipartite system in dissipative environments. *J. Phys. B At. Mol. Opt. Phys.* **2008**, *41*, 205501. [CrossRef]
21. Ikram, M.; Li, F.L.; Zubairy, M. Disentanglement in a two-qubit system subjected to dissipation environments. *Phys. Rev. A* **2007**, *75*, 062336. [CrossRef]
22. Hu, M.L.; Xi, X.Q.; Lian, H.L. Thermal and phase decoherence effects on entanglement dynamics of the quantum spin systems. *Phys. B* **2009**, *404*, 3499–3506. [CrossRef]
23. Orth, P.; Roosen, D.; Hofstetter, W.; Hur, K. Dynamics, synchronization, and quantum phase transitions of two dissipative spins. *Phys. Rev. B* **2010**, *82*, 144423. [CrossRef]
24. Ghasemian, E.; Tavassoly, M.K. Generation of Werner-like states via a two-qubit system plunged in a thermal reservoir and their application in solving binary classification problems. *Sci. Rep.* **2021**, *11*, 3554. [CrossRef] [PubMed]
25. Man, Z.X.; Xia, Y.J.; An, N. Manipulating entanglement of two qubits in a common environment by means of weak measurements and quantum measurement reversals. *Phys. Rev. A* **2012**, *86*, 012325. [CrossRef]
26. Lashin, E.I.; Sadiek, G.; Abdalla, M.S.; Aldufeery, E. Two driven coupled qubits in a time varying magnetic field: Exact approximate solutions. *Appl. Math. Inf. Sci.* **2014**, *8*, 1071, 1071–1084. [CrossRef]
27. Obada, A.; Shaheen, M. Influence of Various Environments on Information and Entanglement Dynamics for Two Interacting Qubits. *J. Russ. Laser Res.* **2015**, *36*, 24–34. [CrossRef]
28. Nourmandipour, A.; Tavassoly, M. Dynamics and protecting of entanglement in two-level systems interacting with a dissipative cavity: The Gardiner–Collett approach. *J. Phys. B At. Mol. Opt. Phys.* **2015**, *48*, 165502. [CrossRef]
29. Zare, F.; Tavassoly, M. Decoherence attenuation in the Tavis-Cummings model via transition frequency modulation with dipole–dipole interaction and multi-photon transitions. *Optik* **2020**, *217*, 164841. [CrossRef]
30. Musavi, S.; Tavassoly, M.; Salimian, S. Entanglement dynamics and population inversion of a two-qubit system in two cavities coupled with optical fiber in the presence of two-photon transition. *Mod. Phys. Lett. A* **2023**, *38*, 2350026. [CrossRef]
31. Nourmandipour, A.; Tavassoly, M.; Bolorizadeh, M. Quantum Zeno and anti-Zeno effects on the entanglement dynamics of qubits dissipating into a common and non-Markovian environment. *J. Opt. Soc. Am. B* **2016**, *33*, 1723–1730. [CrossRef]
32. Yu, T.; Eberly, J. Finite-Time Disentanglement Via Spontaneous Emission. *Phys. Rev. Lett.* **2004**, *93*, 140404. [CrossRef]
33. Yu, T.; Eberly, J. Sudden death of entanglement: Classical noise effects. *Opt. Commun.* **2006**, *264*, 393–397. [CrossRef]
34. Das, S.; Agarwal, G. Decoherence effects in interacting qubits under the influence of various environments. *J. Phys. B At. Mol. Opt. Phys.* **2009**, *42*, 205502. [CrossRef]
35. Shan, C.J.; Liu, J.B.; Cheng, W.W.; Liu, T.K.; Huang, Y.X.; Li, H. Entanglement Dynamics of Two-Qubit System in Different Types of Noisy Channels. *Commun. Theor. Phys.* **2009**, *51*, 1013. [CrossRef]
36. Jiang, L.N.; Ma, J.; Yu, S.Y.; Tan, L.Y.; Ran, Q.W. Entanglement Evolution of the Extended Werner-like State under the Influence of Different Noisy Channels. *Int. J. Theor. Phys.* **2015**, *54*, 440–449. [CrossRef]
37. Rossi, M.; Benedetti, C.; Paris, M. Engineering decoherence for two-qubit systems interacting with a classical environment. *Int. J. Quantum Inf.* **2014**, *12*, 15600035. [CrossRef]
38. Yang, C.; Guo, Y.N.; Peng, H.P.; Lu, Y.B. Dynamics of local quantum uncertainty for a two-qubit system under dephasing noise. *Laser Phys.* **2020**, *30*, 015203. [CrossRef]
39. Chirolli, L.; Burkard, G. Decoherence in solid-state qubits. *Taylor Fr.* **2008**, *57*, 225–285. [CrossRef]

40. Trauzettel, B.; Borhani, M.; Trif, M.; Loss, D. Theory of spin qubits in nanostructures. *J. Phys. Soc. Jpn.* **2008**, *77*, 031012. [CrossRef]
41. Jing, J.; Wu, L.A. Decoherence and control of a qubit in spin baths: An exact master equation study. *Sci. Rep.* **2018**, *8*, 1471. [CrossRef]
42. Malkoc, O.; Stano, P.; Loss, D. Charge-Noise-Induced Dephasing in Silicon Hole-Spin Qubits. *Phys. Rev. Lett.* **2022**, *129*, 247701. [CrossRef]
43. Kakuyanagi, K.; Meno, T.; Saito, S.; Nakano, H.; Semba, K.; Takayanagi, H.; Deppe, F.; Shnirman, A. Dephasing of a Superconducting Flux Qubit. *Phys. Rev. Lett.* **2007**, *98*, 047004. [CrossRef] [PubMed]
44. Anton, S.; Müller, C.; Birenbaum, J.; O'Kelley, S.; Fefferman, A.; Golubev, D.; Hilton, G.; Cho, H.; Irwin, K.; Wellstood, F.; et al. Pure dephasing in flux qubits due to flux noise with spectral density scaling as $1/f^\alpha$. *Phys. Rev. B* **2012**, *85*, 224505. [CrossRef]
45. Spilla, S.; Hassler, F.; Splettstoesser, J. Measurement and dephasing of a flux qubit due to heat currents. *New J. Phys.* **2014**, *16*, 045020. [CrossRef]
46. Hutchings, M.D.; Hertzberg, J.B.; Liu, Y.; Bronn, N.T.; Keefe, G.A.; Brink, M.; Chow, J.M.; Plourde, B.T. Tunable Superconducting Qubits with Flux-Independent Coherence. *Phys. Rev. Appl.* **2017**, *8*, 044003. [CrossRef]
47. Filipp, S.; Goppl, M.; Fink, J.M.; Baur, M.; Bianchetti, R.; Steffen, L.; Wallraff, A. Multimode mediated qubit–qubit coupling and dark-state symmetries in circuit quantum electrodynamics. *Phys. Rev. A* **2011**, *83*, 063827. [CrossRef]
48. Yang, C.P.; Su, Q.P.; Zheng, S.B.; Nori, F. Entangling superconducting qubits in a multi-cavity system. *New J. Phys* **2016**, *18*, 013025. [CrossRef]
49. Scarlino, P.; van Woerkom, D.J.; Mendes, U.C.; Koski, J.V.; Landig, A.J.; Andersen, C.K.; Gasparinetti, S.; Reichl, C.; Wegscheider, W.; Ensslin, K.; et al. Coherent microwave-photon-mediated coupling between a semiconductor and a superconducting qubit. *Nat. Commun.* **2019**, *10*, 3011. [CrossRef]
50. Sedov, D.; Kozin, V.; Iorsh, I. Chiral Waveguide Optomechanics: First Order Quantum Phase Transitions with Z_3 Symmetry Breaking. *Phys. Rev. Lett.* **2020**, *125*, 263606. [CrossRef]
51. Fortunato, E.M.; Viola, L.; Hodges, J.; Teklemariam, G.; Cory, D.G. Implementation of universal control on a decoherence-free qubit. *New J. Phys.* **2002**, *4*, 5. [CrossRef]
52. Singh, H.; Arvind; Kavita, D. Experimentally freezing quantum discord in a dephasing environment using dynamical decoupling. *EPL* **2017**, *118*, 50001. [CrossRef]
53. Blatt, R.; Wineland, D. Entangled states of trapped atomic ions. *Nature* **2008**, *453*, 1008–1015. [CrossRef] [PubMed]
54. Hensen, B.; Bernien, H.; Dréau, A.E.; Reiserer, A.; Kalb, N.; Blok, M.S.; Ruitenberg, J.; Vermeulen, R.L.; Schouten, R.N.; Abellán, C.; et al. Loophole-free Bell inequality violation using electron spins separated by 1.3 kilometres. *Nature* **2015**, *526*, 682–686. [CrossRef] [PubMed]
55. Giustina, M.; Versteegh, M.M; Wengerowsky, S.; Handsteiner, J.; Hochrainer, A.; Phelan, K.; Steinlechner, F.; Kofler, J.; Larsson, J.K.; Abellán, C.; et al. Significant-loophole-free test of Bell's theorem with entangled photons. *Phys. Rev. Lett.* **2015**, *115*, 250401. [CrossRef]
56. Li, M.H.; Wu, C.; Zhang, Y.; Liu, W.Z.; Bai, B.; Liu, Y.; Zhang, W.; Zhao, Q.; Li, H.; Wang, Z.; et al. Test of local realism into the past without detection and locality loopholes. *Phys. Rev. Lett.* **2018**, *121*, 080404. [CrossRef] [PubMed]
57. Rosenfeld, W; Burchardt, D.; Garthoff, R.; Redeker, K.; Ortegel, N.; Rau, M.; Weinfurter, H. Event-Ready Bell Test Using Entangled Atoms Simultaneously Closing Detection and Locality Loopholes. *Phys. Rev. Lett.* **2017**, *119*, 010402. [CrossRef]
58. Storz, S.; Schär, J.; Kulikov, A.; Magnard, P.; Kurpiers, P.; Lütolf, T.; Walter, T.; Copetudo, J.; Reuer, K.; Akin, A.; et al. Loophole-free Bell inequality violation with superconducting circuits. *Nature* **2023**, *617*, 265–270. [CrossRef]
59. Zhang, Y.; Huang, Y.; Li, C.; Guo, G. Experimental preparation of the Werner state via spontaneous parametric down-conversion. *Phys. Rev. A* **2002**, *66*, 062315. [CrossRef]
60. Liu, T.; Wang, C.; Li, J.; Wang, Q. Experimental preparation of an arbitrary tunable Werner state. *EPL* **2017**, *119*, 14002. [CrossRef]
61. Lindblad, G. On the generators of quantum dynamical semigroups. *Commun. Math. Phys.* **1976**, *48*, 119–130. [CrossRef]
62. Sadiek, G.; Almalki, S. Entanglement dynamics in Heisenberg spin chains coupled to a dissipative environment at finite temperature. *Phys. Rev. A* **2016**, *94*, 012341. [CrossRef]
63. Wootters, W.K. Entanglement of Formation of an Arbitrary State of Two Qubits. *Phys. Rev. Lett.* **1998**, *80*, 2245–2248. [CrossRef]
64. Buluta, I.; Ashhab, S.; Nori, F. Natural and artificial atoms for quantum computation. *Rep. Prog. Phys.* **2011**, *74*, 104401. [CrossRef]

Disclaimer/Publisher's Note: The statements, opinions and data contained in all publications are solely those of the individual author(s) and contributor(s) and not of MDPI and/or the editor(s). MDPI and/or the editor(s) disclaim responsibility for any injury to people or property resulting from any ideas, methods, instructions or products referred to in the content.

Article

Interplay between Non-Markovianity of Noise and Dynamics in Quantum Systems

Arzu Kurt

Citation: Kurt, A. Interplay between Non-Markovianity of Noise and Dynamics in Quantum Systems. *Entropy* 2023, 25, 501. https://doi.org/10.3390/e25030501

Academic Editors: Paula I. Villar and Fernando C. Lombardo

Received: 23 February 2023
Revised: 9 March 2023
Accepted: 11 March 2023
Published: 14 March 2023

Copyright: © 2023 by the author. Licensee MDPI, Basel, Switzerland. This article is an open access article distributed under the terms and conditions of the Creative Commons Attribution (CC BY) license (https://creativecommons.org/licenses/by/4.0/).

Department of Physics, Bolu Abant İzzet Baysal University, 14030 Bolu, Türkiye; arzukurt@ibu.edu.tr

Abstract: The non-Markovianity of open quantum system dynamics is often associated with the bidirectional interchange of information between the system and its environment, and it is thought to be a resource for various quantum information tasks. We have investigated the non-Markovianity of the dynamics of a two-state system driven by continuous time random walk-type noise, which can be Markovian or non-Markovian depending on its residence time distribution parameters. Exact analytical expressions for the distinguishability as well as the trace distance and entropy-based non-Markovianity measures are obtained and used to investigate the interplay between the non-Markovianity of the noise and that of dynamics. Our results show that, in many cases, the dynamics are also non-Markovian when the noise is non-Markovian. However, it is possible for Markovian noise to cause non-Markovian dynamics and for non-Markovian noise to cause Markovian dynamics but only for certain parameter values.

Keywords: two-state system; non-Markovianity; continuous time random walk; non-Markovian noise

1. Introduction

Quantum non-Markovianity refers to the existence of memory effects in the dynamics of open quantum systems and has been the subject of many studies with the aim of defining, quantifying, and investigating various schemes to utilize it as a resource for quantum information tasks. Non-Markovianity has been discussed as a possible resource for quantum information tasks such as quantum system control [1], efficient entanglement distribution [2], perfect state transfer of mixed states [3], quantum channel capacity improvement [4], and efficiency of work extraction from the Otto cycle [5]. Miller et al. [6] carried out an optical study of the relation between non-Markovianity and the preservation of quantum coherence and correlations, which are essential resources for quantum metrology applications. Various approaches, from environmental engineering to classical driving to controlling the non-Markovianity of quantum dynamics, have been proposed, analyzed, and experimentally realized in recent years. Most non-Markovianity measures invoke a bidirectional exchange of information between the system and its environment at the root of the memory effects in the dynamics. The seeming contradiction between such an interpretation and the fact that even external classical noise could induce non-Markovian dynamics [7,8] was mostly resolved by showing that random mixing of unitary dynamics might lead to memory effects [9,10]. Representing the quantum environment of a finite-dimensional quantum system using classical stochastic fields has a long history. One of the drawbacks of such an approximation is the effective infinite temperature, which can be resolved by augmenting the master equation with extra terms to restore the correct thermal steady state. Another seemingly difficult task is to account for the lack of feedback from the system to the classical field. Despite these shortcomings, the stochastic Liouville equation (SLE) approach has produced various interesting physical models of open quantum systems [11–16].

There have been several studies on the effect of classical noise on the non-Markovianity of the quantum dynamics of two-state systems. For example, a study by Cialdi et al. investigated the relationship between different classical noises and the non-Markovianity of the dephasing dynamics of a two-level system [17]. The study found that non-Markovianity is influenced by the constituents defining the quantum renewal process, such as the time-continuous part of the dynamics, the type of jumps, and the waiting time. In addition, other studies have explored how to measure and control the transition from Markovian to non-Markovian dynamics in open quantum systems, as well as how to evaluate trace- and capacity-based non-Markovianity. It has been shown that classical environments that exhibit time-correlated random fluctuations can lead to non-Markovian quantum dynamics [18,19]. Costa-Filho et al. investigated the dynamics of a qubit that interacts with a bosonic bath and under the injection of classical stochastic colored noise [20]. The dynamic decoupling of qubits under Gaussian noise and RTN was investigated by Bergli et al. in [21,22]. Cai et al. showed that the environment being non-Markovian noise does not guarantee that the system's dynamics are non-Markovian [23]. When the coupling of the bath to its thermalizing external environment is very strong or on time scales longer than the characteristic microscopic times of the bath, we expect that even fully quantum system-bath models reduce to this case [24]. The addition of non-equilibrium classical noise to dissipative quantum dynamics can be helpful in describing the influence of non-equilibrium environmental degrees of freedom on the transport properties [25]. Goychuk and Hanggi developed a method to average the dynamics of a two-state system driven by non-Markovian discrete noises of the continuous-time random walk type (multi-state renewal processes) [26].

The transition from Markovian to non-Markovian dynamics via tuning of the system-environmental coupling in various quantum systems has been reported [27–31]. The aim of the present study is to provide an answer to the question of whether there is any connection between the non-Markovianity of classical noise and the non-Markovianity of quantum dynamics of a two-state system (TSS) driven by such a noise source. Toward that end, we study the dynamics of a TSS driven by a continuous-time random walk (CTRW)-type stochastic process which is characterized by its residence time distribution (RTD) function. We investigate the effect of biexponential and manifest non-Markovian RTDs. The first one is a simple model of classical non-Markovian noise as a linear combination of two Markovian processes and allows one to study random mixing-induced quantum non-Markovianity, while the latter one can be tuned to study a large number of noise models. We find that exact analytical expressions for the trace distance and entropic measures of non-Markovianity of the dynamics can be obtained for a restricted set of system parameters. It is well known that Markovian classical noise can lead to non-Markovian quantum dynamics. Here, we show that when the driving noise is chosen to be expressively non-Markovian, one can still observe the Markovian quantum dynamics, depending on the noise and system parameters, albeit in a very restricted set. Hence, we show that the existence of non-Markovianity in classical noise does not guarantee quantum non-Markovianity of the dynamics of a TSS driven by that noise.

The outline of this paper is as follows. In Section 2, we describe the TSS and CTRW noise process and the noise averaging procedure that leads to the exact time evolution operator in the Laplace transform domain. The analytical and numerical results of the study for the biased and unbiased TSS for Markovian, as well as the non-Markovian CTRW process, are presented and discussed in Section 3. Section 4 concludes the article with a brief summary of the main findings.

2. Model and Non-Markovianity Measures

The main aim of this section is to introduce the TSS model which will be used to study the effect of the non-Markovianity of the classical noise on the non-Markovianity of the quantum dynamics of the TSS driven by the noise and to summarize the trace distance and entropy-based quantum non-Markovianity measures.

2.1. Model

We consider a two-state system (TSS) with the Hamiltonian

$$H = \frac{1}{2}\hbar\epsilon_0 \sigma_z + \frac{1}{2}\hbar(\Delta_0 + \xi(t))\sigma_x + \frac{1}{2}(E_1 + E_2)\mathcal{I} \qquad (1)$$

where the σ_i values are the Pauli operators, $E_{1,2}$ are the energies of states $|1\rangle$ and $|2\rangle$ of the TSS, Δ_0 is the static tunneling matrix element, $\epsilon_0 = (E_2 - E_1)/\hbar$, and $\mathcal{I}$ is the identity operator. The TSS is driven by two-state non-Markovian noise with amplitudes $\xi(t) = \{\Delta_+, \Delta_-\}$ and stationary-state probabilities $p_\pm^{st} = \langle \tau_\pm \rangle / (\langle \tau_+ \rangle + \langle \tau_- \rangle)$, where $\langle \tau_\pm \rangle$ represents the average residence time of the noise in states $\Delta_\pm$. The stationary autocorrelation function of the noise is defined as $k(t) = \langle \delta\xi(t)\delta\xi(0) \rangle / \langle [\delta\xi]^2 \rangle$, where $\delta\xi(t) = \xi(t) - \langle \xi \rangle_{st}$ and can be expressed in terms of the RTDs in the Laplace space as follows [25,26]:

$$k(s) = \frac{1}{s} - \left(\frac{1}{\langle \tau_+ \rangle} + \frac{1}{\langle \tau_- \rangle}\right) \frac{1}{s^2} \frac{(1-\psi_+(s))(1-\psi_-(s))}{(1-\psi_-(s)\psi_+(s))} \tag{2}$$

where $\psi_\pm(s)$ are Laplace transforms of the residence time distribution of the noise in the Δ_- and Δ_+ states and the autocorrelation time of the noise is defined using $k(t)$ as $\tau_{corr} = \int_0^\infty |k(t)|\, dt$. If $k(t)$ is strictly positive for all t, then τ_{corr} can be obtained from $k(s)$ as $\tau_{corr} = \lim_{s \to 0} k(s)$.

The dynamics of the density matrix $\rho(t)$ of the TSS with the Hamiltonian in Equation (1) can be obtained by expressing it as $\rho(t) = [\mathcal{I} + \sum_i P_i(t)\sigma_i]/2$, where $P_i(t) = \text{Tr}[\rho(t)\sigma_i]$ is

$$\dot{P}(t) = F(t)P(t) \tag{3}$$

where $P(t) = [P_x(t), P_y(t), P_z(t)]^T$ and

$$F[\xi(t)] = \begin{pmatrix} -\epsilon_0 & 0 & 0 \\ \epsilon_0 & 0 & \xi(t) \\ 0 & \xi(t) & 0 \end{pmatrix} \tag{4}$$

The noise propagator $S_\pm(t) = \exp(F[\Delta_\pm])$ for the static values of noise $\xi = \{\Delta_-, \Delta_+\}$ is

$$S_\pm(t) = \sum_k R_\pm^{(k)} \exp\left(i\lambda_\pm^{(k)} t\right) \tag{5}$$

where $\lambda_\pm^0 = 0$, $\lambda_\pm^1 = \Omega_\pm = \sqrt{\epsilon_0^2 + \Delta_\pm^2}$, $\lambda_\pm^2 = -\Omega_\pm$, and

$$R_\pm^{(0)} = \frac{1}{\Omega_\pm^2} \begin{pmatrix} \Delta_\pm^2 & 0 & \epsilon_0 \Delta_\pm \\ 0 & 0 & 0 \\ \epsilon_0 \Delta_\pm & 0 & \epsilon_0^2 \end{pmatrix}$$

$$R_\pm^{(1)} = [R_\pm^{(2)}]^* = \frac{1}{2} \begin{pmatrix} \frac{\epsilon_0^2}{\Omega_\pm^2} & i\frac{\epsilon_0}{\Omega_\pm} & -\frac{\epsilon_0 \Delta_\pm}{\Omega_\pm^2} \\ i\frac{\epsilon_0}{\Omega_\pm} & 1 & i\frac{\Delta_\pm}{\Omega_\pm} \\ -\frac{\epsilon_0 \Delta_\pm}{\Omega_\pm^2} & -i\frac{\Delta_\pm}{\Omega_\pm} & \frac{\Delta_\pm^2}{\Omega_\pm^2} \end{pmatrix} \tag{6}$$

The problem of obtaining the stationary noise average of the propagator in Equation (5) involves both averaging over the initial stationary probabilities. It was shown by Goychuk that this can also be performed exactly in the Laplace space for non-Markovian processes [25]. The noise-averaged propagator can be expressed as follows:

$$\begin{aligned} S(s) = &\, p_+ S_+(s) + p_- S_-(s) - \left(\frac{1}{\tau_+} + \frac{1}{\tau_-}\right) \{C_+ + C_- \\ &[A_+(s)B_-(s) + A_-(s)][I - B_+(s)B_-(s)]^{-1} A_+(s) \\ &[A_-(s)B_+(s) + A_+(s)][I - B_-(s)B_+(s)]^{-1} A_-(s)\} \end{aligned} \tag{7}$$

where

$$\begin{aligned}
S_\pm(s) &= \sum_k \frac{R_\pm^{(k)}}{s - i\lambda_\pm^{(k)}} \\
A_\pm(s) &= \sum_k R_\pm^{(k)} \frac{1 - \psi_\pm\left(s - i\lambda_\pm^{(k)}\right)}{s - i\lambda_\pm^{(k)}} \\
B_\pm(s) &= \sum_k R_\pm^{(k)} \psi_\pm\left(s - i\lambda_\pm^{(k)}\right) \\
C_\pm(s) &= \sum_k R_\pm^{(k)} \frac{1 - \psi_\pm\left(s - i\lambda_\pm^{(k)}\right)}{\left(s - i\lambda_\pm^{(k)}\right)^2}
\end{aligned} \qquad (8)$$

where $\psi(s)$ is the Laplace transform of the distribution of the residence time of the noise.

2.2. Non-Markovianity Measures

Non-Markovianity of random processes has a well-established and widely accepted definition. The non-Markovianity of quantum dynamics, on the other hand, although the subject of an immense number of studies in recent years, has not reached a similar consensus. The trace distance-based measure of non-Markovianity developed in [32,33] quantifies the memory effect in the dynamics with the system's retrieval of information from its environment, which shows up as nonmonotonic behavior in the distinguishability of quantum states. Given two density operators ρ_1 and ρ_2, the trace distance (TD) between them is defined as follows [34]:

$$D(\rho_1, \rho_2) = \mathrm{Tr}\sqrt{(\rho_1 - \rho_2)^\dagger (\rho_1 - \rho_2)} \qquad (9)$$

where Tr stands for the trace operation. TD is bounded from below by $D(\rho_1, \rho_2) = 0$ for $\rho_1 = \rho_2$ and from above by $D(\rho_1, \rho_2) = 1$ if $\rho_1 \perp \rho_2$. As a measure of distinguishability between two quantum states, it can be related to the probability of distinguishing two states with a single measurement [35].

Entropy-based Jensen–Shannon divergence (JSD) between two quantum states is another distinguishability measure used to quantify non-Markovianity [36,37] and is defined as the smoothed version of relative entropy:

$$J(\rho_1, \rho_2) = H\left(\frac{\rho_1 + \rho_2}{2}\right) - \frac{1}{2}(H(\rho_1) + H(\rho_2)) \qquad (10)$$

where $H(.)$ is the von Neumann entropy $H(\rho) = -\mathrm{Tr}\rho \log \rho$. $J(\rho_1, \rho_2)$ has the same bounds as the trace distance in the same limiting cases, but it is not a distance because, contrary to TD, it does not obey the triangle inequality. $\sqrt{J(\rho_1, \rho_2)}$ is shown to be a distance measure [38] and can be used to quantify the non-Markovianity of the quantum dynamics.

The non-Markovianity quantifiers based on a state distinguishability measure $D^d(\rho_1, \rho_2)$ are defined as follows [32,33]:

$$\mathcal{N}^d = \max_{\rho_1(0), \rho_2(0)} \int_{\sigma_d(t) > 0} \sigma_d(t)\, dt \qquad (11)$$

where

$$\sigma_d(t) = \frac{d}{dt} D^d(\rho_1(t), \rho_2(t)) \qquad (12)$$

where the exponent d stands for either the trace distance distinguishability (T) or the Jensen–Shanon entropy divergence (E). Maximization in Equation (11) is carried out over all possible initial states $\rho_{1,2}(0)$. Wissmann et al. [39] showed that $\rho_1(0), \rho_2(0)$, chosen from the antipodal points of the Bloch sphere, maximizes the non-Markovianity measure based on the trace distance for two state systems [32,37]. For the problem studied, both the trace distance and Jensen–Shannon entropy divergence distinguishability measures could be expressed in terms of the population difference $P_z(t)$ and coherences $P_x(t)$ and $P_y(t)$ as follows:

$$D^T = \sqrt{P_x^2 + P_y^2 + P_z^2} \tag{13}$$

$$D^E = \frac{1}{\sqrt{\log 4}}\sqrt{2D^T \operatorname{arctanh}(D^T) + \log(1 - (D^T)^2)} \tag{14}$$

If the chosen distinguishability measure between any two initial states is a monotonic function of time, then the dynamics is said to be Markovian. Otherwise, $\mathcal{N}^d$ quantifies the memory effects in the dynamics.

3. Results and Discussion

We first present the results for TSS, whose state energies were degenerated. When $\epsilon_0 = 0$, the Laplace transformed components of the evolution operator could be expressed in a simple form:

$$S_{yy}(s) = \frac{s(2s^2 + \Delta_-^2 + \Delta_+^2)}{2(s^2 + \Delta_-^2)(s^2 + \Delta_+^2)} + \frac{\Delta^2}{\tau}[\Psi(s) + \Psi^*(s)] \tag{15}$$

$$S_{yz}(s) = -\frac{\Delta_0(s^2 + \Delta_-\Delta_+)}{(s^2 + \Delta_-^2)(s^2 + \Delta_+^2)} - i\frac{\Delta^2}{\tau}[\Psi(s) - \Psi^*(s)] \tag{16}$$

$$S_{zz}(s) = S_{yy}(s), \quad S_{zy}(s) = -S_{yz}(s) \tag{17}$$

where

$$\Psi(s) = \frac{[1 - \psi(s + i\Delta_-)][1 - \psi(s + i\Delta_+)]}{(s + i\Delta_-)^2(s + i\Delta_+)^2[1 - \psi(s + i\Delta_-)\psi(s + i\Delta_+)]} \tag{18}$$

We considered a symmetric two-state discrete noise process such that $\Delta_+ = \Delta = -\Delta_-$ was the amplitude, $\tau_+ = \tau_- = \tau$ was the mean residence time, and $\psi(s) = \psi_+(s) = \psi_-(s)$ was the residence time distribution function of the noise. Since one of the aims of the study was to investigate the relation between the non-Markovianity of the driver noise and the quantum dynamics it created, for the residence time distribution of the noise, we considered two non-Markovian models, namely the bi-exponential and manifest non-Markovian models, which have Markovian-limiting cases.

3.1. Markovian Noise

First, we considered the Markovian noise case, having an RTD $\psi(s) = 1/(1 + s\tau)$ which can be obtained with $\theta = 0, 1$ for the limit of noise with a biexponentially distributed residence time (Equation (27)) or $t_d \to 0$ for the limit of the manifest non-Markovian RTD (Equation (31)), both of which are discussed in Sections 3.2 and 3.3, respectively. For such an RTD, the inverse Laplace transform of the noise propagators in Equations (15)–(17), can be performed exactly to obtain the following:

$$P_y(t) = S(t)\sin(\Delta_0 t + \phi) \tag{19}$$

$$P_z(t) = S(t)\cos(\Delta_0 t + \phi) \tag{20}$$

where the initial values of $P_y(t)$ and $P_z(t)$ are parameterized in terms of ϕ as $P_y(0) = \sin\phi$ and $P_z(0) = \cos\phi$. $S(t)$ in Equations (19) and (20) is the stochastic evolution operator of the Markovian two-state noise:

$$S(t) = e^{-t/\tau}\left[\cosh\left(\sqrt{1 - \Delta^2\tau^2}\, t\right) + \frac{1}{\sqrt{1 - \Delta^2\tau^2}}\sinh\left(\sqrt{1 - \Delta^2\tau^2}\, t\right)\right] \tag{21}$$

The trace distance distinguishability of the dynamics can be calculated with Equation (13) by inserting the population and coherence expressions from Equations (19) and (20) as follows:

$$D(\rho_1, \rho_2) = |S(t)| \qquad (22)$$

One should note that $S(t)$ is a monotonously decreasing function of t for $\Delta \tau < 1$ but displays decaying oscillations when $\Delta \tau > 1$ as the hyperbolic trigonometric functions inside the parentheses transform to ordinary trigonometric functions when $\Delta \tau > 1$. The non-Markovianity measure (Equations (11) and (12)) is defined as the integral of the positive values of the time derivative of $\mathcal{D}$, $\mathcal{N} = 0$ for $\Delta \tau < 1$. Interestingly, the trace distance distinguishability-based non-Markovianity measure for this particular $\mathcal{D}$ and $\Delta \tau > 1$ can be obtained analytically in a simple form as follows:

$$\mathcal{N} = \frac{1}{e^{\frac{\pi}{\sqrt{\Delta^2 \tau^2 - 1}}} - 1} \qquad (23)$$

Here, the non-Markovianity is found to be independent of the static value of the coupling coefficient Δ_0. A similar expression for $\mathcal{N}$ was reported in [18] for a similar Markovian two-state noise. It is also easy to obtain an analytical expression for the Jensen–Shannon entropy divergence for the present case as follows:

$$J(t) = \frac{1}{\log 4}\left\{\log\left[1 - S^2(t)\right] + 2S(t)\operatorname{arctanh}\left[S(t)\right]\right\} \qquad (24)$$

Although it is possible to derive an exact expression for an entropy-based non-Markovianity measure by using Equations (11) and (24), the expression is not compact enough to be helpful in deciphering the relation between $\mathcal{N}^E$ and the noise parameters. Therefore, we display only the calculated entropy-based $\mathcal{N}^E$ along with the one derived from the trace distance distinguishability in Figure 1.

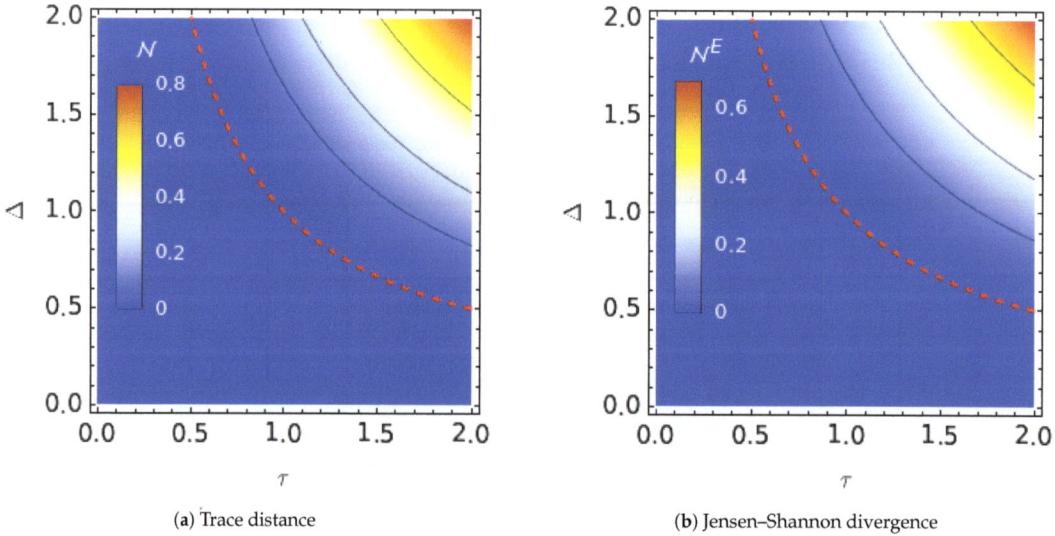

Figure 1. Non-Markovianity of the dynamics for the unbiased TSS as a function of the Markovian noise with an auto-correlation time τ and the amplitude Δ based on the trace distance (a) and Jensen–Shannon divergence distinguishability (b). The red dotted line is the zero contour, while the straight lines denote $\mathcal{N}$ equal to 0.1, 0.25, and 0.5.

The contours of non-Markovianity are plotted in Figure 1 as functions of the mean residence time τ and noise amplitude Δ. As can be seen in Equation (23) and the plot, $\mathcal{N}$ is nonzero as long as the Kubo number of the noise is greater than one, which is known as a slow noise, strong system noise coupling, or strongly colored noise regime [40] Interestingly, both measures were found to signal the same limits ($\Delta\tau > 1$) for the existence of non-Markovianity in the dynamics. Furthermore, even the magnitudes of $\mathcal{N}$ and $\mathcal{N}^E$ were found to be comparable. We observed the same behavior for all the other noise models reported in the following, and for the remainder of the paper, we will report the results only for the trace distance-based measure $\mathcal{N}$.

An interesting dynamics and non-Markovianity behavior was observed if the noise RTD was chosen to have the $\alpha \to 0$ limit of the manifest non-Markovian RTD in Equation (31), which reduced $\psi(s)$ to a form similar to that of Markovian noise with a modified mean residence time. It is easy to perform an exact analytical inverse Laplace transform of the propagator expressions in Equations (15)–(17) for $\psi(s) = 1/(1 + s\tau \tanh(1))$ and find the population difference as follows:

$$P_z(t) = \frac{1}{1+e^2}\left(2\cos(\Delta t) + \left(e^2 - 1\right)S_2(t)\right) \qquad (25)$$

where

$$S_2(t) = e^{-ct/\tau}\left(\cosh(tC/\tau) + \frac{1+e^2}{\sqrt{(1+e^2)^2 - (e^2-1)^2\Delta^2\tau^2}}\sinh(tC/\tau)\right) \qquad (26)$$

where $C = \sqrt{\coth^2 1 - \Delta^2\tau^2}$ and $c = \coth 1$. As t approaches infinity, $S_2(t)$ approaches zero, while $P_z(t)$ exhibits oscillations with an amplitude of $2/(1+e^2)$ and a frequency of Δ. The non-Markovianity of the dynamics, as assessed by both the trace distance and Jensen–Shannon entropy, was found to be unbounded. It is worth noting that the long-term limit of $P_z(t)$ was insensitive to both the noise amplitude Δ and the mean residence time τ. This result contradicts the findings obtained for Markovian noise, for which we found that $\mathcal{N}$ is zero for $\Delta\tau < 1$ and tends toward a finite value for $\Delta\tau > 1$. It should be noted that the $\alpha \to 0$ limit of a manifest non-Markovian process describes a noise with $1/\omega$ as the power spectrum [41] near $\omega = 0$, which is similar to the widely studied $1/f$ noise. Benedetti et al. studied [18] the non-Markovianity of colored $1/f^\alpha$ noise-driven quantum systems and reported finite values for $\mathcal{N}$, in contrast to our findings.

3.2. Biexponentially Distributed Residence Time

The biexponential RTD in the time domain is defined as follows [41]:

$$\psi(t) = \theta\alpha_1\exp(-\alpha_1 t) + (1-\theta)\alpha_2\exp(-\alpha_2 t) \qquad (27)$$

where θ and $(1-\theta)$ are the probabilities of the realization of the transition rates α_1 and α_2, respectively. The mean residence and autocorrelation times of this noise can be expressed as follows:

$$\langle \tau \rangle = \theta/\alpha_1 + (1-\theta)/\alpha_2 \qquad (28)$$

$$\tau_{\text{corr}} = \int_0^\infty |k(t)|\,dt \qquad (29)$$

where $\theta = 0$ and $\theta = 1$ correspond to Markovian noise with mean residence times $1/\alpha_1$ and $1/\alpha_2$, respectively. The two-state noise with biexponential residence time distribution allows one to define a non-Markovianity quantifier, denoted by C_V, which can be tailored by tuning the parameter θ. This quantifier is given by the ratio of the mean autocorrelation time of the non-Markovian noise, $\langle \tau_{\text{corr}} \rangle = \int_0^\infty k(t)dt$, to the autocorrelation time of the Markovian process $\tau_{\text{corr}}^M = \langle \tau \rangle/2$ through the mean residence time $\langle \tau \rangle$ as in Equation (30):

$$C_V^2 = \frac{2}{\langle \tau \rangle} \tau_{corr} \qquad (30)$$

The Laplace-transformed expressions for the noise propagator in Equations (15)–(17) for the biexponential RTD are amenable to be transformed back to the time domain for the unbiased TSS. However, the resulting population, coherence, and trace distance expressions are tedious to display here. On the other hand, for the manifest non-Markovian RTD, the only way to perform the inverse transformation is to use numerically exact inverse Laplace transformation (ILT) methods. We tested the CME [42], Crump [43], Durbin [44], Papoulis [45], Piessens [46], Stehfest [47], Talbot [48], and Weeks numerical ILT algorithms and found that the method based on concentrated matrix exponential (CME) distributions reported in [42] had the best performance in terms of computational cost for a given accuracy. The convergence of the computed quantities as a function of the number of included terms and the working precision was carefully checked, and 300 terms and 64 bit precision were found to be adequate for all the reported calculations to converge to 0.1%.

$\mathcal{N}$ of the TSS dynamics as a function of the noise non-Markovianity parameters C_V is shown in Figure 2a for a noise amplitude $\Delta = 1/4$ with $\Delta_0 = 0, 1$ and $\epsilon_0 = 0, 1$. Remarkably, it was observed that for the four combinations of the site energy difference ϵ_0 and the static coupling Δ_0, the non-Markovianity of the quantum dynamics displayed a broad resonance structure as a function of C_V, which indicates that increasing the non-Markovianity of the classical driving noise beyond a certain threshold would decrease the non-Markovianity of the driven quantum dynamics. Figure 2b shows the trace distance distinguishability at two chosen C_V values and indicates that the main effect of increasing C_V is to increase the dissipation rate of the dynamics. These results indicate that the increasing non-Markovian nature of the driving noise might increase, but it might also decrease the non-Markovianity of the quantum dynamics of the system studied, depending on the magnitude.

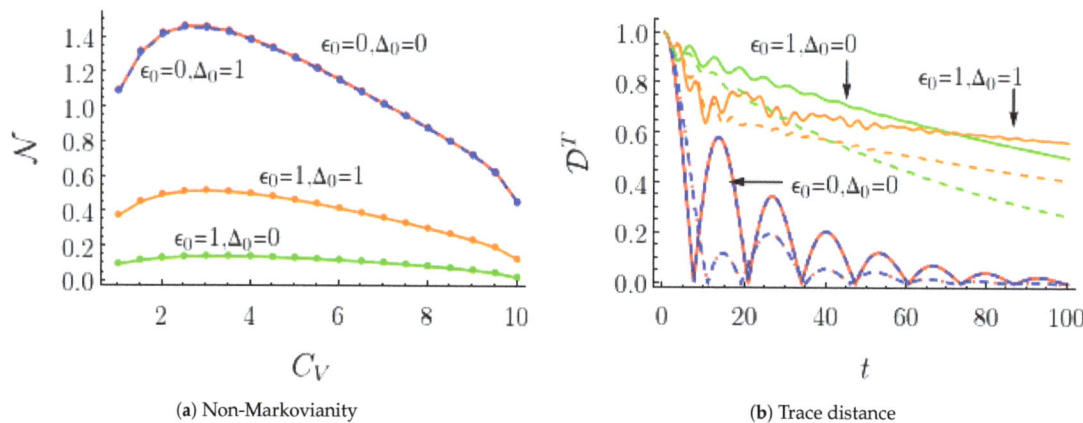

(a) Non-Markovianity   (b) Trace distance

Figure 2. Noise non-Markovianity C_V's dependence on the trace-distance based non-Markovianity measure $\mathcal{N}$ (a) and trace distance distinguishability $\mathcal{D}^T$ (b) for the two-state discrete noise with bi-exponential residence time distribution. The noise parameters were $\Delta = 1/4$, $\alpha_1 = 1/20$, and $\alpha_2 = 1$. θ values were chosen such that C_V ranged from 1 to 10. $\mathcal{N}$ and $\mathcal{D}^T$ for four combinations of TSS transition energy ϵ_0 and electronic coupling Δ_0 values are displayed. Note that for the unbiased case ($\epsilon_0 = 0$), the difference in $\mathcal{N}$ between $\Delta_0 = 0$ and $\Delta_0 = 1$ is minimal and indistinguishable on the plots. The straight (dashed) lines in $\mathcal{D}^T$ plots of (b) were calculated at $C_V = 4$ (10).

3.3. The Manifest Non-Markovian Noise

The other residence time distribution we will investigate is a manifest non-Markovian noise with the RTD defined in the Laplace space as follows [26,41]:

$$\psi(s) = \frac{1}{1 + s\tau g(s)} \tag{31}$$

with

$$g(s) = \frac{\tanh\left[(st_d)^{\alpha/2}\right]}{(st_d)^{\alpha/2}} \tag{32}$$

where τ is the mean residence time of the noise and t_d is another time constant that can be used to control the non-Markovianity of the noise. (At the limit $t_d = 0$, $\psi(t)$ is exponential). The parameter α, which is limited to the range $0 < \alpha < 1$, characterizes the noise-power distribution, where $\psi(s)$ describes noise that shows $1/\omega^{1-\alpha}$ features in its spectrum as $\omega \to 0$ and encompasses various power-law residence time distributions. $\alpha = 1$ describes normal diffusion, while the $0 < \alpha < 1$ case corresponds to subdiffusion with an index α in the transport context [41]. One of the interesting properties of discrete, manifestly non-Markovian noise is that its correlation time is infinite for $\alpha < 1$, which means that the Kubo number is effectively infinite, and no perturbative treatment would produce any reasonably accurate dynamics. The current method based on the Laplace transform is the only way to investigate the dynamics for such residence time distributions. We discussed the two limiting cases, namely $t_d \to 0$ (Markovian) and $\alpha \to 0$ (infinite $\mathcal{C}$), of the manifest non-Markovian RTD above. Here, we present and discuss how the RTD parameters α and t_d affect the trace distance distinguishability and non-Markovianity of the TSS dynamics with different system parameters.

First, we present the trace distance distinguishability along with the associated non-Markovianity $\mathcal{N}$ for the manifestly non-Markovian noise for various t_d and mean residence time τ values in Figure 3 for a biased and unbiased TSS at $\alpha = 0.5$ and $\Delta = 0.5$. As t_d is a rough measure of the non-Markovianity of manifest non-Markovian noise, one can infer, from a comparison of the insets in Figure 3a,c as well as Figure 3c,d, that $\mathcal{N}$ increases with an increasing t_d for both the unbiased and biased TSS. The mean residence time dependence of $\mathcal{N}$ was found to be independent of t_d. $\mathcal{N}$ increased with an increasing τ for all three values considered in this work for the biased as well as the unbiased TSS. Furthermore, $\mathcal{N}$ in the biased case is always found to be lower than that of the unbiased case. Another interesting observation from Figure 3b is that the trace distance distinguishability for the TSS driven by the highly non-Markovian noise tended toward a nonzero constant instead of the expected zero value.

To further delineate the relationship between $\mathcal{N}$ and the noise parameters α and t_d, we present the trace distance-based non-Markovianity measure $\mathcal{N}$ as a function of the exponent α and the t_d time parameter of the noise residence time distribution for the dynamics of the unbiased TSS in Figure 4 in two different combinations of noise amplitude and mean residence time. The mean residence time of the noise is $\tau = 1, 20$ in these graphs, and the amplitude of the noise chosen is $\Delta = 0.1, 0.5$ for the subgraphs. The most important observation from Figure 4 is that the Kubo number was the most important noise parameter that determined the magnitude of the non-Markovianity of the TSS dynamics. The larger Δ led to a larger $\mathcal{N}$ for given α and t_d values. This finding is similar to the one we discussed above for Markovian noise; the existence of non-Markovianity in that case depended on if $\Delta\tau > 1$. For the manifest non-Markovian noise, the dynamics were found to be non-Markovian even for $\Delta\tau < 1$. However, the magnitude of $\mathcal{N}$ still strongly depended on the Kubo number $K = \Delta\tau$. Figure 4 also indicates that $\mathcal{N}$ depends on t_d weakly above a threshold (around $t_d = 15$), and $\mathcal{N}$ increases smoothly with α for a constant t_d in most of the $\alpha - t_d$ plane. It should also be noted that $\mathcal{N}$ can be zero under manifest non-Markovian noise driving as $\alpha \to 1$ when $\Delta \ll 1$. This limit corresponds to white noise with a constant power spectrum at all frequencies.

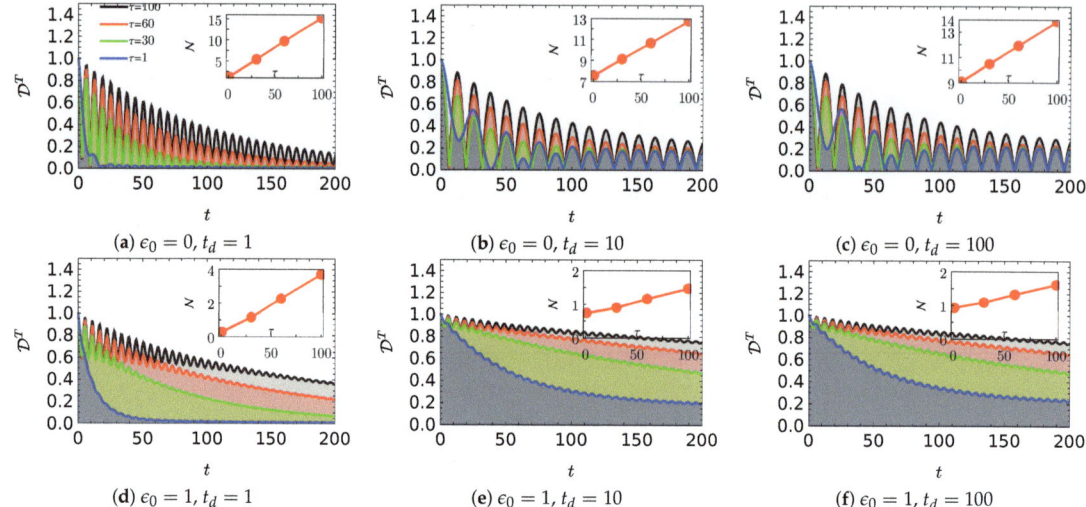

Figure 3. Trace distance as a function of time for the manifestly non-Markovian noise at different t_d parameters and average residence times τ. Insets show the trace distance-based non-Markovianity measure as a function of τ. The other parameters of the noise and the system are $\alpha = 1/2$, $\Delta_0 = 0$, and $\Delta = 1/2$.

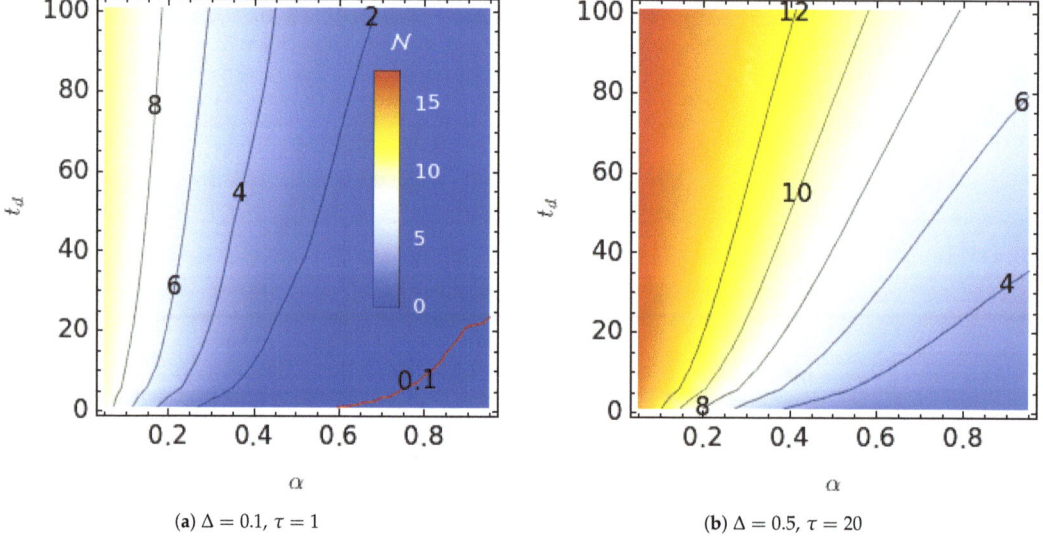

Figure 4. Dependence of α and t_d of trace distance-based non-Markovianity $\mathcal{N}$ on the dynamics of TSS driven with manifest non-Markovian two-state noise at different Kubo numbers: $K = 0.1$ (**a**) and $K = 10$ (**b**). The same color map is used for both plots, and the iso-$\mathcal{N}$ values are shown as the contour labels. The red contour line in (**a**) is the $\mathcal{N} = 0.1$ contour.

4. Conclusions

We studied Jensen–Shannon entropy divergence and trace distance-based measures of non-Markovianity of the dynamics of a two-level system under continuous-time random walk-type stochastic processes with Markovian and non-Markovian residence time distributions to delineate whether there was any connection between the Markovianity of the noise and that of the dynamics. We were able to obtain analytically exact expressions for both measures for the unbiased TSS driven by Markovian CTRW noise. This expression indicates that, above a critical Kubo number for the noise, even Markovian noise can lead to non-Markovian quantum dynamics. The numerical study of a biased TSS with the same external noise was found to be mainly a smearing of the exact boundary between the Markovian and non-Markovian boundary in the noise frequency-noise amplitude or the classical noise-TSS coupling coefficient plane. We used non-Markovian noise with a biexponential distribution as a model of the non-Markovianity produced by random mixing of Markovian dynamics and found that increasing the non-Markovianity of the noise might not lead to increased $\mathcal{N}$ values for the dynamics. We also considered a CTRW with a manifest non-Markovian residence time distribution and showed that the dynamics can be Markovian even for such noise. An interesting finding of this study was obtained at the $\alpha \to 0$ limit of manifest non-Markovian noise. The exact expression obtained for the trace distance at this limit showed that $\mathcal{N}$ was infinite at this limit. As the discussion on the proper definition and measure of the non-Markovianity of quantum dynamics has not been settled yet, the results reported in this study provide a case study for answering the question "does the non-Markovianity of the classical driver determine the non-Markovianity of the driven"?

Funding: This study was supported by the Scientific and Technological Research Council of Türkiye (TUBITAK), project no. 1002-120F011.

Institutional Review Board Statement: It is not relevant for the present study.

Data Availability Statement: Data are available from the author upon reasonable request.

Acknowledgments: The author acknowledges many useful comments from and discussions with Resul Eryiğit.

Conflicts of Interest: The author declares no conflict of interest.

References

1. Reich, D.M.; Katz, N.; Koch, C.P. Exploiting non-Markovianity for quantum control. *Sci. Rep.* **2015**, *5*, 12430. [CrossRef] [PubMed]
2. Xiang, G.-Y.; Hou, Z.-B.; Li, C.-F.; Guo, G.-C.; Breuer, H.-P.; Laine, E.-M.; Piilo, J. Entanglement distribution in optical fibers assisted by nonlocal memory effects. *EPL* **2014**, *107*, 54006. [CrossRef]
3. Laine, E.-M.; Breuer, H.-P.; Piilo, J. Nonlocal memory effects allow perfect teleportation with mixed states. *Sci. Rep.* **2014**, *4*, 4620. [CrossRef] [PubMed]
4. Bylicka, B.; Chruscinski, D.; Maniscalco, S. Non-Markovianity and reservoir memory of quantum channels: A quantum information theory perspective. *Sci. Rep.* **2014**, *4*, 5720. [CrossRef] [PubMed]
5. Thomas, G.; Siddharth, N.; Banerjee, S.; Ghosh, S. Thermodynamics of non-Markovian reservoirs and heat engines. *Phys. Rev. E* **2018**, *97*, 062108. [CrossRef]
6. Miller, M.; Wu, K.-D.; Scalici, M.; Kołodyński, J.; Xiang, G.-Y.; Li, C.-F.; Guo, G.-C.; Streltsov, A. Optimally preserving quantum correlations and coherence with eternally non-Markovian dynamics. *New J. Phys.* **2022**, *24*, 053022. [CrossRef]
7. Pernice, A.; Helm, J.; Strunz, W.T. System–environment correlations and non-Markovian dynamics. *J. Phys. B Atom. Mol. Phys.* **2012**, *45*, 154005. [CrossRef]
8. Megier, N.; Chruscinski, D.; Piilo, J.; Strunz, W. Eternal non-Markovianity: from random unitary to Markov chain realisation. *Sci. Rep.* **2017**, *7*, 6379. [CrossRef]
9. Breuer, H.P.; Amato, G.; Vacchini, B. Mixing-induced quantum non-Markovianity and information flow. *New J. Phys.* **2018**, *20*, 043007. [CrossRef]
10. Chen, X.; Zhang, N.; He, W.E.A. Global correlation and local information flows in controllable non-Markovian open quantum dynamics. *Npj Quantum Inf.* **2022**, *8*, 22. [CrossRef]
11. Haken, H.; Reineker, P. The coupled coherent and incoherent motion of excitons and its influence on the line shape of optical absorption. *Z. Phys.* **1972**, *249*, 253–268. [CrossRef]
12. Haken, H.; Strobl, G. An exactly solvable model for coherent and incoherent exciton motion. *Z. Phys.* **1973**, *262*, 135. [CrossRef]
13. Fox, R.F. Gaussian stochastic processes in physics. *Phys. Rep.* **1978**, *48*, 181. [CrossRef]
14. Kayanuma, Y. Stochastic theory for nonadiabatic level crossing with fluctuating off-diagonal coupling. *J. Phys. Soc. Jpn.* **1985**, *54*, 2047. [CrossRef]
15. Dong, Q.; Torres-Arenas, A.J.; Sun, G.H.; Dong, S.H. Tetrapartite entanglement features of W-Class state in uniform acceleration. *Front. Phys.* **2020**, *15*, 11602. [CrossRef]
16. Shao, J.; Zerbe, C.; Hänggi, P. Suppression of quantum coherence: Noise effect. *Chem. Phys.* **1998**, *235*, 81. [CrossRef]

17. Cialdi, S.; Benedetti, C.; Tamascelli, D.; Olivares, S.; Paris, M.G.A.; Vacchini, B. Experimental investigation of the effect of classical noise on quantum non-Markovian dynamics. *Phys. Rev. A* **2019**, *100*, 052104. [CrossRef]
18. Benedetti, C.; Paris, M.G.A.; Maniscalco, S. Non-markovianity of colored noisy channels. *Phys. Rev. A* **2014**, *89*, 012114. [CrossRef]
19. Benedetti, C.; Buscemi, F.; Bordone, P.; Paris, M.G.A. Non-markovian continuous-time quantum walks on lattices with dynamical noise. *Phys. Rev. A* **2016**, *93*, 042313. [CrossRef]
20. Costa-Filho, J.I.; Lima, R.B.B.; Paiva, R.R.; Soares, P.M.; Morgado, W.A.M.; Franco, R.L.; Soares-Pinto, D.O. Enabling quantum non-Markovian dynamics by injection of classical colored noise. *Phys. Rev. A* **2017**, *95*, 052126. [CrossRef]
21. Bergli, J.; Faoro, L. Exact solution for the dynamical decoupling of a qubit with telegraph noise. *Phys. Rev. B* **2007**, *75*, 054515. [CrossRef]
22. Cywiński, L.; Lutchyn, R.M.; Nave, C.P.; Das Sarma, S. How to enhance dephasing time in superconducting qubits. *Phys. Rev. B* **2008**, *77*, 174509. [CrossRef]
23. Cai, X.; Zheng, Y. Decoherence induced by non-Markovian noise in a nonequilibrium environment. *Phys. Rev. A* **2016**, *94*, 042110. [CrossRef]
24. Cheng, B.; Wang, Q.-H.; Joynt, R. Transfer matrix solution of a model of qubit decoherence due to telegraph noise. *Phys. Rev. A* **2008**, *78*, 022313. [CrossRef]
25. Goychuk, I. Quantum dynamics with non-Markovian fluctuating parameters. *Phys. Rev. E* **2004**, *70*, 016109. [CrossRef]
26. Goychuk, I.; Hänggi, P. Quantum two-state dynamics driven by stationary non-Markovian discrete noise: Exact results. *Chem. Phys.* **2006**, *324*, 160–171. [CrossRef]
27. Liu, B.-H.; Li, L.; Huang, Y.-F.; Li, C.-F.; Guo, G.-C.; Laine, E.-M.; Breuer, H.-P.; Piilo, J. Experimental control of the transition from Markovian to non-Markovian dynamics of open quantum systems. *Nat. Phys.* **2011**, *7*, 931–934. [CrossRef]
28. Bernardes, N.; Carvalho, A.; Monken, C.; Santos, M.F. Environmental correlations and markovian to non-markovian transitions in collisional models. *Phys. Rev. A* **2014**, *90*, 032111. [CrossRef]
29. Brito, F.; Werlang, T. A knob for Markovianity. *New J. Phys.* **2015**, *17*, 072001. [CrossRef]
30. Garrido, N.; Gorin, T.; Pineda, C. Transition from non-Markovian to Markovian dynamics for generic environments. *Phys. Rev. A* **2016**, *93*, 012113. [CrossRef]
31. Chakraborty, S.; Mallick, A.; Mandal, D.; Goyal, S.K.; Ghosh, S. Non-Markovianity of qubit evolution under the action of spin environment. *Sci. Rep.* **2019**, *9*, 2987. [CrossRef]
32. Breuer, H.-P.; Laine, E.-M.; Piilo, J. Measure for the degree of non-Markovian behavior of quantum processes in open systems. *Phys. Rev. Lett.* **2009**, *103*, 210401. [CrossRef]
33. Breuer, H.-P.; Laine, E.-M.; Piilo, J.; Vacchini, B. Colloquium: Non-Markovian dynamics in open quantum systems. *Rev. Mod. Phys.* **2016**, *88*, 021002. [CrossRef]
34. Heinosaari, T.; Ziman, M. *The Mathematical Language of Quantum Theory: From Uncertainty to Entanglement*, 1st ed.; Cambridge University Press: Cambridge, UK, 2011; pp. 159–169.
35. Fuchs, C.A.; Van de Graaf, J. Cryptographic distinguishability measures for quantum-mechanical states. *IEEE Trans. Inf. Theory* **1999**, *45*, 1216. [CrossRef]
36. Majtey, A.P.; Lamberti, P.W.; Prato, D.P. Jensen-Shannon divergence as a measure of distinguishability between mixed quantum states. *Phys. Rev. A* **2005**, *72*, 052310. [CrossRef]
37. Settimo, F.; Breuer, H.-P.; Vacchini, B. Entropic and trace-distance-based measures of non-Markovianity. *Phys. Rev. A* **2022**, *106*, 042212. [CrossRef]
38. Virosztek, D. The metric property of the quantum Jensen-Shannon divergence. *Adv. Math.* **2021**, *380*, 107595. [CrossRef]
39. Wissmann, S.; Karlsson, A.; Laine, E.-M.; Piilo, J.; Breuer, H.-P. Optimal state pairs for non-Markovian quantum dynamics. *Phys. Rev. A* **2012**, *86*, 062108. [CrossRef]
40. Zhou, D.; Lang, A.; Joynt, R. Disentanglement and decoherence from classical non-Markovian noise: random telegraph noise. *Quantum Inf. Process* **2010**, *9*, 727–747. [CrossRef]
41. Goychuk, I.; Hänggi, P. Theory of non-Markovian stochastic resonance. *Phys. Rev. E* **2004**, *70*, 021104. [CrossRef] [PubMed]
42. Horvath, I.; Horvath, G.; Alamosa, S.A.D.; Telek, M. Numerical inverse Laplace transformation using concentrated matrix exponential distributions. *Perform. Eval.* **2019**, *137*, 102067. [CrossRef]
43. Crump, K.S. Numerical inversion of Laplace transforms using a Fourier series approximation. *J. Assoc. Comput. Mach.* **1976**, *23*, 89–96. [CrossRef]
44. Durbin, F. Numerical inversion of Laplace transforms: An effective improvement of Dubner and Abate's method. *Comput. J.* **1973**, *17*, 371–376. [CrossRef]
45. Papoulis, A. A new method of inversion of the Laplace transform. *PIB* **1957**, *XIV*, 405–414. [CrossRef]
46. Piessens, R. A bibliography on numerical inversion of the Laplace transform and applications. *J. Camp. Appl. Math.* **1975**, *1*, 115–126. [CrossRef]
47. Stehfest, H. Algorithm 368: Numerical inversion of Laplace transforms d[5]. *Commun. ACM* **1970**, *13*, 47–49. [CrossRef]
48. Talbot, A. The accurate numerical inversion of Laplace transforms. *IMA J. Appl. Math.* **1970**, *23*, 97–120. [CrossRef]

Disclaimer/Publisher's Note: The statements, opinions and data contained in all publications are solely those of the individual author(s) and contributor(s) and not of MDPI and/or the editor(s). MDPI and/or the editor(s) disclaim responsibility for any injury to people or property resulting from any ideas, methods, instructions or products referred to in the content.

Article

Emulating Non-Hermitian Dynamics in a Finite Non-Dissipative Quantum System

Eloi Flament [1], François Impens [2] and David Guéry-Odelin [1,*]

[1] Laboratoire Collisions, Agrégats, Réactivité, FeRMI, Université de Toulouse, CNRS, UPS, 118 Route de Narbonne, 31062 Toulouse, France; eloi.flament@univ-tlse3.fr

[2] Instituto de Física, Universidade Federal do Rio de Janeiro, Rio de Janeiro 21941-972, RJ, Brazil; impens@if.ufrj.br

* Correspondence: dgo@irsamc.ups-tlse.fr

Abstract: We discuss the emulation of non-Hermitian dynamics during a given time window using a low-dimensional quantum system coupled to a finite set of equidistant discrete states acting as an effective continuum. We first emulate the decay of an unstable state and map the quasi-continuum parameters, enabling the precise approximation of non-Hermitian dynamics. The limitations of this model, including in particular short- and long-time deviations, are extensively discussed. We then consider a driven two-level system and establish criteria for non-Hermitian dynamics emulation with a finite quasi-continuum. We quantitatively analyze the signatures of the finiteness of the effective continuum, addressing the possible emergence of non-Markovian behavior during the time interval considered. Finally, we investigate the emulation of dissipative dynamics using a finite quasi-continuum with a tailored density of states. We show through the example of a two-level system that such a continuum can reproduce non-Hermitian dynamics more efficiently than the usual equidistant quasi-continuum model.

Keywords: open quantum system; quantum simulators; non-Hermitian systems; non-Markovian dynamics

Citation: Flament, E.; Impens, F.; Guéry-Odelin, D. Emulating Non-Hermitian Dynamics in a Finite Non-Dissipative Quantum System. *Entropy* **2023**, 25, 1256. https://doi.org/10.3390/e25091256

Academic Editors: Fernando C. Lombardo and Paula I. Villar

Received: 25 July 2023
Revised: 18 August 2023
Accepted: 21 August 2023
Published: 24 August 2023

Copyright: © 2023 by the authors. Licensee MDPI, Basel, Switzerland. This article is an open access article distributed under the terms and conditions of the Creative Commons Attribution (CC BY) license (https://creativecommons.org/licenses/by/4.0/).

1. Introduction

The decay of unstable states occurs in a wide range of areas of quantum mechanics, including atomic physics, with the limited lifetime of excited electronic states in atoms; condensed matter with various relaxation processes in quantum dot electronic states; in polaron and exciton physics; nuclear physics, with the exponential decay law in radioactivity; and high-energy physics, with the short lifetime of particles such as the Higgs boson. The basic phenomenon underlying these decays is fundamentally the same. It is the irreversible transition from an initial unstable state to a continuum of final states. Such a decay can be derived from first principles. Within the perturbative limit, this problem often offers a first introduction to open quantum systems with Fermi's golden rule. Besides the perturbative limit, the complete resolution of the model reveals three different successive regimes characterized by different decay laws [1–3]: with very short time [4], the decay is quadratic, it is subsequently governed by an exponential law at intermediate time, and eventually exhibits a power law tail at long time scales [5,6]. In general, these studies reveal that a decay can be sensitive to the structure of the environment.

Quantum simulations have become a very important research topic, with various fundamental and technological applications [7]. As any realistic quantum process involves a finite amount of dissipation, a quantum decay emulator appears as an interesting building block for such systems. The simplest model of quantum decay corresponds to the inclusion of a non-Hermitian contribution to the Hamiltonian, which allows emulating non-Hermitian systems. Non-Hermitian dynamics also have their own interest. Since the realization of complex optical PT potentials [8,9], the community has unveiled a very rich phenomenology and numerous applications for effective non-Hermitian systems.

To name a few, we can mention the non-Hermitian skin effect [10,11], non-Hermitian transport [8,12–14], and more generally the intriguing topology of effective non-Hermitian systems [15–21]. Emulating non-Hermitian dynamics can provide access to the above phenomena using different platforms.

Engineering truly non-Hermitian and irreversible quantum dynamics over an arbitrarily long time requires the interaction of the system with an infinite set of states, as in the usual paradigm of infinite discrete quasi-continuum [22,23]. Nevertheless, the emulation of quantum dissipation during a finite time can be sufficient for experimental purposes; for instance, when dissipation is used as an asset to prepare a given quantum state [24]. In this context, simulating dissipative quantum dynamics thanks to coupling with a finite—and ideally minimal—number of ancilla states seems a feasible task. This possibility may have interesting applications in quantum computing, where a smaller number of ancilla states usually corresponds to a simpler setup.

The purpose of this article is to investigate this avenue and provide an emulation of non-Hermitian dynamics for a given time interval with a quasi-continuum made of a *finite* set of ancilla states (see Figure 1). We use the trace distance to quantify the quality of our model, and discuss in detail the minimum number of levels required to obtain an accurate emulation. We also investigate separately the short- and long-term behavior of the associated dynamics. At early times, we compare the quantum evolution of the coupled system with the Zeno effect expected from a genuine continuum. At long times, we observe and characterize quantitatively the emergence of revivals in the presence of the finite continuum, enabling us to set an upper limit for the validity time of this emulation. We connect the appearance of these revivals with adequate measures of non-Markovianity.

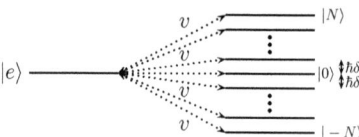

Figure 1. Schematic picture of a quantum system (consisting here of a single discrete state $|e\rangle$) coupled to a finite set of equidistant discrete levels. This model mimics the coupling to a continuum.

We proceed as follows: In Section 2, we provide a brief reminder of the decay for a single discrete level coupled to an infinite continuum. Section 3 presents the considered quasi-continuum model, composed of equidistant energy levels equally coupled to a given state, and discuss its main features. In Section 4, we investigate the same issues for a two-level system whose excited state is coupled to a continuum. We identify a method for defining the minimum size of the discrete continuum using Fourier analysis. In Section 5, we discuss the emergence of non-Makovian evolution at long times and build on the previous sections to design a discrete quasi-continuum with the minimum number of states to reproduce the expected behavior in the strong coupling limit.

2. Decay of a Single Level Coupled to an Effective Continuum

We illustrate our method by first considering a system consisting of a single eigenstate $|e\rangle$ coupled to a large set of independent states $\{|\varphi_f\rangle\}$. This system is the usual paradigm explaining the irreversible exponential decay and Lamb shift undergone by a quantum state coupled to a continuum [23]. We briefly recall below the corresponding derivation in the standard case of an infinite and broad effective continuum consisting of the set of states $\{|\varphi_f\rangle\}$. The quantum system under consideration follows a Hamiltonian given by the sum $H = H_0 + V$ with $H_0 = E_e|e\rangle\langle e| + \sum_f E_f|\varphi_f\rangle\langle\varphi_f|$ the free-system Hamiltonian diagonal in the basis $\{|e\rangle, |\varphi_f\rangle\}$, and with the off-diagonal contribution $V = \sum_f V_{fe}|\varphi_f\rangle\langle e| + \text{h.c.}$ accounting for the coupling between the discrete state and the effective environment.

We search for a solution to a time-dependent Schrödinger of the form:

$$|\psi(t)\rangle = c_e(t)e^{-iE_e t/\hbar}|e\rangle + \sum_f c_f(t)e^{-iE_f t/\hbar}|\varphi_f\rangle. \qquad (1)$$

and subsequently obtain by projection on the eigenstates of H_0 the following integro-differential obeying the coefficient c_e:

$$\dot{c}_e(t) = -\int_0^t dt' K(t-t') c_e(t'), \qquad (2)$$

where the kernel is defined by

$$K(\tau) = \frac{1}{\hbar^2}\sum_f |V_{ef}|^2 e^{i\omega_{ef}\tau}, \qquad (3)$$

with $\omega_{ef} = (E_e - E_f)/\hbar$. Equations (2) and (3) capture the exact quantum dynamics of this system and so far involve no assumptions about the set of final states $\{|\varphi_f\rangle\}$. The function $K(\tau)$ accounts for the memory of the effective environment, resulting in a possibly non-Markovian evolution for the amplitude $c_e(t)$.

We now assume that the effective continuum $\{|\varphi_f\rangle\}$ covers a wide range of frequencies. As a result, the $K(\tau)$ function is expected to peak sharply around $\tau = 0$ when compared to the time-scale of the amplitude evolution; for a genuine continuum with a flat coupling, the sum over all possible final states in Equation (3) would actually yield a Dirac-like distribution. This large timescale separation enables one to pull out the amplitude $c_e(t)$ from the integration of the memory kernel in Equation (2) and to extend the boundary of this integral to infinity. We then obtain a simple closed differential equation for c_e:

$$\dot{c}_e(t) = -\left(\int_0^\infty d\tau K(\tau)\right) c_e(t). \qquad (4)$$

The pre-factor is readily derived within the framework of complex analysis:

$$\int_0^\infty d\tau K(\tau) = i\Delta\omega_e + \frac{\Gamma}{2}, \qquad (5)$$

with

$$\frac{\Gamma}{2} = \frac{2\pi}{\hbar}\sum_f |V_{ef}|^2 \delta(E_e - E_f), \text{ and } \Delta\omega_e = \frac{1}{\hbar}\mathcal{P}\left(\sum_f \frac{|V_{ef}|^2}{E_e - E_f}\right), \qquad (6)$$

where $\mathcal{P}$ denotes the principal value. For the considered coupling to a large set of states, the main effects on the discrete state are therefore an exponential decay of the population at a rate Γ witnessing an irreversible evolution as well as a frequency shift $\Delta\omega_e$, commonly referred to as the Lamb shift. Equation (6) simply expresses Fermi's golden rule for the effective continuum with the density of states $\rho(E) = \sum_f \delta(E - E_f)$. Remarkably, Fermi's golden rule holds, not only for a genuine continuum, but also for a countable set $\{|\varphi_f\rangle\}$ involving only discrete states [23]. Finally, unlike Equation (2), the amplitude $c_e(t)$ at a given time no longer depends on its history; the effective continuum $\{|\varphi_f\rangle\}$ behaves as a Markovian environment. Equations (4) and (5) implicitly define an effective non-Hermitian Hamiltonian $H_{\text{eff}} = \Delta\omega_e - i\frac{\Gamma}{2}$ for this one-level system.

A closer look at Equation (6) reveals the central role played by the density of states $\rho(E)$ of the effective continuum [3,25–27]. Indeed, its properties are responsible for deviations to the exponential law both at short and long times; the existence of an energy threshold ($\rho(E < E_0) = 0$) generates long-time deviations, while the finiteness of the mean energy ($\int \rho(E) E dE < \infty$) explains the short time deviations.

In the same spirit, we examine below how the two characteristics of the quantum evolution discussed above—exponential decay and non-Markovianity—are affected by the use of a finite set as an effective continuum. We restrict our attention to a finite time-interval, as only infinite sets can reproduce these characteristics during arbitrary long times.

3. Coupling of a Single State to a Finite Discretized Continuum

Description of the FQC model. To quantitatively characterize such an irreversible process, we introduce a finite quasi-continuum (FQC) model consisting of a finite set of equidistant energy levels, which are equally coupled to a given state $|e\rangle$ (See Figure 1). This system mimics the decay of an unstable discrete state $|e\rangle$ in a finite time window. In what follows, unless otherwise stated, we always consider FQCs composed of $N_{\text{FQC}} = 2N + 1$ equidistant energy levels symmetrically distributed around the unstable state energy, set by convention to $E = 0$. Here, the total Hilbert space is of dimension $N_{\text{tot}} = N_{\text{sys}} + N_{\text{FQC}} = 2N + 2$. We denote with $\hbar\delta$ the energy gap between two successive FQC states and with $v = |V_{fe}|$ the flat coupling strength between the FQC and the discrete state $|e\rangle$. The expected decay Γ in the limit $N \to +\infty$ is given by Equation (6), which captures the dynamics of an infinite discrete continuum, namely

$$\Gamma = \frac{2\pi}{\hbar^2} \frac{v^2}{\delta}. \tag{7}$$

which corresponds to Fermi's golden rule. In the following, we consider FQCs associated with a fixed common decay rate Γ. We therefore impose $v^2/\delta = Cte$. In our numerical resolution, we implicitly normalize the energies using $\hbar\Gamma$ and the time using Γ^{-1}, which amounts to taking $\hbar = 1$ and $\Gamma = 1$. Our results are valid for arbitrary values of the dissipation rate Γ as long as the dimensionless parameters $\bar{v} = v/(\hbar\Gamma)$, $\bar{t}_f = \Gamma t_f$,... remain identical. The considered FQCs are therefore entirely determined by their size $(2N+1)$ and the coupling strength, v.

The model Hamiltonian in matrix form reads

$$H = \begin{pmatrix} 0 & v & \ldots & v & v & v & v \\ v & -N\hbar\delta & 0 & \ldots & 0 & 0 & 0 \\ v & 0 & -(N-1)\hbar\delta & 0 & \ldots & 0 & 0 \\ v & 0 & \ldots & 0 & \ldots & 0 & 0 \\ v & 0 & . & . & 0 & (N-1)\hbar\delta & 0 \\ v & 0 & 0 & 0 & \ldots & 0 & N\hbar\delta \end{pmatrix}. \tag{8}$$

Examples of FQCs and connection with the Zeno effect. In Figure 2, we compare the evolution of the excited state population for an example of FQC (solid black line) with the exponential decay expected from Fermi's golden rule (dotted line). As expected, we observe a very good agreement, with minor discrepancies at short times (see the inset of Figure 2) and at long times when the population is extremely small. We used a FQC with $N = 15$ and a coupling strength $v = 0.3\,\hbar\Gamma$. In this case, the emulation of quantum decay does not require a very large Hilbert space.

The disagreement at short times corresponds to a quadratic decay of the excited state coupled to a FQC. The initial quadratic profile is directly related to the Zeno effect. This is found by expanding the evolution operator for a short amount of time δt, by writing $|\psi(\delta t)\rangle = e^{-iH\delta t/\hbar}|e\rangle \simeq |e\rangle + |\delta\psi\rangle$ with

$$|\delta\psi\rangle = \left(-\frac{iH}{\hbar}\delta t - \frac{H^2}{2\hbar^2}(\delta t)^2\right)|e\rangle. \tag{9}$$

We infer the initial state population $\pi_e(t) = |\langle e|\psi(t)\rangle|^2$ at early times

$$\pi_e(\delta t) \simeq 1 - \frac{(\delta t)^2}{T_Z^2}, \tag{10}$$

where $T_Z^{-2} = \frac{1}{\hbar^2}(\langle H^2 \rangle_e - \langle H \rangle_e^2) = \frac{1}{\hbar^2}\sum_n \langle e|V|n\rangle\langle n|V|e\rangle = (2N+1)(v/\hbar)^2$. The duration T_Z corresponds to the Zeno time and decreases with the size of the FQC. As T_Z vanishes in the limit $N \to +\infty$, the observed initial quadratic profile witnesses the limited number of states of the FQC. For the parameters $N = 15$ and $v = 0.3\,\hbar\Gamma$, one finds $T_Z \simeq 0.6\,\Gamma^{-1}$, consistent with the inset in Figure 2.

We now provide a second example of FQC, for which the excited state population evolves very differently from the expected exponential decay. We take a FQC with $N = 15$ and $v = 0.45\,\hbar\Gamma$, which corresponds to a larger energy gap between the FQC levels than in the first example, therefore being further away from an ideal continuum. Good agreement is observed up to $t \simeq 5\Gamma^{-1}$, when the population $\pi_e(t)$ grows abruptly (gray dashed line, Figure 2). This revival of the probability distribution in the discrete state reveals the underlying fully coherent dynamics.

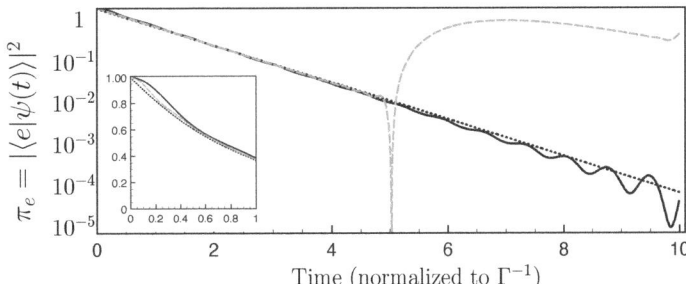

Figure 2. FQC vs. genuine continuum for a single discrete level: Excited state population $\pi_e(t)$ of a discrete level coupled to FQCs obtained from the full quantum dynamics under the Hamiltonian (8) with $N = 15$ and $v = 0.3\,\hbar\Gamma$ (solid black line) and $N = 15$ and $v = 0.45\,\hbar\Gamma$ (gray dashed line) as a function of time (normalized to Γ^{-1}). The dotted line represents the exponential decay expected from a genuine continuum.

Quantitative mapping of successful FQCs for the emulation of a single-state decay. We now proceed to a quantitative mapping of the FQC parameters (N, v) suitable for accurate continuum emulation. In order to capture the accuracy of our model for a given time window, one needs a distance measure between the quantum evolution observed in the presence of an FQC and the genuine continuum. For the single-state quantum system considered here, the density matrix boils down to the excited state population $\pi_e(t)$. We therefore introduce the following distance

$$\mathcal{D}_1(t_f) = \frac{1}{t_f}\int_0^{t_f}|\pi_e(t) - \pi_0(t)|dt. \quad (11)$$

as a figure of merit for the quality of the FQC emulation over the time window $0 \leq t \leq t_f$. $\pi_0(t) = \pi_e(0)e^{-\Gamma t}$ is the exponential decay expected in the large continuum limit. We choose t_f to be larger than several Γ^{-1} to best account for the full decay. In our numerical examples, we systematically use $t_f = 10\Gamma^{-1}$ (unless otherwise specified). The results are summarized in Figure 3a. The good set of parameters for the chosen time interval is provided by the white area. This figure reveals that the quality of the emulation increases with the number of FQC states and decreases with the potential strength v, corresponding to FQCs with a larger energy gap $\hbar\delta$ for a fixed decay rate Γ. In particular, the quality of the emulation drops off sharply above a critical coupling value $v_c \simeq 0.32\,\hbar\Gamma$, which is independent of the number of FQC states. We explain below this abrupt change in terms of quantum interference and revivals of the discrete state population. The dashed gray line of Figure 2 provides an example of the revival of the excited population $\pi_e(t)$ coupled to a FQC with a strength $v \geq v_c$.

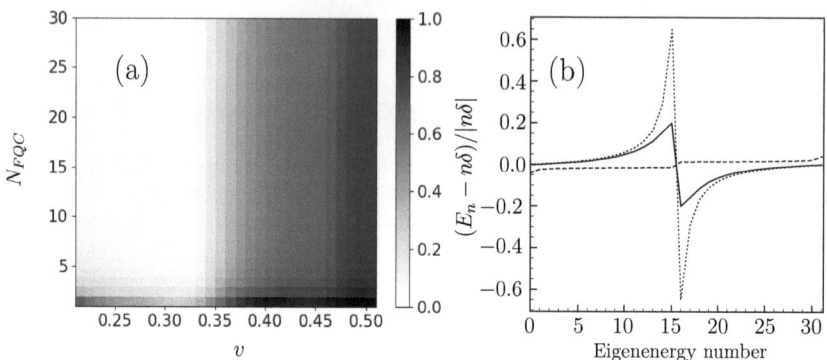

Figure 3. (**a**) Quality of the emulation of a continuum using a FQC: Parameter $\mathcal{D}_1$ (obtained by a numerical resolution of the Shcrödinger equation) as a function of the FQC parameters $\{N, v\}$ (v is given in units of $\hbar\Gamma$). The white zone reveals a very good agreement with the exponential decay expected from a genuine continuum. (**b**) Spectrum analysis of the Hamiltonian (8): Eigenenergies E_n in crescent order normalized by δ for $N = 15$ and $\delta = 5\,\Gamma$ (dotted line), $\delta = 0.5\,\Gamma$ (solid line) and $\delta = 0.05\,\Gamma$ (dashed line). $N_{FQC} = 2N + 1$ is the FQC size.

We now provide a quantitative analysis of the occurrence of such revivals in a given time window. We first look for a necessary condition of revival. For this purpose, we expand the wave function at time t on the eigenbasis:

$$|\psi(t)\rangle = \sum_{n=0}^{N_{\text{tot}}} a_n(0) e^{-iE_n t/\hbar} |\psi_n\rangle, \quad (12)$$

where E_n are the eigenenergies of the total Hamiltonian (8) and with $N_{\text{tot}} = 2N + 2$ the dimension of the total Hilbert space. We denote by T_r the revival time, which necessarily fulfills

$$|||\psi(T_r)\rangle - |\psi(0)\rangle||^2 = 2 \sum_{n=0}^{N_{\text{tot}}} |a_n(0)|^2 (1 - \cos(E_n T_r / \hbar)) \equiv \varepsilon \ll 1. \quad (13)$$

The revivals correspond to a constructive quantum interference occurring at a time T_r determined by the Hamiltonian (8) spectrum. Actually, this spectrum is only marginally affected by the coupling to the discrete state and has a nearly linear dependence of its eigenvalues $E_n \simeq n\hbar\delta$ (see the numerical analysis on Figure 3b). This result is valid for a wide range of energy gaps $\hbar\delta$. The condition (13) requires that for all values of n, $E_n T_r / \hbar = 2\pi k_n$ with k_n an integer. As $E_n \simeq n\hbar\delta$, we find $k_n = n$ and $T_r = 2\pi/\delta$. Figure 4 confirms numerically the predictions of this simple revival model. We have plotted the revival time inferred from the exact resolution of the Schrödinger equations of the model with the Hamiltonian (8) as a function of $1/\delta$.

The above analysis provides a clear criterion for the suitability of the FQC for emulating irreversible dynamics. A necessary condition is the absence of revival during the considered time windows, i.e., $t_f < T_r$. This sets an upper bound on the energy gap, namely $\delta \leq \delta_c = 2\pi/t_f$, or equivalently on the coupling strength $v \leq v_c = \hbar\sqrt{\Gamma/t_f}$, as both quantities are related by Equation (7). For the considered final time $t_f = 10\,\Gamma$, we obtain the value $v_c = 0.316\,\hbar\Gamma$ in very good agreement with the numerical results of Figure 3a. The region $v \geq v_c$ indeed corresponds to the onset of the gray zone, accounting for the degradation in the emulation of dissipative dynamics. In the next Section, we investigate the appropriate choice of the FQC model parameters in the different regimes of a driven two-level system.

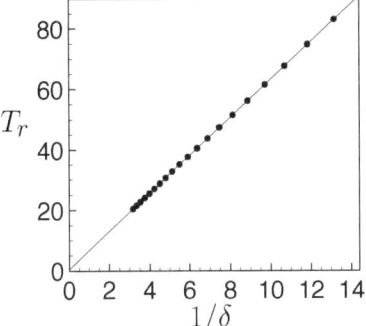

Figure 4. Black disks: Resurgence time T_r as a function of the inverse of the FQC energy gap $\hbar\delta$. T_r is obtained using a numerical resolution of the Schrödinger equation with the Hamiltonian (8). The solid black line represents a linear fit $T_r = a/\delta$ yielding $|a - 2\pi| \leq 10^{-3}$, showing thus an excellent agreement with our prediction for the revival time.

4. Coupling of a Two-Level System to a Finite Discretized Continuum

Model description and equations of motion. In this Section, we consider a two-level atom with a stable ground state $|g\rangle$ and an unstable excited state $|e\rangle$ (see Figure 5), which is the standard model for spontaneous emission in quantum optics [23]. We denote by ω_0 the transition frequency of this two-level system and assume that it is illuminated by a nearly-resonant laser of frequency $\omega_L \simeq \omega_0$. This external field drives the system with a Rabi coupling of frequency Ω_0 between the two atomic levels. The excited state acquires a finite width Γ, due to its coupling with the continuum.

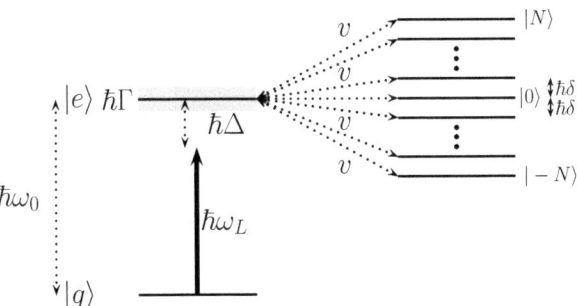

Figure 5. Two-level system driven by a laser pulse with a detuning (Δ), involving a stable ground state $|g\rangle$ and an excited state $|e\rangle$ coupled to a large but finite set of discrete levels. This coupling emulates an unstability and yields an effective linewidth Γ for the transition.

We now consider a $N_{\text{tot}} = 2N + 3$-dimensional Hilbert space encapsulating the two-level quantum system and the FQC. Considering the driving term, the total Hamiltonian is given by

$$H = \begin{pmatrix} 0 & \hbar\Omega_0 & 0 & 0 & \ldots & 0 & 0 \\ \hbar\Omega_0 & \hbar\Delta & v & v & \ldots & v & v \\ 0 & v & -N\hbar\delta & 0 & 0 & \ldots & 0 \\ 0 & v & 0 & -(N-1)\hbar\delta & 0 & \ldots & 0 \\ 0 & . & . & 0 & . & 0 & \ldots \\ 0 & . & . & . & 0 & . & 0 \\ 0 & v & 0 & 0 & \ldots & 0 & N\hbar\delta \end{pmatrix}. \quad (14)$$

on the basis $\{|g\rangle, |e\rangle, |\psi_f\rangle\}$ transformed in the rotating frame with the detuning $\Delta = \omega_0 - \omega_L$. For a given dissipation rate Γ, the system is therefore determined by four independent driving ($\{\Omega_0, \Delta\}$) and FQC ($\{N, \delta\}$, or equivalently $\{N, v\}$ from Equation (7)), parameters. We denote by $|\psi\rangle$ the quantum state of the full Hilbert space. The corresponding density matrix $\rho = |\psi\rangle\langle\psi|$ follows a unitary dynamics $i\hbar \frac{d\rho}{dt} = [H, \rho]$. We now focus on the non-unitary quantum dynamics in the reduced Hilbert space. Specifically, we consider the evolution of the 2×2 density matrix $\rho_r = P_{ge}\rho P_{ge}$, where $P_{ge} = |g\rangle\langle g| + |e\rangle\langle e|$ is the projector on the two-dimensional Hilbert space of the system. The reduced density matrix ρ_r can be obtained by first solving the full unitary dynamics and then applying the projector. In order to highlight the role played by the FQC, the equation of motion for the reduced density matrix can be rewritten in the following form:

$$i\hbar \frac{d\rho_r}{dt} = [H_0, \rho_r] + S_r^{\text{FQC}}. \tag{15}$$

The r.h-s contains the unitary driving of the system Hamiltonian $H_0 = P_{ge} H P_{ge}$, as well as a source term accounting for the interaction with the FQC

$$S_r^{\text{FQC}} = \begin{pmatrix} 0 & \lambda_N \\ \lambda_N^* & \eta_N \end{pmatrix}. \tag{16}$$

where $\lambda_N = v \sum_{i=2}^{2N+2} \rho_{gi}$ and $\eta_N = v \sum_{i=2}^{2N+2} (\rho_{ie} - \rho_{ei})$. This source term drives effective non-unitary dynamics within the considered time interval and depends on the coherence between the FQC levels and the quantum system. The equations above contain no approximation and capture the full quantum dynamics of the two-level system coupled to a FQC.

Non-Hermitian dynamics. Here, we briefly review the equations of motion under an effective non-Hermitian Hamiltonian. Beyond their applications in nanophotonics, effective non-Hermitian Hamiltonians adequately describe the dynamics of open quantum systems in many experimental situations. For instance, this approach has been successfully used to explain the subradiance effects in large atomic clouds [28]. As in Section 2, the effective non-Hermitian Hamiltonian is obtained by deriving differential equations for the two-level system probability amplitudes (c_e, c_g). Using rotating wave-approximation, one finds $H_{\text{eff}} = H_0 + iH_d$ with $H_0 = \hbar\Omega_0(|e\rangle\langle g| + \text{h.c.}) + \hbar\Delta|e\rangle\langle e|$ and $H_d = -\frac{\hbar}{2}\Gamma|e\rangle\langle e|$. The anti-Hermitian contribution iH_d captures the decay towards the continuum. The evolution of the reduced density matrix under the influence of this effective Hamiltonian takes a form analogous to Equation (15)

$$i\hbar \frac{d\tilde{\rho}_r}{dt} = [H_0, \tilde{\rho}_r] + S_r^\infty \tag{17}$$

with a source term $S_r^\infty = i[H_d, \tilde{\rho}_r]_+$ capturing the non-unitary dynamics ($[\ ]_+$ is an anti-commutator). Numerical analysis confirms that S_r^∞ also corresponds to the limit of the FQC source terms S_r^{FQC} (16) within the large quasi-continuum limit $N \to +\infty$. At resonance ($\Delta = 0$), the Schrödinger equation in the presence of H_{eff} boils down to the equation of a damped harmonic oscillator for the probability amplitude c_e

$$\ddot{c}_e + \Gamma\dot{c}_e/2 + \Omega_0^2 c_e = 0. \tag{18}$$

One identifies the three usual dynamical over/critical/under-damping regimes determined by the ratio Ω_0/Γ (see the black dashed lines in Figure 6).

Example of successful FQC-emulated dynamics. In Figure 6, we investigate the suitability of a FQC with parameters $\{N, v\} = \{30, 0.3\,\hbar\Gamma\}$ for the emulation of non-Hermitian dynamics in these different regimes. We obtain the evolution of the excited state population $\pi_e(t)$ coupled to this FQC using a numerical resolution of the Schrödinger equation with the Hamiltonian (14), and compare it to the evolution under the non-Hermitian dynamics given by Equation (18). Excellent agreement is observed for the three distinct regimes, covering a wide range of Ω_0/Γ values. We investigate below how to determine the minimal number of levels of an adequate FQC.

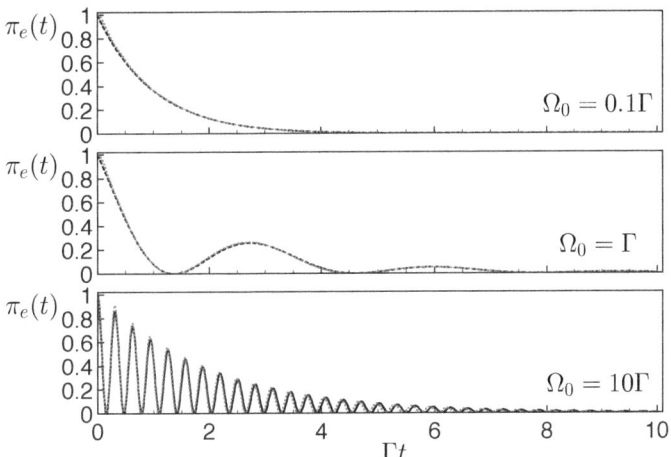

Figure 6. Non-Hermitian vs. FQC dynamics for the two-level system in the following regimes: over-damping ($\Omega_0 = 0.1\Gamma$, upper panel), critical ($\Omega_0 = 1\Gamma$, middle panel), and under-damping ($\Omega_0 = 0.1\Gamma$, lower panel) in the presence of an FQC with $N_{FQC} = 2N + 1 = 61$ levels and $v = 0.3\hbar\Gamma$ (full quantum dynamics, gray dotted line) or from the non-Hermitian dynamics (Equation (18), black dashed line). Both lines are superimposed, showing the excellent emulation of non-Hermitian dynamics with the considered FQCs.

Quantitative mapping of successful FQCs for the emulation of two-level non-Hermitian dynamics. Before proceeding to a more systematic analysis of the suitability of the FQC, we introduce a quantitative measure for the accuracy of FQC-emulated dynamics. Specifically, in the considered two-level system, we take the trace distance [29] between the reduced density matrices evolved respectively under the influence of a FQC (unitary evolution with H (14) followed by projection with P_{eg}) and following non-Hermitian dynamics (Equation (17)). This distance is defined for two density matrices ρ and σ by

$$T(\rho, \sigma) = \frac{1}{2}\text{Tr}\left(\sqrt{(\rho - \sigma)^\dagger (\rho - \sigma)}\right). \tag{19}$$

In order to obtain a quantitative estimate of the fidelity over the whole considered interval, we use the mean trace distance over the considered time window:

$$\mathcal{D}_2(t_f) = \frac{1}{t_f}\int_0^{t_f} T(\rho_e(t), \sigma(t))dt. \tag{20}$$

This definition in terms of trace distance coincides with the measure $\mathcal{D}_1$ introduced in Equation (11) in the one-dimensional case.

As in Section 3, we proceed to a systematic study of the appropriate FQC parameters (N, v) for the emulation of non-Hermitian dynamics. We here separately consider the three different regimes evidenced by Equation (18) and we use the mean trace distance (20) between the respective density matrices evolving in the presence of a FQC (ρ_r) or following non-Hermitian dynamics ($\tilde{\rho}_r$). The results are summarized in Figure 7a–c for the different ratios Ω_0/Γ corresponding to the three distinct regimes of non-Hermitian dynamics. In order to avoid the revival effect discussed in Section 3, we take a slightly shorter time interval $t_f = 8\Gamma^{-1}$. A comparison between the mappings presented Figures 3a and 7a–c reveals very different characteristics in the FQC emulation for the one- and two-level systems. For the one-level system, successful FQC emulation only requires the absence of revivals, associated with a condition $v \leq v_c$ independent of the FQC size N. Differently, we see for the two-level case that the number $2N + 1$ of FQC states has a critical influence on the fidelity of the FQC-emulated dynamics. These figures reveal an abrupt transition when the parameter N falls below a critical value $N(v)$, depending on the coupling strength v for a given ratio Ω_0/Γ. This raises the question of how to choose suitable FQC parameters.

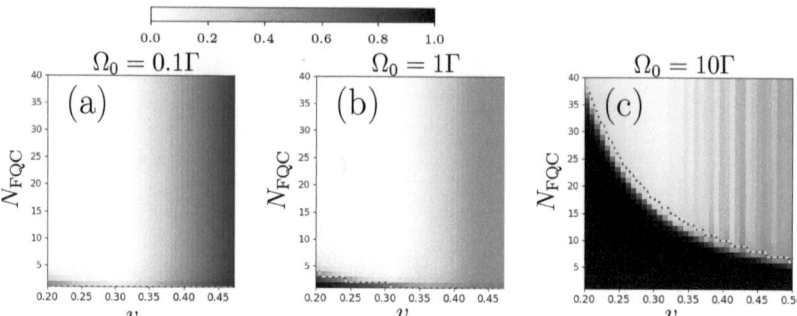

Figure 7. Quality of the emulation of non-Hermitian dynamics with a FQC: 2D plots of the mean trace distance (normalized to its maximum value) between the non-Hermitian model and the dynamics in a FQC model with parameters N_{tot}, v (in units of $\hbar\Gamma$) for the respective ratios $\Omega_0/\Gamma = 0.1$ (a), 1 (b), 10 (c). The dotted black–white line corresponds to the number $N_{max}(v)$.

Suitability criteria for FQC. Here, we determine the subset of FQC states that are significantly populated during the time evolution. Intuitively, this set should form the minimal FQC which accurately captures dissipative quantum dynamics. As can be seen below, the populated modes essentially depend on the Rabi frequency Ω_0 and dissipation rate Γ.

This situation is reminiscent of the dynamical Casimir effect (DCE), in which a continuum of vacuum electromagnetic modes becomes gradually populated under the harmonic motion of a moving mirror (See Ref. [30] for a review). In the DCE, the mirror oscillation at a frequency Ω_m induces the emission of photons of frequencies $\omega \leq \Omega_m$ in initially unpopulated electromagnetic modes. A similar effect is observed with a moving two-level atom [31,32] in the vacuum field. We find below that our FQC model with a Rabi driving reproduces these features, with the emergence of sidebands at the Rabi frequency in the FQC population. As in the DCE, the external drive provides energy to the system, which eventually leaks into the continuum.

To analyze this effect, we introduce the expansion

$$|\psi(t)\rangle = c_e(t)|e\rangle + c_g(t)|g\rangle + \sum_{p=-N}^{N} c_p(t)|p\rangle \qquad (21)$$

into the Schrödinger equation. A projection on the kth state of the FQC yields a differential equation for the coefficient $c_k(t)$ driven by the excited state probability amplitude $c_e(t)$. This equation is formally solved as

$$c_k(t) = -\frac{iv}{\hbar} \int_0^t c_e(t') e^{ik\delta t'} dt'. \qquad (22)$$

In the long-time limit, the coefficient $c_k(t)$ tends towards the Laplace transform of the excited state amplitude at the frequency $k\delta$ (up to a constant factor). In order to estimate the occupation probability $|c_k(t)|^2$ at time $t < t_f$, we use the probability amplitude $\tilde{c}_e(t)$ given by non-Hermitian dynamics (Equation (18)). The latter is indeed an excellent approximation of the excited state probability $c_e(t)$ in coupling to a sufficiently large FQC (see Figure 6). We find

$$c_k(\infty) = \frac{v\Omega_0}{\hbar\sqrt{\Delta_0}} \left[\frac{-1}{-\frac{\Gamma}{4} + \frac{i\sqrt{\Delta_0}}{2} + ik\delta} + \frac{1}{-\frac{\Gamma}{4} - \frac{i\sqrt{\Delta_0}}{2} + ik\delta} \right] \quad (23)$$

with $\Delta_0 = 4\Omega_0 - \frac{\Gamma^2}{4}$. Figure 8 shows the probability occupations $|c_k(t_f)|^2 \simeq |c_k(\infty)|^2$. These distributions exhibit two sidebands centered about k values, such that $|k|\delta \simeq \Omega_0$, symmetrically distributed around $k = 0$ for our choice of $\Delta = 0$. A similar generation of sidebands is observed for the dynamical Casimir effect [30]. These occupancy probabilities actually determine the number of relevant FQC states and the size of the minimal appropriate FQC. Indeed, we have indicated in Figure 7a–c the maximum occupancy number $N_{\max}(v)$ as a function of the coupling strength v. This quantity is defined as $|c_{N_{\max}(v)}(t_f)|^2 = \max_n\{|c_n(t_f)|^2\}$ for the considered coupling strength v and Rabi frequency Ω_0. As the occupation peak approximately corresponds to the Rabi frequency, we expect $N_{\max}(v) \simeq \Omega_0/\delta = \Omega_0 \hbar^2\Gamma/(2\pi v^2)$ from Equation (7). In Figure 7a–c, the line representing the maximum occupancy number $N_{\max}(v)$ is almost superimposed on the interface between the suitable and unsuitable FQCs (white/grey zones, respectively). This confirms that the suitable FQCs are those that host all the significantly populated levels. The population of each Fourier components is represented for different Ω_0/Γ ratios in Figure 8: in the weak coupling limit $\Omega_0 \ll \Gamma$, $N_{\max}(v)$ is mainly determined by the dissipation rate Γ, while in the strong coupling limit $\Omega_0 \gg \Gamma$, it scales linearly with the Rabi frequency Ω_0 (for $v = 0.3$, $N_{\max}(v) \simeq 1.77\Omega_0$).

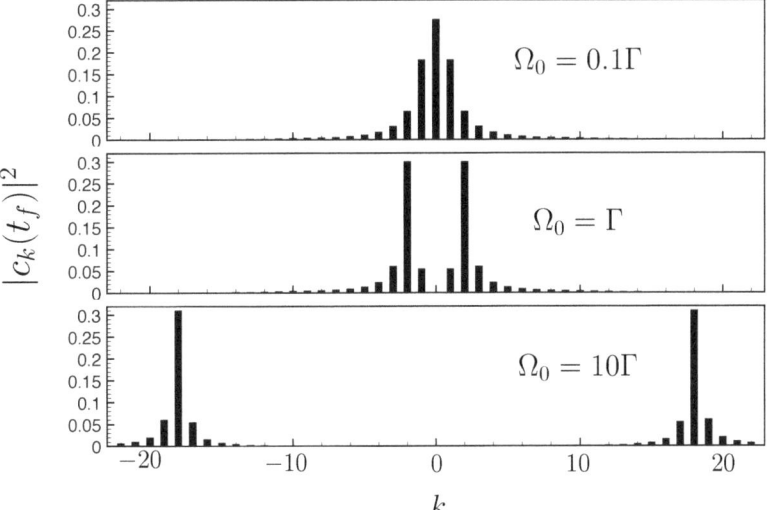

Figure 8. Distribution of populations: Occupation probability $|c_k(t_f)|^2$ at the final time t_f as a function of the level k of the FQC for $\Omega_0 = 0.1\,\Gamma$ (lower panel), $\Omega_0 = \Gamma$ (middle panel) and $\Omega_0 = 10\,\Gamma$ (upper panel). Parameters: $2N + 1 = 45$ levels, and with $v = 0.3\,\hbar\Gamma$.

5. Non-Markovian Dynamics and Adaptive Quasi-Continuum

In this Section, we analyze quantitatively the finiteness-related effects in the evolution of an FQC-coupled quantum system. First, we establish the connection between the presence of revivals (discussed in Section 3) and a measure of non-Markovianity applied to the FQC-emulated dynamics. Second, we show that the FQC structure of equidistant energy levels induces a mismatch of the effective Rabi frequency and decay rates when compared to the equivalent parameters in the non-Hermitian model. Solving this issue suggests an adaptative structure of FQCs, discussed below, capable of reproducing non-Hermitian dynamics with a considerably reduced number of states.

5.1. Revivals and Non-Markovianity

Revivals in the excited state probability $\pi_e(t)$, discussed in Section 3 for the single-level system, also occur in the two-level FQC emulated dynamics for large values of coupling strength v. Such revivals are indeed a symptom of non-Markovian dynamics in the FQC; their exact form depends on the initial quantum state and therefore reveals a memory effect in the quantum evolution. Despite the successful emulation of dissipative dynamics over a given time interval, these revivals show that some information about the initial state has been transmitted and stored in the FQC. The revival appears as a kind of constructive interference effect when information about the initial state, stored in the FQC, returns to the system. Non-Hermitian dynamics (17) are Markovian, and so the emergence of non-Markovianity reveals a discrepancy between the FQC-emulated system and the ideal irreversible case.

These considerations suggest quantitatively studying the non-Markovianity of the FQC-emulated dynamics. We proceed by using the measure from Ref. [33], summarized below for convenience. This measure uses the trace distance $T(\rho, \sigma)$ (19), which has a direct interpretation in terms of the distinguishability of the associated quantum states. Indeed, if we consider an emitter which randomly prepares one of the two quantum states $\{\rho, \sigma\}$ with equal probability, the probability of an observer successfully identifying the correct quantum state through a measurement is simply $\frac{1}{2}(1 + T(\rho, \sigma))$. Markovian processes correspond to a decreasing trace distance for any set of states following the quantum evolution associated with the process. In this case, no information likely to improve the dinstinguability of the states $\{\rho(t), \sigma(t)\}$ is acquired by the system during the evolution. The unitary evolution operator of a closed quantum system, and more generally complete positive trace-preserving maps, fall into this category. Conversely, non-Markovian quantum processes are those that exhibit at least a temporary positive variation of the trace distance for some pair of initial states. This increase witnesses a flow of information from the environment back to the system.

To obtain a quantitative measure, one introduces the rate of variation of the trace distance for a given quantum process

$$\sigma_{\rho_1^0, \rho_2^0}(t) = \frac{d}{dt} T(\rho_1(t), \rho_2(t)). \tag{24}$$

where $\rho_{1,2}(t)$ are two density matrices undergoing the quantum process under consideration and therefore following the same evolution operator/dynamic equation, but with distinct initial conditions $\rho_{1,2}(0) = \rho_{1,2}^0$. Quantum processes with $\sigma_{\rho_1^0, \rho_2^0}(t) > 0$ correspond to an increasing trace distance, and therefore a flow of information from the environment to the system. The non-Markovianity measure is given by [33]

$$\Sigma(t) = \max\nolimits_{\rho_1^0, \rho_2^0} \int_0^t dt'\, \Theta(\sigma_{\rho_1^0, \rho_2^0}(t'))\, \sigma_{\rho_1^0, \rho_2^0}(t') \tag{25}$$

The Heaviside function $\Theta(x)$ (s.t. $\Theta(x) = 1$ if $x \geq 0$ and $\Theta(x) = 0$ for $x < 0$) guarantees that only time intervals with an increasing trace distance effectively contribute to the integral. The quantity $\Sigma(t)$ is obtained by considering all possible initial quantum states $\rho_i^0 = |\psi_i\rangle\langle\psi_i|$ (with $|\psi_i\rangle$ a generic quantum state of the full Hilbert space), and the considered evolution corresponds to $\rho(t) = P_{eg} e^{-iHt/\hbar} \rho^0 e^{iHt/\hbar} P_{eg}$, where H is the Hamiltonian (14) and P_{eg} the projection operator introduced earlier for the two-dimensional subspace. In Figure 9b, we have plotted the evolution of the non-Markovianity $\Sigma(t)$ as a function of time for a given FQC, to be compared with the time evolution of the excited-state population $\pi_e(t)$ in Figure 9a. We deliberately chose a time window during which a revival was observed. Figure 9a,b reveal that a sharp increase in the non-Markovianity $\Sigma(t)$ occurs at the onset of the probability revival. The non-Makovianity $\Sigma(t)$ thus provides another determination of the time window over which the FQC dynamics accurately emulate an irreversible non-Hermitian evolution.

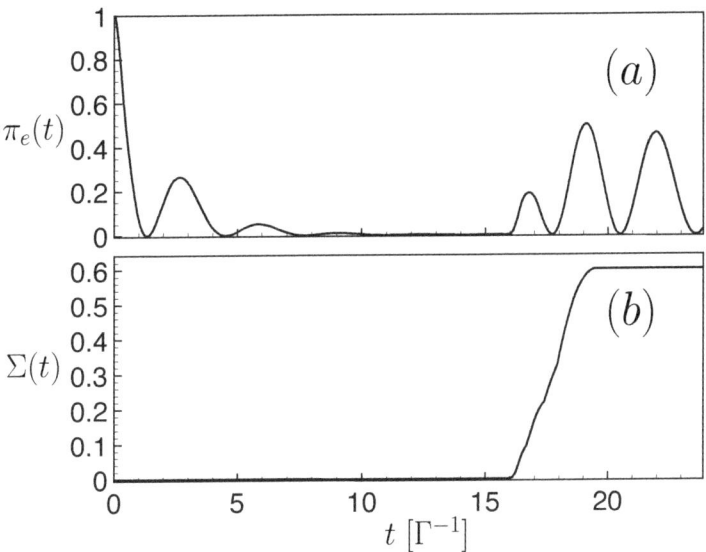

Figure 9. Revivals and non-Markovianity: (**a**) Time evolution of the population in the unstable state $\pi_e(t)$. (**b**) Measure of the non-markovianity $\Sigma(t)$ (see text) of a two-level system composed of $2N+3 = 53$ levels as a function of time. Parameters: $\Omega_0 = 1\,\Gamma$, $v = 0.25\,\hbar\Gamma$. The non-Markovianity measure is estimated within a 1% accuracy by sampling over a set of 256 initial states.

5.2. Adaptive FQC

The study carried out in Section 4 reveals the minimal size of suitable FQCs scales with the Rabi frequency Ω_0. Here, we go one step further and propose adapting the FQC's structure depending on the Rabi coupling Ω_0. We no longer consider exclusively flat FQCs with equidistant levels around the excited state energy. Instead, we study adaptive FQCs with an enhanced density of states around the occupation peaks depicted in Figure 8. As seen below, such adaptive FQCs yield an optimized emulation of non-Hermitian dynamics.

We begin by investigating the influence of the discrete FQC structure on the emulated quantum dynamics. Figure 7c exhibits a slightly gray zone associated with a slight mismatch between the FQC evolution and the non-Hermitian dynamics. This is the regime we wish to investigate. For this purpose, we consider non-Hermitian dynamics (18) in the strong coupling regime $\Omega_0 \gg \Gamma$. The corresponding excited-state population reads

$$\pi_e(t) = e^{-\Gamma t/2}\cos^2(\Omega t), \qquad (26)$$

with $\Omega = \Omega_0(1 - (\Gamma/4\Omega_0)^2)^{1/2}$. An effective Rabi frequency ($\tilde{\Omega}$) and dissipation rate ($\tilde{\Gamma}$) for the FQC dynamics are obtained by fitting the excited-state population $\pi_e(t)$ with the form (26) of exact non-Hermitian dynamics. The corresponding results are represented as a solid gray line in Figure 10a,b for different FQCs.

The discrepancy between the FQC model and the ideal non-Hermitian case can be explained through a closer examination of the integration Kernel (2), or more precisely its equivalent for the two-level case. To reproduce non-Hermitian dynamics, the integration Kernel must take a form analogous to Equation (4). In this case, the frequency mismatch $\Delta\Omega = \tilde{\Omega} - \Omega$ cannot be attributed to a Lamb shift effect, as the principal part of the kernel in Equation (6) cancels out in the presence of a symmetric FQC with a homogeneous coupling constant. The slight frequency shift is therefore a signature of the non-Markovianity of the FQC dynamics, i.e., of the residual error committed by replacing Equation (2) with Equation (4). The corresponding approximations, namely of short kernel memory and the extension of the integral in Equation (4) to infinity, are indeed jeopardized by the discrete FQC structure. Intuitively, the discrete states of zero-energy (i.e., of energy close to the unstable excited state) can increase the error. We also note that these central states are not significantly populated in the FQC dynamics: the highly populated levels correspond to peak population sidebands centered on $\pm\Omega_0/\delta$.

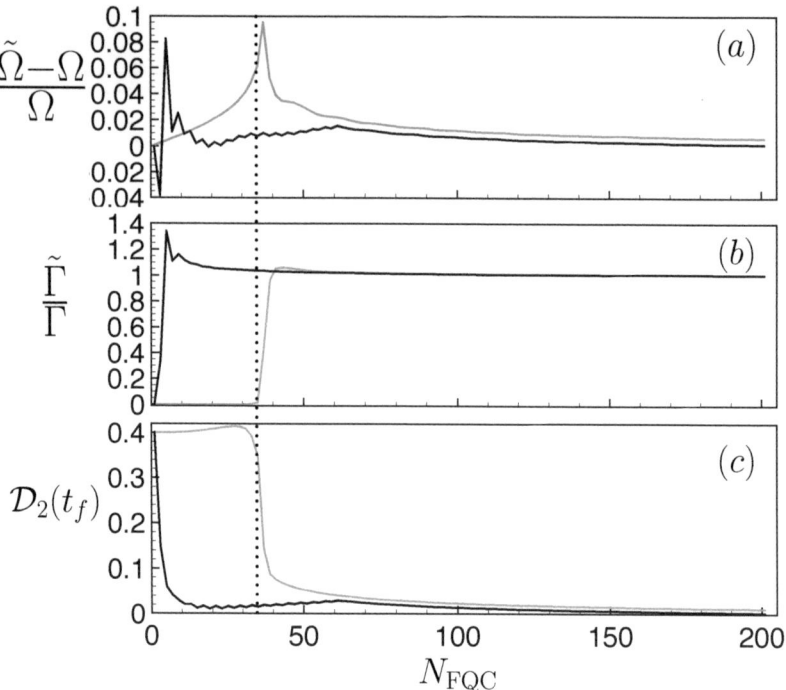

Figure 10. Mismatch between the effective Rabi frequency (**a**), damping rates (**b**), and mean trace distance (**c**) as a function of the size N_{FQC} obtained by comparing the FQC model with the non-Hamiltonian model. We have plotted the effective parameters obtained from a flat FQC made of equidistant levels (gray solid line) and for an FQC with an adaptative structure with removed central states (black solid line). The dotted line corresponds to $N_{FQC} = 35$, considered in Figure 11. Parameter $v = 0.3\,\hbar\Gamma$.

These observations raise the question of the relevant optimal FQC structure in this regime. From Figure 3b, the central FQC state eigenenergies undergo the largest shift from the linear dispersion relation expected from an ideal continuum. Furthermore, Figure 8c shows that, in the strong driving regime ($\Omega_0 \gg \Gamma$), the final population of these states is very small. These considerations suggest that the central components of the FQC play a minor role, or even a deleterious role.

To confirm this intuition, we studied a different FQC model obtained from the former FQC, by removing the states close to the $E = 0$ energy while preserving the symmetry of the distribution. The corresponding results are shown in Figure 10a–c (solid black line). For large FQC sizes, both the flat and adaptive FQC provide a good emulation of non-Hermitian dynamics, although the latter had the same error in the $\tilde{\Omega}, \tilde{\Gamma}, \mathcal{D}_2$ parameters with a much smaller size. For small sizes $N_{FQC} \leq 30$, the spectrum of the regular flat FQC is too narrow to include the highly populated Rabi sidebands of Figure 8c. Consequently, small regular FQCs produce a negligible effective damping rate $\tilde{\Gamma}$. On the other hand, by construction, the adaptive FQC contains states nearby these sidebands. Thus, even small adaptive FQCs $N_{FQC} \simeq 10$ already give an effective damping rate $\tilde{\Gamma}$ close to the appropriate value. At intermediate sizes ($10 \leq N_{FQC} \leq 50$), adaptive FQCs also outperform regular FQCs: a strong improvement is observed in the agreement between the effective Rabi frequency $\tilde{\Omega}$ and damping rate $\tilde{\Gamma}$ with their non-Hermitian counterparts Ω, Γ, as well as a significant reduction in the mean trace distance compared to exact non-Hermitian dynamics. We conclude that, for the damped Rabi dynamics considered, adaptive FQCs with a tailored distribution (involving mostly states close to the Rabi frequency sidebands $\pm \Omega_0 / \delta$ and presenting a hole in the central zone near the unstable state energy ($E = 0$)) provide a higher fidelity to non-Hermitian dynamics with constant resources, i.e., with the same number of states and for an identical time window.

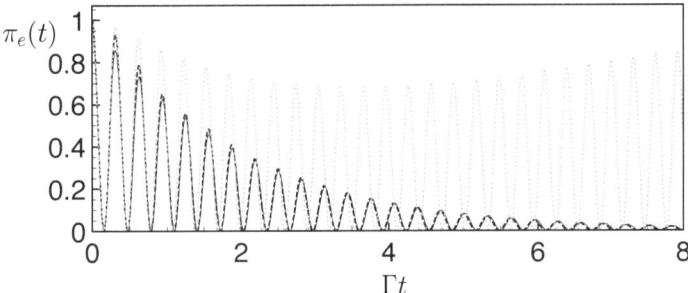

Figure 11. FQC-emulated dynamics in the strong coupling regime ($\Omega_0 = 10\Gamma$): excited population as a function of time for a flat FQC (dotted gray line) made of $N_{FQC} = 2N + 1 = 35$ equidistant levels and for an adaptative FQC (dashed black line) made of $N_{FQC} = 2N = 34$ levels. For both FQCs, we used the parameters $v = 0.3\,\hbar\Gamma$. The solid gray line represent the exact evolution expected from the non-Hermitian dynamics (Equation (26)), and is almost superimposed onto the adaptative FQC results.

Figure 11 provides an example, where both kinds of FQC (flat vs. adaptive) produce very different qualitative behaviors, while having a very similar number of states. While the coupling to a regular equidistant FQC cannot account for the damping of the Rabi oscillation, the quantum system coupled to the adaptive FQCs yields an excellent agreement with the predicted non-Hermitian dynamics (Equation (26)).

6. Conclusions

In conclusion, we discussed the emulation of non-Hermitian quantum dynamics during a given time window with a finite quasi-continuum composed of discrete states. We specifically considered the exponential decay of an unstable state, and the Rabi driving of a two-level quantum system exhibiting an unstable state. We characterized the short- and long-time deviations of the FQC-emulated system compared to the exact non-Hermitian case. Short-time deviations can be interpreted in terms of the Zeno effect, while the long-term deviations correspond to a probability revival that can be quantified using a measure of non-Markovianity. We provided a criterion for the adequacy of the discrete FQCs considered by evaluating the occupancy probabilities of the quasi-continuum states. There is a trade-off between using FQCs involving a large number of states and achieving high accuracy in emulating non-Hermitian dynamics. We showed that, in the strong coupling regime, this trade-off can be significantly improved by considering FQCs with an adapted density of states. This study is potentially relevant for many body systems, where a given subsystem can be coupled to a large set of states corresponding to the surrounding bodies [7]. Quantum dots coupled to nano-wires are a promising platform for implementing low-dimensional systems coupled to FQCs [34,35]. This work also paves the way for the emulation of non-Hermitian dynamics with a finite set of states. A long-term goal is to integrate a tunable dissipation within quantum simulators [7]. Different methods have been investigated to reach this goal, relying on the Zeno effect [36–39], atom losses [40,41], and multichromatic Floquet [42], to name a few.

Author Contributions: Conceptualization, D.G.-O.; Formal analysis, D.G.-O., E.F.; Investigation, E.F.; Methodology, D.G.-O., E.F., F.I.; Validation, D.G.-O., F.I.; Supervision, D.G.-O.;writing—original draft preparation, D.G.-O., F.I.; writing—review and editing, F.I., D.G.-O. All authors have read and agreed to the published version of the manuscript.

Funding: This research was funded by the Brazilian agencies CNPq (310265/2020-7), CAPES and FAPERJ (210.296/2019), by the CAPES-PRINT Program and by INCT-IQ (465469/2014-0).

Data Availability Statement: Data is contained within the article or supplementary material.

Conflicts of Interest: The authors declare no conflict of interest.

References

1. Khalfin, S.A. Contribution to the Decay Theory of a Quasi-Stationary State. *Sov. J. Exp. Theor. Phys.* **1958**, *6*, 1053.
2. Chiu, C.B.; Sudarshan, E.C.G.; Misra, B. Time evolution of unstable quantum states and a resolution of Zeno's paradox. *Phys. Rev. D* **1977**, *16*, 520–529. [CrossRef]
3. Greenland, P.T. Seeking non-exponential decay. *Nature* **1988**, *335*, 298. [CrossRef]
4. Wilkinson, S.R.; Bharucha, C.F.; Fischer, M.C.; Madison, K.W.; Morrow, P.R.; Niu, Q.; Sundaram, B.; Raizen, M.G. Experimental evidence for non-exponential decay in quantum tunnelling. *Nature* **1997**, *387*, 575–577. [CrossRef]
5. Rothe, C.; Hintschich, S.I.; Monkman, A.P. Violation of the Exponential-Decay Law at Long Times. *Phys. Rev. Lett.* **2006**, *96*, 163601. [CrossRef]
6. Torrontegui, E.; Muga, J.G.; Martorell, J.; Sprung, D.W.L. Enhanced observability of quantum postexponential decay using distant detectors. *Phys. Rev. A* **2009**, *80*, 012703. [CrossRef]
7. Altman, E.; Brown, K.R.; Carleo, G.; Carr, L.D.; Demler, E.; Chin, C.; DeMarco, B.; Economou, S.E.; Eriksson, M.A.; Fu, K.M.C.; et al. Quantum Simulators: Architectures and Opportunities. *PRX Quantum* **2021**, *2*, 017003. [CrossRef]
8. Guo, A.; Salamo, G.J.; Duchesne, D.; Morandotti, R.; Volatier-Ravat, M.; Aimez, V.; Siviloglou, G.A.; Christodoulides, D.N. Observation of $\mathcal{PT}$-Symmetry Breaking in Complex Optical Potentials. *Phys. Rev. Lett.* **2009**, *103*, 093902. [CrossRef] [PubMed]
9. Wan, W.; Chong, Y.; Ge, L.; Noh, H.; Stone, A.D.; Cao, H. Time-Reversed Lasing and Interferometric Control of Absorption. *Science* **2011**, *331*, 889. [CrossRef]
10. Weidemann, S.; Kremer, M.; Helbig, T.; Hofmann, T.; Stegmaier, A.; Greiter, M.; Thomale, R.; Szameit, A. Topological funneling of light. *Science* **2020**, *368*, 311. [CrossRef]
11. Okuma, N.; Kawabata, K.; Shiozaki, K.; Sato, M. Topological Origin of Non-Hermitian Skin Effects. *Phys. Rev. Lett.* **2020**, *124*, 086801. [CrossRef] [PubMed]
12. Zhang, K.L.; Yang, X.M.; Song, Z. Quantum transport in non-Hermitian impurity arrays. *Phys. Rev. B* **2019**, *100*, 024305. [CrossRef]
13. Damanet, F.; Mascarenhas, E.; Pekker, D.; Daley, A.J. Controlling Quantum Transport via Dissipation Engineering. *Phys. Rev. Lett.* **2019**, *123*, 180402. [CrossRef] [PubMed]
14. Shu, C.; Zhang, K.; Su, K. Loss-induced universal one-way transport in periodically driven systems. *arXiv* **2023**, arXiv:2306.10000.
15. Yao, S.; Wang, Z. Edge States and Topological Invariants of Non-Hermitian Systems. *Phys. Rev. Lett.* **2018**, *121*, 086803. [CrossRef]
16. Kunst, F.K.; Edvardsson, E.; Budich, J.C.; Bergholtz, E.J. Biorthogonal Bulk-Boundary Correspondence in Non-Hermitian Systems. *Phys. Rev. Lett.* **2018**, *121*, 026808. [CrossRef]
17. Martinez Alvarez, V.M.; Barrios Vargas, J.E.; Foa Torres, L.E.F. Non-Hermitian robust edge states in one dimension: Anomalous localization and eigenspace condensation at exceptional points. *Phys. Rev. B* **2018**, *97*, 121401. [CrossRef]
18. Lee, C.H.; Thomale, R. Anatomy of skin modes and topology in non-Hermitian systems. *Phys. Rev. B* **2019**, *99*, 201103. [CrossRef]
19. Yokomizo, K.; Murakami, S. Non-Bloch Band Theory of Non-Hermitian Systems. *Phys. Rev. Lett.* **2019**, *123*, 066404. [CrossRef]
20. Zhang, K.; Yang, Z.; Fang, C. Correspondence between Winding Numbers and Skin Modes in Non-Hermitian Systems. *Phys. Rev. Lett.* **2020**, *125*, 126402. [CrossRef]
21. Yang, Z.; Zhang, K.; Fang, C.; Hu, J. Non-Hermitian Bulk-Boundary Correspondence and Auxiliary Generalized Brillouin Zone Theory. *Phys. Rev. Lett.* **2020**, *125*, 226402. [CrossRef]
22. Stey, G.; Gibberd, R. Decay of quantum states in some exactly soluble models. *Physica* **1972**, *60*, 1–26. [CrossRef]
23. Tannoudji, C.C.; Grynberg, G.; Dupont-Roe, J. *Atom-Photon Interactions: Basic Processes and Applications*; John Wiley and Sons: New York, NY, USA, 1992.
24. Verstraete, F.; Wolf, M.M.; Cirac, J.I. Quantum computation and quantum-state engineering driven by dissipation. *Nat. Phys.* **2009**, *5*, 633. [CrossRef]
25. Winter, R.G. Evolution of a Quasi-Stationary State. *Phys. Rev.* **1961**, *123*, 1503–1507. [CrossRef]
26. Fonda, L.; Ghirardi, G.C.; Rimini, A. Decay theory of unstable quantum systems. *Rep. Prog. Phys.* **1978**, *41*, 587. [CrossRef]
27. Peres, A. Nonexponential decay law. *Ann. Phys.* **1980**, *129*, 33–46. [CrossRef]
28. Bienaimé, T.; Piovella, N.; Kaiser, R. Controlled Dicke Subradiance from a Large Cloud of Two-Level Systems. *Phys. Rev. Lett.* **2012**, *108*, 123602. [CrossRef] [PubMed]
29. Nielsen, M.A.; Chuang, I.L. *Quantum Computation and Quantum Information*; Cambridge University Press: Cambridge, UK, 2000.

30. Dalvit, D.A.R.; Maia Neto, P.A.; Mazzitelli, F.D. Fluctuations, Dissipation and the Dynamical Casimir Effect. *Lect. Notes Phys.* **2011**, *834*, 287.
31. e Souza, R.D.M.; Impens, F.; Neto, P.A.M. Microscopic dynamical Casimir effect. *Phys. Rev. A* **2018**, *97*, 032514. [CrossRef]
32. Impens, F.; e Souza, R.D.M.; Matos, G.C.; Neto, P.A.M. Dynamical Casimir effects with atoms: From the emission of photon pairs to geometric phases. *Europhys. Lett.* **2022**, *138*, 30001. [CrossRef]
33. Breuer, H.P.; Laine, E.M.; Piilo, J. Measure for the Degree of Non-Markovian Behavior of Quantum Processes in Open Systems. *Phys. Rev. Lett.* **2009**, *103*, 210401. [CrossRef] [PubMed]
34. Ricco, L.S.; Kozin, V.K.; Seridonio, A.C.; Shelykh, I.A. Reshaping the Jaynes-Cummings ladder with Majorana bound states. *Phys. Rev. A* **2022**, *106*, 023702. [CrossRef]
35. Ricco, L.S.; Kozin, V.K.; Seridonio, A.C.; Shelykh, I.A. Accessing the degree of Majorana nonlocality in a quantum dot-optical microcavity system. *Sci. Rep.* **2022**, *12*, 1983. [CrossRef]
36. Syassen, N.; Bauer, D.M.; Lettner, M.; Volz, T.; Dietze, D.; García-Ripoll, J.J.; Cirac, J.I.; Rempe, G.; Dürr, S. Strong Dissipation Inhibits Losses and Induces Correlations in Cold Molecular Gases. *Science* **2008**, *320*, 1329–1331. [CrossRef] [PubMed]
37. Barontini, G.; Labouvie, R.; Stubenrauch, F.; Vogler, A.; Guarrera, V.; Ott, H. Controlling the Dynamics of an Open Many-Body Quantum System with Localized Dissipation. *Phys. Rev. Lett.* **2013**, *110*, 035302. [CrossRef]
38. Zhu, B.; Gadway, B.; Foss-Feig, M.; Schachenmayer, J.; Wall, M.L.; Hazzard, K.R.A.; Yan, B.; Moses, S.A.; Covey, J.P.; Jin, D.S.; et al. Suppressing the Loss of Ultracold Molecules Via the Continuous Quantum Zeno Effect. *Phys. Rev. Lett.* **2014**, *112*, 070404. [CrossRef]
39. Tomita, T.; Nakajima, S.; Danshita, I.; Takasu, Y.; Takahashi, Y. Observation of the Mott insulator to superfluid crossover of a driven-dissipative Bose-Hubbard system. *Sci. Adv.* **2017**, *3*, e1701513,
40. Rauer, B.; Grišins, P.; Mazets, I.E.; Schweigler, T.; Rohringer, W.; Geiger, R.; Langen, T.; Schmiedmayer, J. Cooling of a One-Dimensional Bose Gas. *Phys. Rev. Lett.* **2016**, *116*, 030402. [CrossRef]
41. Schemmer, M.; Bouchoule, I. Cooling a Bose Gas by Three-Body Losses. *Phys. Rev. Lett.* **2018**, *121*, 200401. [CrossRef]
42. Impens, F.; Guéry-Odelin, D. Multichromatic Floquet engineering of quantum dissipation. *arXiv* **2023**, arXiv:2306.01676.

Disclaimer/Publisher's Note: The statements, opinions and data contained in all publications are solely those of the individual author(s) and contributor(s) and not of MDPI and/or the editor(s). MDPI and/or the editor(s) disclaim responsibility for any injury to people or property resulting from any ideas, methods, instructions or products referred to in the content.

Article

Control of the von Neumann Entropy for an Open Two-Qubit System Using Coherent and Incoherent Drives [†]

Oleg V. Morzhin *[ID] and Alexander N. Pechen *[ID]

Department of Mathematical Methods for Quantum Technologies & Steklov International Mathematical Center, Steklov Mathematical Institute of Russian Academy of Sciences, 8 Gubkina Str., 119991 Moscow, Russia
* Correspondence: morzhin.oleg@yandex.ru (O.V.M.); apechen@gmail.com (A.N.P.)
[†] Dedicated to the 120th anniversary of the birth of John von Neumann (28 December 1903).

Abstract: This article is devoted to developing an approach for manipulating the von Neumann entropy $S(\rho(t))$ of an open two-qubit system with coherent control and incoherent control inducing time-dependent decoherence rates. The following goals are considered: (a) minimizing or maximizing the final entropy $S(\rho(T))$; (b) steering $S(\rho(T))$ to a given target value; (c) steering $S(\rho(T))$ to a target value and satisfying the pointwise state constraint $S(\rho(t)) \leq \overline{S}$ for a given $\overline{S}$; (d) keeping $S(\rho(t))$ constant at a given time interval. Under the Markovian dynamics determined by a Gorini–Kossakowski–Sudarshan–Lindblad type master equation, which contains coherent and incoherent controls, one- and two-step gradient projection methods and genetic algorithm have been adapted, taking into account the specifics of the objective functionals. The corresponding numerical results are provided and discussed.

Keywords: quantum control; von Neumann entropy; quantum thermodynamics; open quantum system; coherent control; incoherent control; optimization methods; two-qubit system

Citation: Morzhin, O.V.; Pechen, A.N. Control of the von Neumann Entropy for an Open Two-Qubit System Using Coherent and Incoherent Drives. *Entropy* **2024**, *26*, 36. https://doi.org/10.3390/e26010036

Academic Editors: Fernando C. Lombardo and Paula I. Villar

Received: 24 November 2023
Revised: 20 December 2023
Accepted: 23 December 2023
Published: 29 December 2023
Corrected: 2 September 2024

Copyright: © 2023 by the authors. Licensee MDPI, Basel, Switzerland. This article is an open access article distributed under the terms and conditions of the Creative Commons Attribution (CC BY) license (https://creativecommons.org/licenses/by/4.0/).

1. Introduction

The theory of (optimal) control of quantum systems (atoms, molecules, etc.) is important for developing quantum technologies [1–20]. Modeling of control problems for quantum systems is based on various quantum mechanical equations with Markovian or non-Markovian dynamics, e.g., the Schrödinger, von Neumann, Gorini–Kossakowski–Sudarshan–Lindblad (GKSL) equations, and various objective functionals to be minimized or maximized. In practical applications, often the controlled quantum system is open, i.e., interacting with its environment, and this environment is considered as an obstacle for controlling the system. However, in some cases, one can use the environment as a useful control resource, such as, for example, in the *incoherent control* approach [21,22], where the spectral, generally time-dependent and non-equilibrium density of incoherent photons is used as a control function jointly with the *coherent control* via lasers to manipulate such a quantum system dynamics. Following this approach, various types and aspects of optimal control problems for one- and two-qubit systems were analyzed [23–28].

One particularly important class of quantum control objectives includes thermodynamic quantities and entropy of the quantum system. Properties of the von Neumann entropy in general are discussed, e.g., in [29–33]. The von Neumann entropy appears in various applied aspects of quantum theory, has applications in quantum communication and statistical physics [34–37], or even in cross-linguistic comparisons of language networks [38]. The von Neumann entropy of reduced density matrices of a bipartite quantum system provides a good measure of entanglement. It appears in various thermodynamic quantities, such as Helmholtz free energy, can serve as a degree of purity of a quantum state, etc. The system–bath interaction can play a crucial role in the emergence of the laws of thermodynamics from quantum consideration [39]. The control of dissipative quantum systems, which changes entropy of the quantum state, has been studied in various works. In particular, an analytical solution for the optimal control of a quantum dissipative three-level system leading to the decrease in entropy was provided [40]. Entropy production for controlled Markovian evolution was studied in [41]. The von Neumann entropy and Rényi entropy changes for the laser cooling of molecules were investigated [42]. A detailed study of entropy changing control targets is explored in [43], when the external drive influences not only the primary system but also the dissipation induced by the environment. Similar to the control of entropy is the state-to-state control between two Gibbs states, which is used to accelerate thermalization and cool for an open system [44]. The effects of the population decay, leading to the reduction of entropy, in a two-level Markovian dissipative system were considered in [45]. Reference [46] considers entanglement entropy maximization for the Lipkin–Meshkov–Glick model operating with $N = 50$ spins and the subsystem with $L = N/2$ spins using the free-gradient chopped random basis (CRAB) ansatz. Non-Markovian regimes can also be effective, e.g., for quantum battery and heat machines [47]. Reference [48] considers a stochastic master equation with a finite-dimensional measurement-based quantum feedback control and linear entropy. Reference [49] considers an open four-level atomic system and analyzes coherent control for the von Neumann entropy (total and reduced versions) via quantum interference. In [50–52], a controller design approach for a closed quantum system described by the Scrödinger equation in terms of the von Neumann/Shannon entropy was proposed. Reference [53] considers the spatial control of entropy for a three-level ladder-type atomic system that interacts with optical laser fields and an incoherent pumping field.

Reference [54] provides the formulation and analysis of control objectives describing optimization of thermodynamic quantities of the form $\langle O \rangle - \beta^{-1} S(\rho(T))$, where O is some quantum observable (e.g., energy with Hamiltonian H), β is inverse temperature, and $S(\rho(T))$ is some concave type of entropy, e.g., the von Neumann entropy, of an open quantum system density matrix at the final time T. The system evolution was considered as driven by some coherent and/or incoherent controls, including Markovian and non-Markovian cases, particularly the cases of master equations with coherent and incoherent controls [21]. The objective was expressed as a Mayer-type functional determined by the final state $\rho(T)$ ($\hat\rho_f$ at the final time t_f in the notations of [54]). The applied control $c = (u, n)$ (note that in [54] this most general combination of coherent and incoherent controls was denoted by symbol u, which in the present work denotes only coherent control, whereas the combination of controls here we denote by c) directs the evolution of the system from the initial state to the final state and specifies the value of the objective which depends, through $\rho(T)$, on the control c. A specific important example of such an objective is Helmholtz free energy, which corresponds to $O = H$. In the case of trivial observable $O = \text{const} \cdot \mathbb{I}$, the objective is reduced to the entropic form and differs from the entropy by a non-essential for the optimization constant term. Based on this objective and Reference [54], we define below several other control problems involving entropy.

The entropy of a quantum state was introduced by L. Landau to describe states of composite quantum systems [55], which is related to using entropy as a measure of entanglement, and by J. von Neumann to describe the thermodynamic properties of quantum systems [56]. This provides the motivation to introduce control problems focused on steering and maintaining the von Neumann entropy of system states. Objectives of forms (4)–(8) serve as examples of naturally extending problems related to maximizing or minimizing quantities involving entropy to controlling their behavior over a certain time range. Such a natural extension, in general, can include (but is not limited to) the following:

- Control the behavior of thermodynamic quantities, such as Helmholtz free energy, not only at the final time instant but over some time range;
- Control of the degree of entanglement of a bipartite system over time;
- To not only maximize or minimize but rather control the rate of entropy production.

The basic task for all these problems is to manipulate entropy over a given time range which, including optimization methods, we consider in this work.

In general, quantum (open-loop) control, both for closed and open quantum systems, various types of optimization tools are used:
- For infinite-dimensional optimization, e.g., the Pontryagin maximum principle (PMP) [20,57,58], Krotov-type methods ([24,59,60], [19], § 16.2.2, [61], pp. 253–259), one- and two-step gradient projection methods (GPM-1, GPM-2) [23,24,28], etc.;
- For finite-dimensional optimization under various classes of parameterized controls, e.g., gradient ascent pulse engineering (GRAPE)-type methods (e.g., [25–27], Section 3, [62]) (GRAPE-type methods operate with piecewise-constant controls, matrix exponentials, and gradients), CRAB ansatz [46,63] (coherent control is considered in terms of sine, cosine, etc.), genetic algorithm (GA) [21,64,65], dual annealing [24], etc.

In this article, we develop an approach for (open loop) control of the von Neumann entropy for open quantum systems driven by simultaneous coherent and incoherent controls. For such a system, we study control objectives based on the von Neumann entropy of the system states:

$$S(\rho(t)) = -\text{Tr}(\rho(t) \log \rho(t)) = - \sum_{\lambda_i(t) \neq 0} \lambda_i(t) \log \lambda_i(t), \tag{1}$$

where log denotes the natural matrix logarithm and $\lambda_i(t)$ are eigenvalues of $\rho(t)$. For the initial time $t = 0$ and final time $t = T$, we consider, correspondingly, $S(\rho_0)$ and $S(\rho(T))$. The approach is based on using bounded coherent and incoherent controls to manipulate the von Neumann entropy. Since the control of the entropy requires, in general, changing the degree of purity of the system density matrix, it requires the ability to generate a given non-unitary dynamics. For this, the combination of coherent and incoherent controls introduced in [21] makes a suitable tool.

To achieve these goals, we formulate the corresponding objective functionals. These functionals contain either differentiable or non-differentiable forms. For the differentiable cases, both for the objective functionals of the Mayer and Mayer–Bolza types, we develop the one- and two-step GPMs for piecewise continuous controls based on deriving gradients of the objective functionals and the corresponding adjoint systems. For the non-differentiable cases, piecewise linear controls are considered instead, and finite-dimensional optimization is performed using GA. Moreover, various forms of regularization in controls are provided.

The structure of the article is the following. In Section 2, we briefly outline the incoherent control approach. In Section 3, the objective functionals involving entropy for the described above problems are defined. In Section 4, we consider—as an example—an open two-qubit system whose dynamics are determined by a GKSL-type master equation, which contains coherent and incoherent controls. Section 5 describes the optimization approaches. Section 6 provides and discusses the analytical and numerical results. Conclusions Section 7 resumes the article.

2. Incoherent Control and Time-Dependent Decoherence Rates

The idea of incoherent control is to consider the environment as a useful resource for manipulating quantum systems. There are various approaches to using the environment as a control. We exploit the idea proposed and developed for generic quantum systems in [21,22]. In this approach, the state of the environment is used as a control. Usually, the state of the environment is considered as the Gibbs (thermal) state with some temperature. However, the state of the environment can be a more general non-thermal non-equilibrium state. If the environment consists of photons, which is one of the most typical physical examples of the environment, its more general non-equilibrium state at some time instant t is characterized by the distribution $n_{\mathbf{k},\alpha}(t)$ of photons in momenta $\mathbf{k}$ and polarization α. Moreover, this state and, hence, this distribution can evolve with time. Non-thermal distributions for photons are relatively easy to generate, so that it is a physical and technically possible way of control. In this work, we neglect polarization and directional dependence so that, here, the control is the distribution of photons only in frequency ω and time, $n_\omega(t)$. In the most general consideration, polarization and directional selectivity can be taken into account for the control.

A time-evolving distribution of photons induced generally time-dependent decoherence rates of the system, which is immersed in this photonic environment, so that under certain approximations, the master equation for the system density matrix can be considered as

$$\frac{d\rho(t)}{dt} = \mathcal{L}_t^{u,n}(\rho(t)) := -i[H_t^{u,n}, \rho(t)] + \varepsilon \underbrace{\sum_k \gamma_k(t) \mathcal{D}_k(\rho(t))}_{\mathcal{D}_t^n(\rho(t))}, \quad \rho(0) = \rho_0, \quad t \in [0, T]. \quad (2)$$

Here, both Markovian and non-Markovian cases can be included. The general formulation below is performed for both Markovian and non-Markovian cases, while only the Markovian case is explicitly analyzed. In [21], the dissipators $\mathcal{D}_k$ corresponding to the weak coupling and low-density limits in the theory of open quantum systems were explicitly considered. In general, other regimes, e.g., the ultrastrong coupling and the strong-decoherence limits [66,67], or weakly damped quantum systems in various regimes [68], can be considered as well. For the weak coupling limit case, the decoherence rate for the transition between system states $|i\rangle$ and $|j\rangle$ with transition frequency $\omega_{ij} = E_j - E_i$ (here, E_i is the energy of the system state $|i\rangle$) were considered in [21] as

$$\gamma_{ij}(t) = \pi \int \delta(\omega_{ij} - \omega_\mathbf{k}) |g(\mathbf{k})|^2 (n_{\omega_{ij}}(t) + \kappa_{ij}) d\mathbf{k}, \quad i,j = 1, \ldots, N.$$

Here, $\kappa_{ij} = 1$ for $i > j$ and $\kappa_{ij} = 0$; otherwise, $\omega_\mathbf{k}$ is the dispersion law for the bath (e.g., $\omega = |\mathbf{k}|c$ for photons, where $\mathbf{k}$ denotes photon momentum, c denotes the speed of light), and $g(\mathbf{k})$ describes the coupling of the system to the $\mathbf{k}$-th mode of the photonic reservoir. For $i > j$, the summand $\kappa_{ij} = 1$ describes spontaneous emission and γ_{ij} determines the rate of both spontaneous and induced emissions between levels i and j. For $i < j$, γ_{ij} determines the rate of induced absorption. These decoherence rates appear in (2), where $k = (i,j)$ is multi-indexed.

Such incoherent control appears to be rich enough to approximately generate, when combined with fast coherent control, arbitrary density matrices of generic quantum systems within the scheme proposed in [22]. Hence, it can approximately realize the strongest possible degree of quantum state control—controllability of open quantum systems in the set of all density matrices. This scheme has several important features. (1) It was obtained for a physical class of dissipators $\mathcal{D}_k$ known in the weak coupling limit. (2) It was obtained for generic quantum systems of an arbitrary dimension and for almost all values of the system parameters. (3) A simple explicit analytic solution for incoherent control was obtained. (4) The control scheme is robust to variations of the initial state—the optimal control steers simultaneously *all* initial states into the target state, thereby physically realizing all-to-one Kraus maps theoretically exploited for quantum control in [69] and recently experimentally for an open single qubit in [70]. In [22], coherent and incoherent controls were separated in time (first coherent control, followed by incoherent) and were applied to the system on different time scales determined by the parameters of the system. Incoherent control was applied on a time scale slower than coherent control. When coherent and incoherent controls are applied simultaneously, such a difference in time scales may lead to bounds on variations of incoherent control, considering that incoherent control should be varied slowly compared to coherent control. In the analysis below, Equation (12) is used to take into account such bounds on variations of the incoherent control. To shorten the incoherent control time scale, the first stage of the incoherent control scheme proposed in [22] was further modified for a two-level system in [27], significantly reducing the control time scale. Such an incoherent control can be technically implemented, e.g., as it was done for controlling multi-species atomic and molecular systems with $Gd_2O_2S:Er^{3+}$ (6%) samples [71].

3. Control Objective Functionals Involving Entropy

In this section, we define control objective functionals, describing various problems involving entropy including both Markovian and non-Markovian cases.

Fixing T, ρ_0, control $c = (u, n)$, ε, and so on, one solves the initial problem (2) with the initial condition $\rho(0) = \rho_0$ to find the corresponding solution ρ, a matrix function defined at $[0, T]$. For each state $\rho(t)$, consider its von Neumann entropy $S(\rho(t))$. Using this standard notion of the von Neumann entropy, we formulate below several objective functionals based on the following objective functional for minimizing or maximizing the von Neumann entropy as considered in [54].

- *Minimizing or maximizing* the von Neumann entropy, or more general thermodynamic quantities (O is a Hermitian observable, for example, the Hamiltonian of the system, in this case, it is Helmholtz free energy) at a final time, as defined in [54]:

$$J_O(c) = \langle O \rangle - \frac{1}{\beta} S(\rho(T)) \to \inf / \sup, \quad \beta > 0. \tag{3}$$

Case $O = 0$ corresponds to the minimization or maximization of the entropy itself. Based on this objective, one can define the problem of keeping the thermodynamic observable invariant at the whole time range, steering the entropy to a given target level, making it follow a predefined trajectory, etc.

- For the problem of *keeping* the required invariant $S(\rho(t)) \equiv S(\rho_0)$ at the whole time range $[0, T]$, we consider

$$J_1(c) = (S(\rho(T)) - S(\rho_0))^2 + P \int_0^T (S(\rho(t)) - S(\rho_0))^2 dt \to \inf, \tag{4}$$

where the penalty coefficient $P > 0$ and the final time T are fixed. Although one can expect such a case that making the integral close to zero does not provide $S(\rho(t)) \approx S(\rho_0)$ at the whole $[0, T]$; however, (4) is of interest, because, first, it can be useful and, second, it is appropriate for the described below gradient approach (GPMs). Moreover, as a variant, one can formulate the problem

$$J_2(c) = \max_{\{t_1 > 0, \dots, t_k, \dots, t_M = T\}} |S(\rho(t_k)) - S(\rho_0)| \to \inf, \tag{5}$$

which is considered below together with piecewise linear controls and GA.

- For the problem of *steering* the von Neumann entropy to a given target value S_{tar}, we consider

$$J_3(c) = (S(\rho(T)) - S_{\text{tar}})^2 \to \inf, \quad S_{\text{tar}} \neq S(\rho_0), \tag{6}$$

where T is fixed, as necessary for the considered GPMs. In extension, one can analyze a series of such steering problems for various values T and look for such an approximately minimal T for which the required value S_{tar} is reached.

- In addition to the steering problem with J_3, we consider the pointwise state constraint $S(\rho(t)) \leq \overline{S}$ for a given $\overline{S} > S(\rho_0)$ at the whole $[0, T]$ by adding to J_3 the integral term, taking into account the constraint:

$$J_4(c) = (S(\rho(T)) - S_{\text{tar}})^2 + P \int_0^T (\max\{S(\rho(t)) - \overline{S}, 0\})^2 dt \to \inf, \quad P > 0. \tag{7}$$

Here, the final time T and the penalty coefficient $P > 0$ are fixed. Moreover, as a variant, one can consider non-fixed T and take into account the state constraint as follows:

$$J_5(c,T) = |S(\rho(T)) - S_{\text{tar}}|$$
$$+ P \max_{\{t_1 > 0, \ldots, t_k, \ldots, t_M = T\}} \left(\max\{S(\rho(t_k)) - \overline{S}, 0\} \right) \to \inf, \quad P > 0, \quad (8)$$

where T is considered free at a given range $[T_1, T_2]$. As for J_2, we consider J_5 for piecewise linear controls and perform finite-dimensional optimization using GA.

For the objective functionals $J_1(c)$, $J_3(c)$, $J_4(c)$, below the GPM-1 and GPM-2 are formulated for the class of bounded piecewise continuous controls. For a unified description of the GPMs for these three optimal control problems, we use the following notation:

$$\Phi(c) \text{ is } J_1(c) \text{ or } J_3(c) \text{ or } J_4(c),$$

$$F(\rho) = \begin{cases} (S(\rho) - S(\rho_0))^2, & \text{if } J_1 \text{ is used,} \\ (S(\rho) - S_{\text{tar}})^2, & \text{if } J_3 \text{ or } J_4 \text{ is used,} \end{cases} \quad (9)$$

$$g(\rho) = \begin{cases} 0, & \text{if } J_3 \text{ is used,} \\ (S(\rho) - S(\rho_0))^2, & \text{if } J_1 \text{ is used,} \\ (\max\{S(\rho) - \overline{S}, 0\})^2, & \text{if } J_4 \text{ is used.} \end{cases} \quad (10)$$

The objective functionals $J_2(c)$, $J_5(c,T)$, as it is noted above, we consider with piecewise linear controls. Such a control c is determined by control parameters corresponding to a set of nodes at $[0, T]$. For example, one can consider a uniform grid $\{t_1 = 0, \ldots, t_s, \ldots, t_N = T\}$ with the step $\Delta t = T/N$ and the representation

$$u(t) = u^s + (t - t_s)(u^{s+1} - u^s)/\Delta t, \quad n_j(t) = n_j^s + (t - t_s)(n_j^{s+1} - n_j^s)/\Delta t, \quad j = 1,2$$

that allows introducing the vector of parameters,

$$\mathbf{a} = (a_i)_{i=1}^{3N} = (u^1, \ldots, u^N, n_1^1, \ldots, n_1^N, n_2^1, \ldots, n_2^N),$$

satisfying the constraints $|u^s| \leq u_{\max}$, $n_j^s \in [0, n_{\max}]$ for $j = 1,2$ and $s = 1, \ldots, N$, and defining such controls u, n_1, n_2. Moreover, as we show below, it can be useful to define a more sophisticated class of controls by defining c as piecewise linear at a subset of $[0, T]$ and setting constant (zero) for other times; in such a way, c is defined not only by $\mathbf{a}$. Anyway, we have deal with finite-dimensional optimization, where $J_2(c)$, $J_5(c,T)$ are represented by the corresponding objective functions $q_2(\mathbf{a})$ and $q_5(\mathbf{a},T)$ to be minimized. Moreover, for these objective functions, one can decide to add regularization in controls, e.g., for J_5, as follows:

$$q_5(\mathbf{a}, T; \gamma) = q_5(\mathbf{a}, T) + \gamma_u \max_{1 \leq s \leq N}\{|u^s|\} + \gamma_n \left(\max_{1 \leq s \leq N}\{n_1^s\} + \max_{1 \leq s \leq N}\{n_2^s\} \right) \to \inf, \quad (11)$$

where the coefficients γ_u, $\gamma_n \geq 0$. Moreover, as a variant, for the parameters, which represent incoherent controls, consider the inequality constraints $|n_j^{s+1} - n_j^s| \leq \delta_n^j$, $s = 1, \ldots, N-1$, $j = 1,2$, where the largest allowed jumps $\delta_n^j > 0$, $j = 1,2$ are predefined, and taking into account these constraints. E.g., for $J_2(c)$ and $q_2(\mathbf{a})$, consider

$$q_2(\mathbf{a}; \gamma) = q_2(\mathbf{a}) + \gamma_u \max_{1 \leq s \leq N}\{|u^s|\}$$
$$+ \gamma_n \sum_{j=1}^{2} \max\{\max_{1 \leq s \leq N-1}\{|n_j^{s+1} - n_j^s| - \delta_n^j, 0\}\} \to \inf. \quad (12)$$

This equation is used to take into account possible bounds on variations of the incoherent control.

For the objectives, for which GPMs are used below, e.g., for $J_3(c)$, one can add the following regularization term (like to [24], p. 14):

$$R(c; \gamma) = \int_0^T \left(\gamma_u u^2(t) + \gamma_n (n_1(t) + n_2(t)) \right) dt, \quad \gamma_u, \gamma_n \geq 0. \tag{13}$$

4. Markovian Two-Qubit System

As in [23,24], consider, as a particular case for (2), an open two-qubit system whose dynamics are determined by a GKSL-type master equation which contains coherent and incoherent controls and $H_t^{u,n} = H_S + H_{c(t)}$. Here, we deal with the following:

- The system state $\rho(t) : \mathcal{H} \to \mathcal{H}$ as a 4×4 density matrix (positive semi-definite, $\rho(t) \geq 0$, with unit trace, $\mathrm{Tr}\rho(t) = 1$) and a given initial density matrix ρ_0;
- Scalar coherent control u, vector incoherent control $n = (n_1, n_2)$, and the corresponding vector control $c = (u, n)$ considered in this work, in general, as piecewise continuous functions on $[0, T]$;
- H_S being the free Hamiltonian defined below;
- The controlled Hamiltonian $H_{c(t)} = \varepsilon H_{\mathrm{eff}, n(t)} + H_{u(t)}$, consisting of the effective Hamiltonian $H_{\mathrm{eff}, n(t)}$, which represents the Lamb shift and depends on $n(t)$, and of the Hamiltonian $H_{u(t)} = Vu(t)$, which describes interaction of the system with $u(t)$ and contains a Hermitian matrix V specified below as in [24];
- $\mathcal{D}_t^n$ being the controlled superoperator of dissipation, where we consider a special form of a Lindblad superoperator known in the weak coupling limit (see [21], etc.);
- The parameter $\varepsilon > 0$ describing the coupling strength between the system and the environment;
- The system of units with the Planck constant $\hbar = 1$.

The following detailed forms of the Hamiltonians are considered:

$$H_S = H_{S,1} + H_{S,2}, \quad H_{S,j} = \frac{\omega_j}{2} W_j, \quad W_1 := \sigma_z \otimes \mathbb{I}_2, \quad W_2 := \mathbb{I}_2 \otimes \sigma_z, \tag{14}$$

$$H_{\mathrm{eff}, n(t)} = \sum_{j=1}^{2} H_{\mathrm{eff}, n_j(t)}, \quad H_{\mathrm{eff}, n_j(t)} = \Lambda_j W_j n_j(t), \tag{15}$$

$$H_{u(t)} = Vu(t), \quad V = Q_1 \otimes \mathbb{I}_2 + \mathbb{I}_2 \otimes Q_2, \tag{16}$$

$$Q_j = \sum_{\alpha = x,y,z} \lambda_\alpha^j \sigma_\alpha = \sin \theta_j \cos \varphi_j \sigma_x + \sin \theta_j \sin \varphi_j \sigma_y + \cos \theta_j \sigma_z, \tag{17}$$

where $j = 1, 2$. Here $\sigma_x = \begin{pmatrix} 0 & 1 \\ 1 & 0 \end{pmatrix}$, $\sigma_y = \begin{pmatrix} 0 & -i \\ i & 0 \end{pmatrix}$, and $\sigma_z = \begin{pmatrix} 1 & 0 \\ 0 & -1 \end{pmatrix}$ are the X, Y, and Z Pauli matrices. The free Hamiltonian $H_{S,j}$ contains the transition frequency ω_j of the jth qubit. The effective Hamiltonian $H_{\mathrm{eff}, n(t)}$ represents the Lamb shift which describes shifts in transition frequencies of the qubits under the influence of the environment. The coefficients $\Lambda_j > 0$, $j = 1, 2$ together with $n_j(t)$ describe the influence of the environment on the Lamb shift. In $H_{u(t)}$, the unit vectors $\lambda^j := (\lambda_x^j, \lambda_y^j, \lambda_z^j) \in \mathbb{R}^3$, $j = 1, 2$. Physically, the Hamiltonian can describe either a pair of two-level atoms in electric fields polarized along the directions $\lambda^j := (\lambda_x^j, \lambda_y^j, \lambda_z^j) \in \mathbb{R}^3$, $j = 1, 2$, or two particles with spin $1/2$ in magnetic fields along the directions λ^j. In this model, the qubits independently interact with the coherent controls of the same intensity but with different directions determined by vectors λ^j, so that the interaction Hamiltonian V is the sum of two terms. In [23], in addition to this form, the case when coherent control induces interaction between the qubits was also considered. In contrast to [24], and this work, the articles [23] consider only the case where $Q_1 = Q_2 = \sigma_x$, i.e., in the present terms, $\theta_j = \pi/2$ and $\varphi_j = 0$, $j = 1, 2$.

As in [23,24], consider the following two-qubit superoperator of dissipation:

$$\mathcal{D}_t^n(\rho(t)) = \mathcal{D}_{n(t),1}(\rho(t)) + \mathcal{D}_{n(t),2}(\rho(t)), \qquad (18)$$

$$\mathcal{D}_{n(t),j}(\rho(t)) = \Omega_j(n_j(t)+1)\left(2\sigma_j^- \rho \sigma_j^+ - \sigma_j^+ \sigma_j^- \rho - \rho \sigma_j^+ \sigma_j^-\right)$$
$$+ \Omega_j n_j(t)\left(2\sigma_j^+ \rho \sigma_j^- - \sigma_j^- \sigma_j^+ \rho - \rho \sigma_j^- \sigma_j^+\right), \qquad j=1,2. \qquad (19)$$

The coefficients $\Omega_j > 0$, $j = 1,2$ are determined by the system–environment microscopic interaction. The matrices $\sigma_j^\pm$ are

$$\sigma_1^\pm = \sigma^\pm \otimes \mathbb{I}_2, \qquad \sigma_2^\pm = \mathbb{I}_2 \otimes \sigma^\pm \qquad \text{with} \qquad \sigma^+ = \begin{pmatrix} 0 & 0 \\ 1 & 0 \end{pmatrix}, \quad \sigma^- = \begin{pmatrix} 0 & 1 \\ 0 & 0 \end{pmatrix}. \qquad (20)$$

Incoherent control n has the physical meaning of the density of particles of the system environment and, therefore, should be non-negative. Moreover, we consider the parallelepipedal constraints:

$$c(t) = (u(t), n_1(t), n_2(t)) \in [-u_{\max}, u_{\max}] \times [0, n_{\max}]^2 = Q, \quad \text{for all} \quad t \in [0, T], \qquad (21)$$

where $u_{\max}$, $n_{\max} > 0$. The parameters ε, ω_1, ω_2, Λ_1, Λ_2, θ_1, θ_2, φ_1, φ_2, Ω_1, Ω_2, $u_{\max}$, $n_{\max}$ are considered fixed when we formulate the optimal control problems, while modifying some of them alters the quantum dynamics, i.e., one can vary them for a deeper analysis.

In this article, the two-qubit system is considered, in general, with piecewise continuous controls. The described below GPMs operate in theory with such controls, and the performed computer implementations of GPMs use piecewise linear interpolation for controls. For the non-differentiable objectives, we consider piecewise linear controls that, in contrast to piecewise constant controls used in the GRAPE-type method in [25], is another way of parameterization of controls.

For such a Markovian two-qubit system, the corresponding evolution equation for real-valued states was obtained in [23] and has the form

$$\frac{dx(t)}{dt} = (A + B_u u(t) + B_{n_1} n_1(t) + B_{n_2} n_2(t)) x(t), \quad x(0) = x_{\rho_0}, \qquad (22)$$

obtained using the parameterization of the system density matrix,

$$\rho = \begin{pmatrix} \rho_{1,1} & \rho_{1,2} & \rho_{1,3} & \rho_{1,4} \\ \rho_{1,2}^* & \rho_{2,2} & \rho_{2,3} & \rho_{2,4} \\ \rho_{1,3}^* & \rho_{2,3}^* & \rho_{3,3} & \rho_{3,4} \\ \rho_{1,4}^* & \rho_{2,4}^* & \rho_{3,4}^* & \rho_{4,4} \end{pmatrix} = \begin{pmatrix} x_1 & x_2 + ix_3 & x_4 + ix_5 & x_6 + ix_7 \\ x_2 - ix_3 & x_8 & x_9 + ix_{10} & x_{11} + ix_{12} \\ x_4 - ix_5 & x_9 - ix_{10} & x_{13} & x_{14} + ix_{15} \\ x_6 - ix_7 & x_{11} - ix_{12} & x_{14} - ix_{15} & x_{16} \end{pmatrix}. \qquad (23)$$

To analyze the dynamics of each qubit separately, we consider the reduced density matrices $\rho^j \in \mathbb{C}^{2\times 2}$, $j = 1,2$, and the corresponding Bloch vectors for the two qubits

$$\rho^1 = \mathrm{Tr}_{\mathcal{H}_2} \rho = \sum_{k=1}^{2} (\mathbb{I}_2 \otimes \langle k|) \rho (\mathbb{I}_2 \otimes |k\rangle), \qquad \rho^2 = \mathrm{Tr}_{\mathcal{H}_1} \rho = \sum_{k=1}^{2} (\langle k| \otimes \mathbb{I}_2) \rho (|k\rangle \otimes \mathbb{I}_2), \qquad (24)$$

where $|k\rangle$ are basis vectors in $\mathcal{H}_1$ and $\mathcal{H}_2$. Because the density matrix of a qubit can be bijectively mapped to the Bloch ball (in $\mathbb{R}^3$, this ball is centered in the point $(0,0,0)$ and has the unit radius), consider Bloch vectors $r^j = (r_x^j, r_y^j, r_z^j)$ where $r_\alpha^j = \mathrm{Tr}(\rho^j \sigma_\alpha)$, $\alpha \in \{x,y,z\}$, $|r^j| \leq 1$, $j = 1,2$. In terms of parameterization (23), one has:

$$r^1 = (2(x_4 + x_{11}), \quad -2(x_5 + x_{12}), \quad x_1 + x_8 - x_{13} - x_{16}), \qquad (25)$$
$$r^2 = (2(x_2 + x_{14}), \quad -2(x_3 + x_{15}), \quad x_1 - x_8 + x_{13} - x_{16}). \qquad (26)$$

Reduced density matrices are $\rho^j = \frac{1}{2}\begin{pmatrix} 1+r_z^j & r_x^j - ir_y^j \\ r_x^j + ir_y^j & 1-r_z^j \end{pmatrix}$, $j = 1, 2$. Further, for density matrices $\rho^1(t)$ and $\rho^2(t)$ vs $t \in [0, T]$, we consider their von Neumann entropies, i.e., $S(\rho^j(t)) = -\text{Tr}(\rho^j(t) \log \rho^j(t))$, $j = 1, 2$, and the sum $S(\rho^1(t)) + S(\rho^2(t))$. The behavior of these quantities in the numerical experiments is shown below in Figures 1b and 2c,f,i.

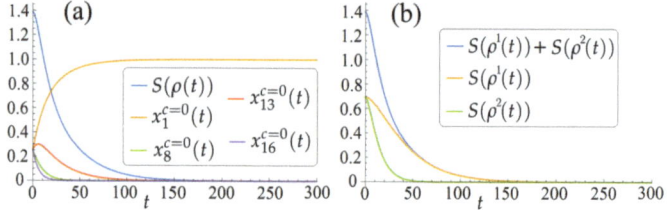

Figure 1. For the initial state $\rho_0 = \frac{1}{4}\mathbb{I}_4$ and the control $c = 0$: (**a**) the von Neumann entropy $S(\rho(t))$ and $x_j^{c=0}(t)$, $j = 1, 8, 13, 16$, i.e., the diagonal elements of the diagonal $\rho(t)$, vs $t \in [0, T = 300]$ (in this case, the entropy steers from the largest value $\log 4 \approx 1.39$ to zero, indicating the system's state purification and minimization of $S(\rho(T))$; (**b**) the entropies $S(\rho^1(t))$ and $S(\rho^2(t))$ for the first and second qubits, correspondingly, and the sum $S(\rho^1(t)) + S(\rho^2(t))$, vs $t \in [0, T = 300]$, steer to zero.

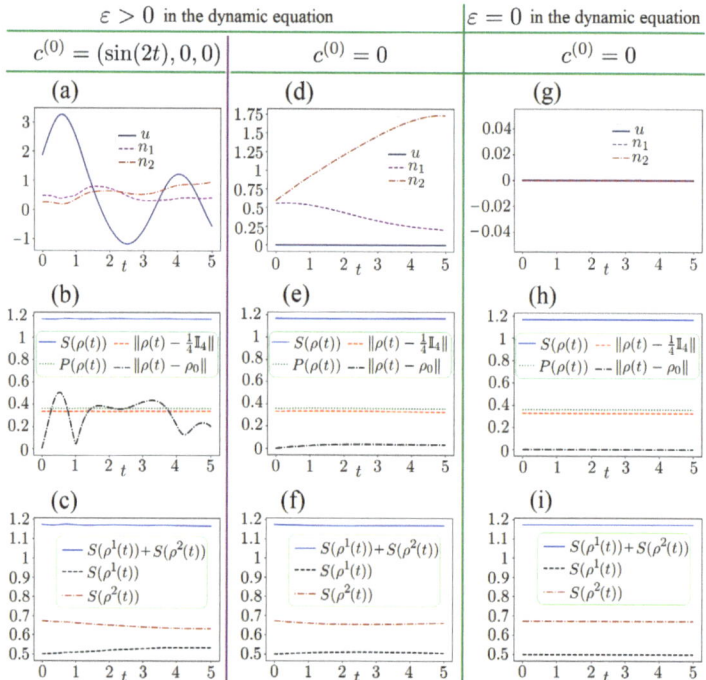

Figure 2. For the problem of keeping the invariant $S(\rho(t)) \equiv S(\rho_0)$ at the whole $[0, T = 5]$. Problem (4) and GPM-2 are used: (1) the subfigures (**a**–**c**) shows the results for $\varepsilon = 0.1$ and $c^{(0)} = (\sin(2t), 0, 0)$; (2) the subfigures (**d**–**f**) shows the results for $\varepsilon = 0.1$ and $c^{(0)} = 0$; (3) the subfigures (**g**–**i**) shows the results for $\varepsilon = 0$ (i.e., without taking into account the Lamb shift and the dissipator) and $c^{(0)} = 0$. The subfigures (**a**,**d**,**g**) show the obtained controls; for these controls, the subfigures (**b**,**e**,**h**) and (**c**,**f**,**i**) show, correspondingly, the two-qubit system characteristics ($S(\rho(t))$, etc.) vs t and the entropies $S(\rho^1(t))$, $S(\rho^2(t))$, their sums vs t. In the cases related to the subfigures (**c**,**f**), we see that for each qubit its entropy is not constant.

5. Numerical Optimization Tools: Markovian Two-Qubit Case

5.1. Gradient-Based Optimization Approach for the Problems with J_1, J_3, J_4

5.1.1. Pontryagin Function and Krotov Lagrangian

According to the theory of optimal control (e.g., [72]), for the unified optimal control problem with $\Phi(c)$ representing J_1, J_3, J_4, the Pontryagin function is

$$h(\chi, \rho, c) = \langle \chi, -i[H_c, \rho] + \varepsilon \mathcal{D}_n(\rho) \rangle - P g(\rho) - \gamma_u u^2 - \gamma_n (n_1 + n_2)$$
$$= \langle \mathcal{K}^c(\chi, \rho), c \rangle - \gamma_u u^2 - \gamma_n (n_1 + n_2) + \overline{h}(\chi, \rho),$$

where χ and ρ are 4×4 density matrices; $c = (u, n_1, n_2) \in \mathbb{R}^3$; the functions

$$\mathcal{K}^c = (\mathcal{K}^u, \mathcal{K}^{n_1}, \mathcal{K}^{n_2}), \quad \mathcal{K}^u(\chi, \rho) = \langle \chi, -i[V, \rho] \rangle, \quad (27)$$

$$\mathcal{K}^{n_j}(\chi, \rho) = \left\langle \chi, -i[\Lambda_j W_j, \rho] + \varepsilon \Omega_j \left(2\sigma_j^- \rho \sigma_j^+ + 2\sigma_j^+ \rho \sigma_j^- - \{\mathbb{I}_4, \rho\} \right) \right\rangle, \quad j = 1, 2 \quad (28)$$

(the 4×4 identity matrix $\mathbb{I}_4$ appears in $\mathcal{K}^{n_j}$ since $\sigma_j^+ \sigma_j^- + \sigma_j^- \sigma_j^+ = \mathbb{I}_4$), $j = 1, 2$; the term

$$\overline{h}(\chi, \rho) = \left\langle \chi, -i[H_S, \rho] + \varepsilon \sum_{j=1}^{2} \Omega_j (2\sigma_j^- \rho \sigma_j^+ - \{\sigma_j^+ \sigma_j^-, \rho\}) \right\rangle - P g(\rho).$$

As the Introduction notes, various Krotov-type iterative methods are used in quantum optimal control. In this article, we do not use any Krotov-type method, but we use the Krotov Lagrangian, which is the following for the unified problem:

$$L(c, \rho) = G(\rho(T)) - \int_0^T R(t, \rho(t), c(t)) dt,$$
$$G(\rho(T)) = F(\rho(T)) + \langle \chi(T), \rho(T) \rangle - \langle \chi(0), \rho_0 \rangle,$$
$$R(t, \rho, c) = \left\langle \chi(t), -i[H_c, \rho] + \varepsilon \mathcal{L}_n^D(\rho) \right\rangle + \langle \dot{\chi}(t), \rho \rangle - P g(\rho) - \gamma_u u^2 - \gamma_n (n_1 + n_2).$$

The function χ is defined in the next subsection as the solution of the adjoint system also defined below. For each admissible control c, the values of the Krotov Lagrangian and $\Phi(c)$ coincide, as in the general V.F. Krotov theory [61].

5.1.2. Unified Adjoint System and Gradient

Consider the increment of L at admissible controls $c, c^{(k)}$ (for the further consideration, we introduce $k \geq 0$ as an iteration index):

$$L(c, \rho) - L(c^{(k)}, \rho^{(k)}) = G(\rho(T)) - G(\rho^{(k)}(T))$$
$$- \int_0^T (R(t, \rho(t), c(t)) \quad R(t, \rho^{(k)}(t), c^{(k)}(t))) dt, \quad (29)$$

where the control process $(c^{(k)}, \rho^{(k)})$ is known.

By analogy with [61] (pp. 239–240) in the theory of optimal control, here for the increment (29), we consider the first-order Taylor expansions for G, R. At admissible controls c, $c^{(k)}$, this gives the representation

$$\Phi(c) - \Phi(c^{(k)}) = \left\langle \frac{d}{d\rho} G(\rho^{(k)}(T)), \rho(T) - \rho^{(k)}(T) \right\rangle$$
$$- \int_0^T \left\langle \frac{\partial}{\partial \rho} R(t, \rho^{(k)}(t), c^{(k)}(t)), \rho(t) - \rho^{(k)}(t) \right\rangle dt$$
$$- \int_0^T \left\langle \frac{\partial}{\partial c} R(t, \rho^{(k)}(t), c^{(k)}(t)), c(t) - c^{(k)}(t) \right\rangle_{E^3} dt + r.$$

Here, the notations with the derivatives mean that we initially find these derivatives with respect to ρ or (ρ, c), and after that, we substitute $\rho = \rho^{(k)}(T)$, etc.; r is the corresponding residual. Setting the derivatives $\frac{d}{d\rho}G(\rho^{(k)}(T))$ and $\frac{\partial}{\partial\rho}R(t, \rho^{(k)}(t), c^{(k)}(t))$ to be zero gives the *adjoint system* which defines the function $\chi^{(k)}$ as detailed below. As the result, the increment formula for the unified objective $\Phi(c)$ has the form

$$\Phi(c) - \Phi(c^{(k)}) = -\int_0^T \left\langle \frac{\partial}{\partial c} R(t, \rho^{(k)}(t), c^{(k)}(t)), c(t) - c^{(k)}(t) \right\rangle_{E^3} dt + r, \qquad (30)$$

$$\frac{\partial}{\partial c} R(t, \rho^{(k)}(t), c^{(k)}(t)) = \frac{\partial}{\partial c} h(\chi^{(k)}(t), \rho^{(k)}(t), c^{(k)}(t)) = \mathcal{K}^c(\chi^{(k)}(t), \rho^{(k)}(t))$$
$$- \gamma_u (u^{(k)}(t))^2 - \gamma_n (n_1^{(k)}(t) + n_2^{(k)}(t)).$$

The differentiation of the unified function F is needed to obtain the condition for the final co-state $\chi^{(k)}(T)$, i.e., the transversality condition; for differentiation of F, it is needed to consider the various forms of F shown in (9). For differentiation of R, it is needed to consider the various forms of $g(\rho)$ shown in (10). Using the matrix differential calculus (e.g., [73]), for the problems the following derivatives are found:

$$\frac{dS(\rho)}{d\rho} = -\log\rho - \mathbb{I}_{\dim\mathcal{H}=4},$$

$$\frac{dF(\rho)}{d\rho} = \begin{cases} \frac{d}{d\rho}(S(\rho) - S(\rho_0))^2, & \text{if } J_1 \text{ is used,} \\ \frac{d}{d\rho}(S(\rho) - \overline{S})^2, & \text{if } J_3 \text{ or } J_4 \text{ is used,} \end{cases}$$

$$= -2(\log\rho + \mathbb{I}_4) \begin{cases} S(\rho) - S(\rho_0), & \text{if } J_1 \text{ is used,} \\ S(\rho) - \overline{S}, & \text{if } J_3 \text{ or } J_4 \text{ is used,} \end{cases}$$

$$\frac{dg(\rho)}{d\rho} = \begin{cases} 0, & \text{if } J_3 \text{ is used,} \\ \frac{d}{d\rho}(S(\rho) - S(\rho_0))^2, & \text{if } J_1 \text{ is used,} \\ \frac{d}{d\rho}(\max\{S(\rho - \overline{S}, 0\})^2, & \text{if } J_4 \text{ is used} \end{cases}$$

$$= -2(\log\rho + \mathbb{I}_4) \begin{cases} 0, & \text{if } J_3 \text{ is used,} \\ S(\rho) - S(\rho_0), & \text{if } J_1 \text{ is used,} \\ \max\{S(\rho) - \overline{S}, 0\}, & \text{if } J_4 \text{ is used.} \end{cases}$$

To compute the derivative $\frac{\partial}{\partial \rho} R(t, \rho^{(k)}(t), c^{(k)}(t))$, one needs to operate with the right-hand side of the system (2) and take into account the corresponding properties such that the anti-commutativity property of commutator and cyclic permutation of matrices under trace. In this regard, and using the given formulas above for $\frac{dg(\rho)}{d\rho}$, we, as a result, obtain the adjoint system shown below in Proposition 1. This adjoint system contains the following superoperator acting on $\chi^{(k)}(t)$ (this superoperator is the same as derived in [24]):

$$\mathcal{D}^\dagger_{n^{(k)}(t)}(\chi^{(k)}(t)) = \sum_{j=1}^{2} \left[\Omega_j \left(n_j^{(k)}(t) + 1 \right) \left(2\sigma_j^+ \chi^{(k)}(t) \sigma_j^- - \{\sigma_j^+ \sigma_j^-, \chi^{(k)}(t)\} \right) \right.$$
$$\left. + \Omega_j n_j^{(k)}(t) \left(2\sigma_j^- \chi^{(k)}(t) \sigma_j^+ - \{\sigma_j^- \sigma_j^+, \chi^{(k)}(t)\} \right) \right],$$

where "†" reflects that $\langle \chi^{(k)}(t), T_1(\rho(t)) - T_1(\rho^{(k)}(t)) \rangle = \langle T_1^\dagger(\chi^{(k)}(t)), \rho(t) - \rho^{(k)}(t) \rangle$ and also $\langle \chi^{(k)}(t), T_2(\rho(t)) - T_2(\rho^{(k)}(t)) \rangle = \langle T_2^\dagger(\chi^{(k)}(t)), \rho(t) - \rho^{(k)}(t) \rangle$ with the operators

$T_1 := 2\sigma_j^- \cdot \sigma_j^+$, $T_2 := 2\sigma_j^+ \cdot \sigma_j^-$ and $(\sigma_j^+)^\top = \sigma_j^-$, $(\sigma_j^-)^\top = \sigma_j^+$; note that $\sigma_j^+\sigma_j^-$, $\sigma_j^-\sigma_j^+$ are Hermitian.

Proposition 1. *(Adjoint system). For the Markovian two-qubit case of the system (2) and the unified objective functional $\Phi(c)$ containing the unified terminant $F(\rho(T))$ and integrand $g(\rho(t))$, the adjoint system has the following form:*

$$\frac{d\chi^{(k)}(t)}{dt} = -i[H_{c^{(k)}(t)}, \chi^{(k)}(t)] - \varepsilon \mathcal{D}^\dagger_{n^{(k)}(t)}(\chi^{(k)}(t)) - P\frac{dg(\rho^{(k)}(t))}{d\rho}, \quad (31)$$

$$\chi^{(k)}(T) = -\frac{dF(\rho^{(k)}(T))}{d\rho}. \quad (32)$$

If the adjoint system is used with taking into account one of the two pointwise state constraints, then the system depends on $\rho^{(k)}$. Anyway, the adjoint system is linear in co-state $\chi^{(k)}$. This system is solved backward in time. In view of (30) and like the formula (3.16) in [24], we consider the gradient of the unified objective.

Proposition 2. *(Gradient). For the Markovian two-qubit case of the system (2) and the unified objective functional $\Phi(c)$, the corresponding gradient at a given admissible control $c^{(k)}$ has the form*

$$\operatorname{grad}\Phi(c^{(k)})(t) = \Big(-\mathcal{K}^u(\chi^{(k)}(t), \rho^{(k)}(t)) + 2\gamma_u u^{(k)}(t),$$
$$-\mathcal{K}^{n_j}(\chi^{(k)}(t), \rho^{(k)}(t)) + \gamma_n, \quad j = 1, 2 \Big), \quad t \in [0, T]. \quad (33)$$

Here $\rho^{(k)}$ is the solution of the Markovian case of the system (2) with control $c^{(k)}$, while $\chi^{(k)}$ is the solution of the adjoint system (31), (32) with the control process $(\rho^{(k)}, c^{(k)})$; the vector function $\mathcal{K}^c(\chi, \rho)$ defined in (27), (28) is used with these solutions.

In general, the formula (33) for the unified gradient reminds us, e.g., of the gradient formula (2.5.29) given in Reference [74] on the theory of optimal control with real-valued states.

5.1.3. Projection Form of the PMP

Following the projection form of the PMP known in the theory of optimal control (e.g., see [75]) and also its use in quantum control [58], below such a projection form of the PMP is formulated.

Proposition 3. *(Projection form of the differential version of the PMP for the unified problem with the objective $\Phi(c)$). For the Markovian two-qubit case of the system (2) and the unified objective functional $\Phi(c)$ with piecewise continuous controls satisfying (21) for a fixed final time $T > 0$, if an admissible control $\widehat{c} = (\widehat{u}, \widehat{n}_1, \widehat{n}_2)$ is a local minimum point of $\Phi(c)$ to be minimized, then for $\widehat{c}$ there exist such the solutions $\widehat{\rho}$ and $\widehat{\chi}$ that the pointwise condition*

$$\widehat{c}(t) = \operatorname{Pr}_Q\big(\widehat{c}(t) - \alpha \operatorname{grad}\Phi(\widehat{c})(t)\big), \quad t \in [0, T], \quad \alpha > 0, \quad (34)$$

holds and, in detail, has the form

$$\widehat{u}(t) = \begin{cases} -u_{\max}, & \widetilde{u}(t;\alpha) < -u_{\max}, \\ u_{\max}, & \widetilde{u}(t;\alpha) > u_{\max}, \\ \widetilde{u}(t;\alpha), & |\widetilde{u}(t;\alpha)| \le u_{\max}, \end{cases}$$

$$\text{where} \quad \widetilde{u}(t;\alpha) = \widehat{u}(t) + \alpha(\mathcal{K}^u(\widehat{\chi}(t), \widehat{\rho}(t)) - 2\gamma_u \widehat{u}(t)),$$

$$\widehat{n}_j(t) = \begin{cases} 0, & \widetilde{n}_j(t;\alpha) < 0, \\ n_{\max}, & \widetilde{n}_j(t;\alpha) > n_{\max}, \\ \widetilde{n}_j(t;\alpha), & \widetilde{n}_j(t;\alpha) \in [0, n_{\max}], \end{cases}$$

where $\widehat{n}_j(t;\alpha) = \widehat{n}_j(t) + \alpha(\mathcal{K}^{n_j}(\widehat{\chi}(t), \widehat{\rho}(t)) - \gamma_n)$, $j = 1, 2$.

5.1.4. One- and Two-Step Gradient Projection Methods

In the theory of optimal control, there are various forms of GPM-1 operating with control functions (e.g., see in [76–78]). In quantum control, for example, work [28] exploits GPM-1, which uses two algorithmic parameters (coefficient α for the gradient of the considered in that article objective functional and parameter $\theta \in [0,1]$ of the convex combination between the given control $c^{(k)}$ and depending on the α projection form for constructing $c^{(k+1)}$) and a scheme of one-dimensional optimization with respect to θ at each iteration, to search for the best variation of $c^{(k)}$ in the sense of the best decreasing objective. In contrast to [28], this article considers GPM-1 without the aforementioned convex combination and with a fixed α at the whole set of iterations. The considered GPM-2 is based on the heavy-ball method (see the works [79,80]), its projection version [81,82] and the recent papers [23,24], where the corresponding GPM-2 adaptations are used for quantum control.

For the unified optimal control problem and a given admissible initial guess $c^{(0)}$, consider the following GPMs iterative processes operating in the functional space of controls.

- GPM-1. The iteration process in the vector form is as follows and is reminiscent of (34):

$$c^{(k+1)}(t) = \text{Pr}_Q\big(c^{(k)}(t) - \alpha\,\text{grad}\,\Phi(c^{(k)})(t)\big), \quad \alpha > 0, \quad k \geq 0. \qquad (35)$$

In detail, we have

$$u^{(k+1)}(t) = \begin{cases} -u_{\max}, & u^{(k)}(t;\alpha) < -u_{\max}, \\ u_{\max}, & u^{(k)}(t;\alpha) > u_{\max}, \\ u^{(k)}(t;\alpha), & |u^{(k)}(t;\alpha)| \leq u_{\max}, \end{cases}$$

where $u^{(k)}(t;\alpha) = u^{(k)}(t) + \alpha(\mathcal{K}^u(\chi^{(k)}(t), \rho^{(k)}(t)) - 2\gamma_u u^{(k)}(t))$,

$$n_j^{(k+1)}(t) = \begin{cases} 0, & n_j^{(k)}(t;\alpha) < 0, \\ n_{\max}, & n_j^{(k)}(t;\alpha) > n_{\max}, \\ n_j^{(k)}(t;\alpha), & n_j^{(k)}(t;\alpha) \in [0, n_{\max}], \end{cases}$$

where $n_j^{(k)}(t;\alpha) = n_j^{(k)}(t) + \alpha(\mathcal{K}^{n_j}(\chi^{(k)}(t), \rho^{(k)}(t)) - \gamma_n)$, $j = 1, 2$;

- GPM-2. The iteration process in the vector form is as follows:

$$c^{(k+1)}(t) = \text{Pr}_Q\big(c^{(k)}(t) - \alpha\,\text{grad}\,J(c^{(k)})(t) \\ + \beta(c^{(k)}(t) - c^{(k-1)}(t))\big), \quad \alpha, \beta > 0, \quad k \geq 1, \qquad (36)$$

where $c^{(1)}$ is obtained using GPM-1 for a given initial guess $c^{(0)}$.

Here, the algorithmic parameters $\alpha, \beta > 0$ are fixed for all iterations. One may consider this, on the one hand, as a drawback, because we do not try to effectively variate these parameters, and, on the other hand, as a simpler case for the analysis. Moreover, here, relying on the various known computational facts about the heavy-ball method (e.g., see [83,84]), we take $\beta \in (0,1)$ and more likely $\beta = 0.8, 0.9$ in GPM-2, but not $\beta = 10$, etc. `TensorFlow MomentumOptimizer` [84] under the setting `use_nesterov = False` represents the heavy-ball method, where the parameter is 0.9 by default.

5.2. Zeroth-Order Stochastic Optimization for the Problems with J_2, J_5

GA belongs to zeroth-order stochastic tools, such as differential evolution, simulated annealing, particle-swarm optimization, sparrow search algorithm, etc., whose stochastic behavior models try to find a global minimizer of an objective function without its gradient due to these behavior models. In this article, the GA implementation [85] has been adjusted for the problems with the objectives J_2, J_5.

When a GA realization works with large $u_{\max}$, $n_{\max}$, then one can expect that the algorithm may miss a closer-to-optimal point, which is in a smaller subdomain. Because of the stochastic nature of GA, one can expect that, for the same optimization problem, the results of different trials of the GA may differ significantly even with the same deterministic settings (mutation probability, etc.). That is why one can perform—for the same optimization problem—several trials of the GA and then select the lowest computed value of the objective over the trials. However, e.g., if we consider the keeping problem (5) with regularization in controls and consider J_2 as sufficiently close to zero, and the profiles in the computed controls are acceptable, then it is not needed to perform more trials of the GA, because we know that zero is the lower bound for J_2.

6. Analytical and Numerical Analysis: Markovian Two-Qubit Case

In the numerical experiments, the following values of the system parameters are used:

$$\omega_1 = 1, \quad \omega_2 = 0.5, \quad \Lambda_1 = 0.3, \quad \Lambda_2 = 0.5, \quad \Omega_1 = 0.2, \quad \Omega_2 = 0.6, \quad \varepsilon = 0.1,$$
$$\varphi_1 = \pi/4, \quad \varphi_2 = \pi/3, \quad \theta_1 = \pi/3, \quad \theta_2 = \pi/4. \tag{37}$$

(except for Case 3 in Section 6.2, where for comparison, we set $\varepsilon = 0$). All the parameters are expressed in the relative units of free oscillation of the first qubit, which has period $T_1 = 2\pi$. Free oscillations of the second qubit have period $T_2 = 2T_1$. The decoherence rate is by the order of magnitude smaller than the oscillations of the first qubit. The difference between the qubit's free transition frequencies may occur twice, for example, in superconducting qubits. The system-environment coupling is determined by the parameter ε. This parameter specifies the (uncontrolled) decoherence rate, i.e., the rate of decoherence when $u = 0$ and $n \equiv 0$). Generally, the decoherence rate is several orders of magnitude smaller than the rate of free dynamics. In this study, we focus on cases where the decoherence rate is an order of magnitude slower than the free dynamics.

In the computer realizations (in Python) of GPM-1 and GPM-2, piecewise linear interpolation of controls u, n_1, n_2 is used at a uniform grid introduced over $[0, T]$ with M subintervals, i.e., with $M + 1$ time instances. To solve the considered ODEs, `solve_ivp` from `SciPy` is used.

6.1. Results on the von Neumann Entropy under Zero Coherent and Incoherent Controls

If one takes $c = 0$, then (22) becomes $\dfrac{dx^{c=0}}{dt} = Ax^{c=0}$, $x^{c=0}(0) = x_{\rho_0}$ whose solution is $x^{c=0}(t) = e^{At} x_{\rho_0}$. For the parameterized initial density matrix $\rho_0 = \mathrm{diag}(a_1, a_2, a_3, a_4)$ (s.t. $a_j \geq 0$, $j = 1, 2, 3, 4$, $\sum_{j=1}^{4} a_j = 1$) and the corresponding initial state $x_{\rho_0} = (a_1, \text{six zeros}, a_2, \text{four zeros}, a_3, 0, 0, a_4)$, as Reference [23] shows, system (22) for $c = 0$ has the following exact solution:

$$x_1^{c=0}(t) = a_1 + a_2 - a_2 e^{-2\varepsilon\Omega_2 t} + e^{-2\varepsilon(\Omega_1+\Omega_2)t}(e^{2\varepsilon\Omega_1 t} - 1)(a_3 e^{2\varepsilon\Omega_2 t} + a_4(e^{2\varepsilon\Omega_2 t} - 1)),$$
$$x_8^{c=0}(t) = e^{-2\varepsilon\Omega_2 t}(a_2 + a_4 - a_4 e^{-2\varepsilon\Omega_1 t}), \quad x_{13}^{c=0}(t) = e^{-2\varepsilon\Omega_1 t}(a_3 + a_4 - a_4 e^{-2\varepsilon\Omega_2 t}),$$
$$x_{16}^{c=0}(t) = a_4 e^{-2\varepsilon(\Omega_1+\Omega_2)t}, \quad x_j^{c=0}(t) = 0, \quad j \in \overline{1,16} \setminus \{1, 8, 13, 16\}, \quad t \geq 0. \tag{38}$$

The corresponding density matrix ρ is diagonal. Then the final von Neumann entropy is

$$S(\rho(T)) = -\sum_{x_j^{c=0}(T) \neq 0, \, j=1,8,13,16} x_j^{c=0}(T) \log x_j^{c=0}(T). \tag{39}$$

Using (25), (26), we obtain for the Bloch vectors:

$$r^1(t) = \left(r_x^1(t), r_y^1(t), r_z^1(t)\right) = \left(0, \, 0, \, x_1^{c=0}(t) + x_8^{c=0}(t) - x_{13}^{c=0}(t) - x_{16}^{c=0}(t)\right),$$

$$r^2(t) = \left(r_x^2(t), r_y^2(t), r_z^2(t)\right) = \left(0, 0, x_1^{c=0}(t) - x_8^{c=0}(t) + x_{13}^{c=0}(t) - x_{16}^{c=0}(t)\right).$$

Thus, the jth reduced density matrix is also diagonal, $\rho^j(t) = \frac{1}{2}\begin{pmatrix} 1+r_z^j(t) & 0 \\ 0 & 1-r_z^j(t) \end{pmatrix}$, and we have $S(\rho^j(t)) = \begin{cases} -\frac{1+r_z^j(t)}{2}\log\frac{1+r_z^j(t)}{2} - \frac{1-r_z^j(t)}{2}\log\frac{1-r_z^j(t)}{2}, & \text{if } r_z^j(t) \notin \{\pm 1\}, \\ 0, & \text{if otherwise.} \end{cases}$

Case 1: $\rho_0 = \frac{1}{4}\mathbb{I}_4$ ($a_1 = a_2 = a_3 = a_4 = \frac{1}{4}$), i.e., the completely mixed quantum state whose von Neumann entropy is the largest among 4×4 density matrices. Using (39), for (37) and $T = 50, 200, 250$, we obtain, correspondingly, $S(\rho(T)) \approx 0.2571, 0.0016, 0.0003$. For a sufficiently large T, this steering allows the purification of the system states with good quality. This corresponds to the problem of minimizing the objective functional $J_0(c) = S(\rho(T)) \to \inf$ that relates to (3). We see that in the considered case, the purification goal is achieved using the system-free evolution, i.e., without any non-trivial control c. Figure 1 shows $x_j^{c=0}(t), j = 1, 8, 13, 16$, and $S(\rho(t))$ computed via (39) vs $t \in [0, T = 300]$. We see that approximately $x_1^{c=0}$ steers to 1, while $x_8^{c=0}, x_{13}^{c=0}$, and $x_{16}^{c=0}$ steer to zero. This means that the system approximately steers to the pure state $\rho = \text{diag}(1, 0, 0, 0)$.

Case 2: $\rho_0 = \text{diag}(\frac{1}{2}, \frac{3}{10}, \frac{1}{10}, \frac{1}{10})$, i.e., a mixed quantum state. If we take Formula (38) with $\varepsilon = 0$, then we have $x_1^{c=0}(t) \equiv \frac{1}{2}, x_8^{c=0}(t) \equiv \frac{3}{10}, x_{13}^{c=0}(t) \equiv \frac{1}{10}$, and $x_{16}^{c=0}(t) \equiv \frac{1}{10}$ for any $t \geq 0$. For any time, this particular dynamic system does not leave the state ρ_0 (x_{ρ_0})—this is a *singular point* of the system vector field. This analytical finding relates with one of the considered below cases for the keeping problem (and with the right column of the subfigures in Figure 2) analyzed in the next subsection.

6.2. The Problem of Keeping the Initial Entropy $S(\rho_0)$

Consider the initial state $\rho_0 = \text{diag}(\frac{1}{2}, \frac{3}{10}, \frac{1}{10}, \frac{1}{10})$ with $S(\rho_0) \approx 1.168$ and the problem of keeping the von Neumann entropy $S(\rho(t))$ at the level $S(\rho_0)$ at the whole $[0, T = 5]$.

6.2.1. Using the Problem (4) and GPM

Set the coefficient $P = 0.1$ in (4). Set the bounds $u_{\max} = 30$, $n_{\max} = 10$ in (21). The regularization (13) is not used in each of the described below three cases. We use GPM-2 (see the iteration formula (36)) with the gradient of the corresponding functional, parameters $\alpha = 3$, $\beta = 0.9$ fixed for the whole number of iterations. For comparison, GPM-1 (see the iteration formula (35)) with the same α is used. With respect to the both terms of the objective J_1, we use the following stopping criterion for GPMs:

$$\left((S(\rho^{(k)}(T)) - S(\rho_0))^2 \leq \varepsilon_{\text{stop},1}\right) \;\&\; \left(\frac{1}{P}\int_0^T (S(\rho(t)) - S_{\rho_0})^2 dt \leq \varepsilon_{\text{stop},2}\right). \quad (40)$$

Set $\varepsilon_{\text{stop},1} = 10^{-6}$ and $\varepsilon_{\text{stop},2} = 10^{-5}$.

Consider the following three cases: (1) $\varepsilon = 0.1$ and $c^{(0)} = (\sin(2t), 0, 0)$; (2) $\varepsilon = 0.1$ and $c^{(0)} = 0$; (3) $\varepsilon = 0$ and $c^{(0)} = 0$. For the GPM computer implementations, we consider piecewise linear interpolation for u, n_1, n_2 at the uniform time grid with $M = 10^3$ subintervals.

Case 1 ($\varepsilon = 0.1$ and $c^{(0)} = (\sin(2t), 0, 0)$). GPM-2 at the cost of 132 iterations reaches (40). For this case, consider the left column of the subfigures in Figure 2. We see that all the computed controls u, n_1, n_2 are non-zero here. We see that the graphs of $S(\rho(t))$ (blue solid), degree of purity $P(\rho(t)) = \text{Tr}\rho^2(t)$, and the Hilbert–Schmidt distance $\|\rho(t) - \frac{1}{4}\mathbb{I}_4\| = [\text{Tr}((\rho(t) - \frac{1}{4}\mathbb{I}_4)^2)]^{1/2}$ vs $t \in [0, T]$ are close to the constants that relate to the idea of the keeping problem. At the same time, the graph of $\|\rho(t) - \rho_0\|$ is far from constant and shows that this (approximate) keeping relates to sufficiently different distances between the system states and ρ_0 at various time instances. For comparison, GPM-1 is used for the same $c^{(0)}$. Let the largest allowed number of iterations be 500 for this method. At the cost of 500 iterations, GPM-1 does not reach the stopping criterion (40), but the terminal part of J_1 is near 3×10^7 (rather less than $\varepsilon_{\text{stop},1} = 10^{-6}$) and $\frac{1}{T} \int_0^T (S(\rho(t)) - S_{\rho_0})^2 dt \approx 0.0005$. Thus, both GPM-1 and GPM-2 work good here, but GPM-2 reaches the criterion at the cost of 132 iterations.

Case 2 ($\varepsilon = 0.1$ and $c^{(0)} = (0,0,0)$). Only the initial guess is different, i.e., we use the same values (37), etc., the same other settings in GPM-2. At the cost of 253 iterations, GPM-2 reaches (40). The resulting control c contains the control $u = 0$, while both the obtained controls n_1, n_2 are non-trivial. The middle column of the subfigures in Figure 2 shows the obtained results. Thus, in this keeping problem, it is sufficient to adjust only n_1, n_2 under $u = 0$. Moreover, note that for $c^{(0)} = 0$, its component $u^{(0)} = 0$ is singular in the sense that the corresponding switching function $\mathcal{K}^u(\chi^{(0)}(t), \rho^{(0)}(t)) \equiv 0$ at the whole $[0, T]$.

Case 3 ($\varepsilon = 0.1$ and $c^{(0)} = (0,0,0)$). In contrast to the previous case, here we do not take into account the Lamb shift and the dissipator. The right column of the subfigures in Figure 2 shows that, in this case, the system dynamics achieve the goal of keeping $S(\rho(t))$ at the level $S(\rho_0)$ at the whole $[0, T = 5]$.

6.2.2. Using the Problem (5) and Genetic Algorithm

Further, the keeping problem is considered as minimizing the objective J_2 in the class of piecewise linear controls via the GA. Here, the class of piecewise linear controls u, n_1, n_2 is defined at the uniform grid introduced at $[0, T = 5]$ with only $M = 10$ subintervals (compare with $M = 10^3$ used for interpolation of controls in the GPM computer realization). Thus, here, we consider $3(M + 1) = 33$ control parameters. Consider $u_{\max} = n_{\max} = 4$ and use the regularization (12) with $\gamma_u = 0$, $\gamma_n = 0.01$, $\delta_{n_1} = \delta_{n_2} = 1$. For GA, we set the allowed number of iterations to 350. Figure 3 shows the results obtained due to some GA trial that started from an automatically generated initial point. In this case, we obtain $J_2 = q_2 \approx 0.005$, satisfying the regularization requirements for incoherent controls in (12) with the largest allowed jumps $\delta_{n_1} = \delta_{n_2} = 1$. All the resulting controls u, n_1, n_2 are non-trivial here.

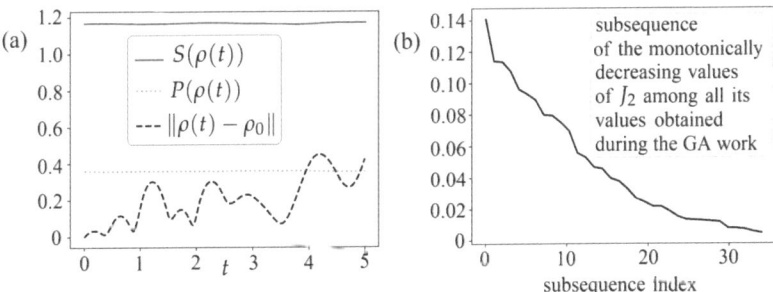

Figure 3. For the problem of keeping the invariant $S(\rho(t)) \equiv S(\rho_0)$ at the whole $[0, T = 5]$. Considering piecewise linear controls (with $M = 10$ subintervals) relates to the GA finite-dimensional optimization. At the resulting controls computed with some GA trial: (**a**) $S(\rho(t))$, $P(\rho(t))$, and $\|\rho(t) - \rho_0\|$ vs $t \in [0, T = 5]$; (**b**) the subsequence of the monotonically decreasing values of J_2 among all its values computed during the GA work.

6.3. The Problem of Steering the von Neumann Entropy to a Predefined Value

Consider the steering problem as only the terminal problem, i.e., we use the objective J_3 and (6). As with objective J_1, we also consider the system with the values in Equation (37), setting bounds $u_{\max} = 30$, $n_{\max} = 10$. We set the initial state $\rho_0 = \text{diag}(0, \frac{1}{2}, 0, \frac{1}{2})$ with $S(\rho_0) = \log 2 \approx 0.7$ and the target value $S_{\text{tar}} = 0.4$. Set $T = 40$. With respect to the regularization (13), we consider two cases: with and without this regularization. GPM-2 is used with $\alpha = 3$ and $\beta = 0.9$. Piecewise linear interpolation for controls is used with $M = 10^3$ equal subintervals. We take $c^{(0)} = 0.5$. The stopping criterion is $J_3(c^{(k)}) \leq \varepsilon_{\text{stop}} = 10^{-6}$.

Case 1: Without the regularization (13). GPM-2, at the cost of 42 iterations, meets the stopping criterion. The obtained results are shown in Figure 4a,b,c. We see that all the resulting controls are non-trivial.

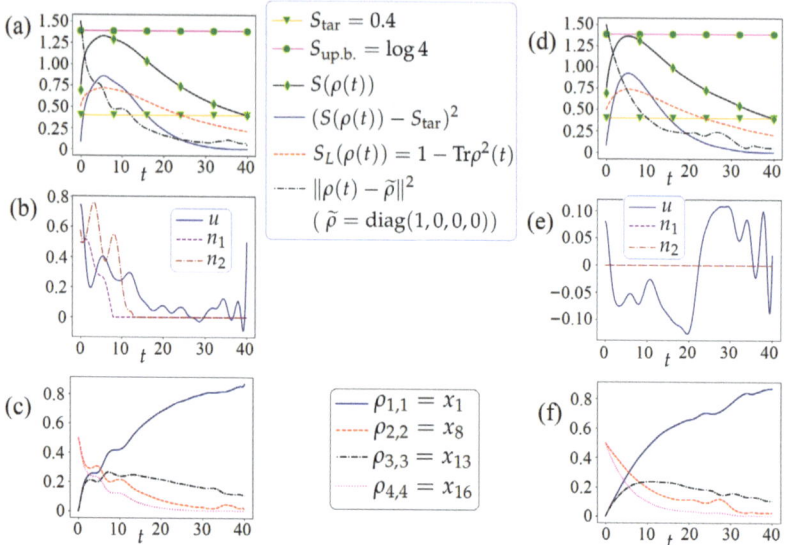

Figure 4. For the problem of steering the von Neumann entropy to the predefined value $S_{\text{tar}} = 0.4$ from the initial value $S(\rho_0) \approx 0.7$: without (see the subfigures (**a**–**c**)) and with (see the subfigures (**d**–**f**)) the regularization (13). Here, $S_{\text{up.b.}} = \log 4$ is the von Neumann entropy upper bound, $S_L(\rho(t)) = 1 - P(\rho(t))$ is the linear entropy.

Case 2: With the regularization (13). Set $\gamma_u = \gamma_n = 10^{-3}$. GPM-2, at the cost of 34 iterations, meets the stopping criterion. The obtained results are shown in Figure 4d,e,f. We see that only coherent control is computed as non-trivial. Thus, for the considered steering problem, it is sufficient to adjust only non-trivial coherent control.

6.4. The Steering Problem for the von Neumann Entropy under the Pointwise Constraint for This Entropy

In view of the graphs of $S(\rho(t))$ vs t in Figure 4a,d, we introduce and try to satisfy the pointwise constraint $S(\rho(t)) \leq \overline{S} = 1, t \in [0, T = 40]$, in addition to the requirement to reach the value $S_{\text{tar}} = 0.4$.

Consider both the problems (7) and (8) and, correspondingly, GPM and GA.

6.4.1. Using the Problem (7) and GPM

Consider the objective J_4 and the problem (7). With respect to both terms of the objective J_4, we use the following stopping criterion for the GPMs:

$$\left((S(\rho^{(k)}(T)) - S_{\text{tar}})^2 \leq \varepsilon_{\text{stop},1}\right) \& \left(\frac{1}{P}\int_0^T (\max\{S(\rho^{(k)}(t)) - \overline{S}, 0\})^2 dt \leq \varepsilon_{\text{stop},2}\right). \quad (41)$$

Set $\varepsilon_{\text{stop},1} = 10^{-6}$ and $\varepsilon_{\text{stop},2} = 10^{-3}$. We take the penalty coefficient $P = 0.05$ in J_4. The regularization (13) is not used here. We set the bounds $u_{\max} = 30$, $n_{\max} = 10$ in (21).

GPM-2 with $\alpha = 3$, $\beta = 0.9$ at the cost of 39 iterations provides reaching (41). The results are shown in Figure 5a,b.

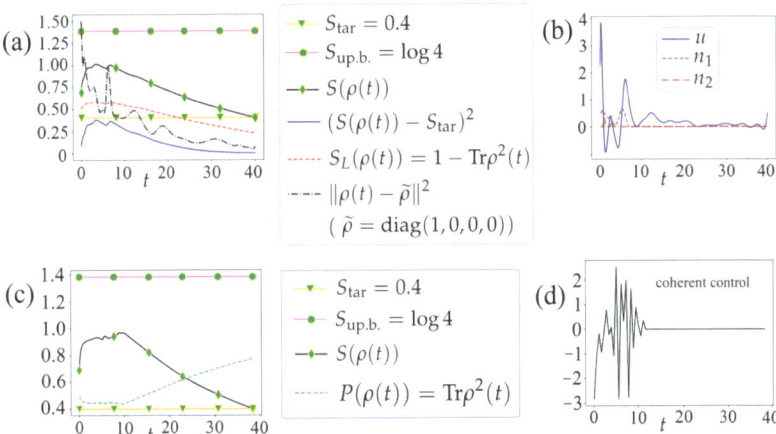

Figure 5. For the problem of steering, the von Neumann entropy to the predefined value $S_{\text{tar}} = 0.4$ from the initial value $S(\rho_0) \approx 0.7$ under the state constraint $S(\rho(t)) \leq \overline{S} = 1$: (1) with respect to the problem (7) (without the regularization (13)) and using GPM-2 (subfigures (**a**,**b**)); (2) with respect to the problem (8) (with the described in the main text special class of controls) and the regularized objective (11) (with $\gamma_u = 0.1$, $\gamma_n = 0$) and using the GA (subfigures (**c**,**d**)). We see that, for approximate steering, it is appropriate to adjust only coherent control under the zero incoherent controls.

6.4.2. Using the Problem (8) and Genetic Algorithm

Consider the problem of steering the von Neumann entropy under the pointwise constraint on $S(\rho(t))$ as minimizing J_5. Here, taking into account the structure of the resulting controls obtained via GPM-2, and shown in Figure 5b, we construct the following special class of piecewise linear controls. Let both incoherent controls be zero throughout the interval $[0, T = 40]$, while coherent control is zero at $(0.3T, T]$, and is a piecewise linear function at $[0, 0.3T]$, which is determined at the uniform grid with $M = 20$ subintervals taken at $[0, 0.3T]$. Consider the bound $u_{\max} = 4$ and penalty factor $P = 0.5$. In this optimization problem, T is not fixed and is considered as a control parameter varied at the range $[T_1, T_2] = [38, 40]$. Thus, here the objective function g_5 depends on $M + 1 = 21$ control parameters, which determine coherent control, and T. Moreover, the regularization in the control parameters according to (11) is used with $\gamma_u = 0.1$, $\gamma_n = 0$. The upper bound for the number of iterations of the GA is set at 200. The results of certain GA trials are shown in Figure 5c,d. The resulting value $|S(\rho(T)) - S_{\text{tar}}| \approx 6 \times 10^{-5}$ and the computed pointwise max-max term in J_5 is zero. Thus, we see that, for approximate steering, it is appropriate to adjust only coherent control under the zero incoherent controls here.

7. Conclusions

In this article, we consider the general problem of controlling the von Neumann entropy of quantum systems either at some final time or over some time interval. The example of the two-qubit system is considered in detail with the following control goals: (1) minimizing or maximizing the final entropy $S(\rho(T))$; (2) steering $S(\rho(T))$ to a given target value; (3) steering $S(\rho(T))$ to a target value and satisfying the pointwise state constraint $S(\rho(t)) \leq \overline{S}$ for a given $\overline{S}$; (4) keeping $S(\rho(t))$ constant at a given time interval. Under the Markovian two-qubit dynamics determined by a GKSL-type master equation with coherent and incoherent controls: (1) for the differentiable cases and piecewise continuous controls, one- and two-step gradient projection methods have been adapted by deriving the corresponding adjoint systems and gradients for the objective functionals; (2) for the non-differentiable cases and piecewise linear controls, a finite-dimensional optimization with the genetic algorithm has been performed. The numerical experiments conducted with these optimization tools demonstrate their appropriateness for the problems considered and enable the identification of various structures in the resulting controls. A more detailed analysis of the entropy involving objective functionals, taking into account the Hilbert–Schmidt distances and the reduced density matrices (24), is an open direction for future research.

Author Contributions: Conceptualization, discussion, methodology, O.V.M. and A.N.P.; software, numerical experiments, visualization, O.V.M.; writing, O.V.M. and A.N.P. All authors have read and agreed to the published version of the manuscript.

Funding: This work was performed at the Steklov International Mathematical Center and supported by the Ministry of Science and Higher Education of the Russian Federation (agreement no. 075-15-2022-265).

Institutional Review Board Statement: Not applicable.

Data Availability Statement: All the input data used during this study and all the computed data during this study are written or shown in the figures in this article.

Conflicts of Interest: The authors declare no conflict of interest.

Abbreviations

The following abbreviations are used in this manuscript:

GKSL	Gorini–Kossakowski–Sudarshan–Lindblad
PMP	Pontryagin's maximum principle
GPM-1, GPM-2	one- and two-step gradient projection methods
GA	genetic algorithm

References

1. Dong, D.; Petersen, I.R. *Learning and Robust Control in Quantum Technology*; Springer: Cham, Switzerland, 2023. [CrossRef]
2. Kuprov, I. Spin: From Basic Symmetries to Quantum Optimal Control; Springer: Cham, Switzerland, 2023. [CrossRef]
3. Koch, C.P.; Boscain, U.; Calarco, T.; Dirr, G.; Filipp, S.; Glaser, S.J.; Kosloff, R.; Montangero, S.; Schulte-Herbrüggen, T.; Sugny, D.; et al. Quantum optimal control in quantum technologies. Strategic report on current status, visions and goals for research in Europe. *EPJ Quantum Technol.* **2022**, *9*, 19. [CrossRef]
4. Kurizki, G.; Kofman, A.G. *Thermodynamics and Control of Open Quantum Systems*; Cambridge University Press: Cambridge, UK, 2022. [CrossRef]
5. D'Alessandro, D. *Introduction to Quantum Control and Dynamics*, 2nd ed.; Chapman & Hall: Boca Raton, Fl, USA, 2021. [CrossRef]
6. Kwon, S.; Tomonaga, A.; Bhai, G.L.; Devitt, S.J.; Tsai, J.-S. Gate-based superconducting quantum computing. *J. Appl. Phys.* **2021**, *129*, 041102. [CrossRef]
7. Bai, S.-Y.; Chen, C.; Wu, H.; An, J.-H. Quantum control in open and periodically driven systems. *Adv. Phys. X* **2021**, *6*, 1870559. [CrossRef]
8. Acín, A.; Bloch, I.; Buhrman, H.; Calarco, T.; Eichler, C.; Eisert, J.; Esteve, D.; Gisin, N.; Glaser, S.J.; Jelezko, F.; et al. The quantum technologies roadmap: A European community view. *New J. Phys.* **2018**, *20*, 080201. [CrossRef]
9. Koch, C.P. Controlling open quantum systems: Tools, achievements, and limitations. *J. Phys. Condens. Matter* **2016**, *28*, 213001. [CrossRef]
10. Dong, W.; Wu, R.; Yuan, X.; Li, C.; Tarn, T.-J. The modelling of quantum control systems. *Sci. Bull.* **2015**, *60*, 1493–1508. [CrossRef]
11. Cong, S. *Control of Quantum Systems: Theory and Methods*; John Wiley & Sons: Hoboken, NJ, USA, 2014.
12. Altafini, C.; Ticozzi, F. Modeling and control of quantum systems: An introduction. *IEEE Trans. Automat. Control* **2012**, *57*, 1898–1917. [CrossRef]
13. Bonnard, B.; Sugny, D. *Optimal Control with Applications in Space and Quantum Dynamics*; AIMS: Springfield, MA, USA, **2012**.
14. Gough, J.E. Principles and applications of quantum control engineering. *Philos. Trans. R. Soc. A* **2012**, *370*, 5241–5258. [CrossRef]
15. Shapiro, M.; Brumer, P. *Quantum Control of Molecular Processes*, 2nd revised ed.; Enlarged Edition; Wiley–VCH Verlag: Weinheim, Germany, **2012**. [CrossRef]
16. Brif, C.; Chakrabarti, R.; Rabitz, H. Control of quantum phenomena: Past, present and future. *New J. Phys.* **2010**, *12*, 075008. [CrossRef]

17. Fradkov, A.L. *Cybernetical Physics: From Control of Chaos to Quantum Control*; Springer: Berlin/Heidelberg, Germany, 2007. [CrossRef]
18. Letokhov, V. *Laser Control of Atoms and Molecules*; Oxford University Press: Oxford, UK, 2007.
19. Tannor, D.J. *Introduction to Quantum Mechanics: A Time Dependent Perspective*; University Science Books: Sausilito, CA, USA, 2007.
20. Butkovskiy, A.G.; Samoilenko, Y.I. *Control of Quantum–Mechanical Processes and Systems*; Translated from the Edition Published in Russian in 1984; Kluwer Academic Publishers: Dordrecht, The Netherland, 1990.
21. Pechen, A.; Rabitz, H. Teaching the environment to control quantum systems. *Phys. Rev. A* **2006**, *73*, 062102. [CrossRef]
22. Pechen, A. Engineering arbitrary pure and mixed quantum states. *Phys. Rev. A* **2011**, *84*, 042106. [CrossRef]
23. Morzhin, O.V.; Pechen, A.N. Optimal state manipulation for a two-qubit system driven by coherent and incoherent controls. *Quantum Inf. Process.* **2023**, *22*, 241. [CrossRef]
24. Morzhin, O.V.; Pechen, A.N. Krotov type optimization of coherent and incoherent controls for open two-qubit systems. *Bull. Irkutsk State Univ. Ser. Math.* **2023**, *45*, 3–23. [CrossRef]
25. Petruhanov, V.N.; Pechen, A.N. Optimal control for state preparation in two-qubit open quantum systems driven by coherent and incoherent controls via GRAPE approach. *Int. J. Mod. Phys. B* **2022**, *37*, 2243017. [CrossRef]
26. Petruhanov, V.N.; Pechen, A.N. GRAPE optimization for open quantum systems with time-dependent decoherence rates driven by coherent and incoherent controls. *J. Phys. A Math. Theor.* **2023**, *56*, 305303. [CrossRef]
27. Morzhin, O.V.; Pechen, A.N. On optimization of coherent and incoherent controls for two-level quantum systems. *Izv. Math.* **2023**, *87*, 1024–1050. [CrossRef]
28. Morzhin, O.V.; Pechen, A.N. Minimal time generation of density matrices for a two-level quantum system driven by coherent and incoherent controls. *Int. J. Theor. Phys.* **2021**, *60*, 576–584. [CrossRef]
29. Holevo, A.S. *Quantum Systems, Channels, Information: A Mathematical Introduction*, 2nd revised ed.; Expanded Edition; De Gruyter: Berlin, Germany; Boston, MA, USA, 2019. [CrossRef]
30. Wilde, M.M. *Quantum Information Theory*, 2nd ed.; Cambridge University Press: Cambridge, UK, 2017.
31. Shirokov, M.E. Continuity of the von Neumann entropy. *Commun. Math. Phys.* **2010**, *296*, 625–654. [CrossRef]
32. Nielsen, M.; Chuang, I. *Quantum Computation and Quantum Information*, 10th anniversary ed.; Cambridge University Press: Cambridge, UK, 2010. [CrossRef]
33. Petz, D. Entropy, von Neumann and the von Neumann entropy. In *John von Neumann and the Foundations of Quantum Physics*; Rédei, M., Stöltzner, M., Eds.; Springer: Dordrecht, The Netherland, 2001; pp. 83–96. [CrossRef]
34. Ohya, M.; Petz, D. *Quantum Entropy and Its Use*; Springer: Berlin/Heidelberg, Germany, 1993.
35. Ohya, M.; Volovich, I. *Mathematical Foundations of Quantum Information and Computation and Its Applications to Nano- and Bio-Systems*; Springer: Dordrecht, The Netherland, 2011. [CrossRef]
36. Ohya, M.; Watanabe, N. Quantum entropy and its applications to quantum communication and statistical physics. *Entropy* **2010**, *12*, 1194–1245. [CrossRef]
37. Bracken, P. Classical and quantum integrability: A formulation that admits quantum chaos. In *A Collection of Papers on Chaos Theory and Its Applications*; Bracken, P., Uzunov, D.I., Eds.; IntechOpen: London, UK, 2020. [CrossRef]
38. Vera, J.; Fuentealba, D.; Lopez, M.; Ponce, H.; Zariquiey, R. On the von Neumann entropy of language networks: Applications to cross-linguistic comparisons. *EPL* **2021**, *136*, 68003. [CrossRef]
39. Kosloff, R. Quantum thermodynamics: A dynamical viewpoint. *Entropy* **2013**, *15*, 2100–2128. [CrossRef]
40. Sklarz, S.E.; Tannor, D.J.; Khaneja, N. Optimal control of quantum dissipative dynamics: Analytic solution for cooling the three-level Λ system. *Phys. Rev. A* **2004**, *69*, 053408. [CrossRef]
41. Pavon, M.; Ticozzi, F. On entropy production for controlled Markovian evolution. *J. Math. Phys.* **2006**, *47*, 063301. [CrossRef]
42. Bartana, A.; Kosloff, R.; Tannor, D.J. Laser cooling of molecules by dynamically trapped states. *Chem. Phys.* **2001**, *267*, 195–207. [CrossRef]
43. Kallush, S.; Dann, R.; Kosloff, R. Controlling the uncontrollable: Quantum control of open system dynamics. *Sci. Adv.* **2022**, *8*, eadd0828. [CrossRef] [PubMed]
44. Dann, R.; Tobalina, A.; Kosloff, R. Fast route to equilibration. *Phys. Rev. A* **2020**, *101*, 052102. [CrossRef]
45. Ohtsuki, Y.; Mikami, S.; Ajiki, T.; Tannor, D.J. Optimal control for maximally creating and maintaining a superposition state of a two-level system under the influence of Markovian decoherence. *J. Chin. Chem. Soc.* **2023**, *70*, 328–340. [CrossRef]
46. Caneva, T.; Calarco, T.; Montangero, S. Chopped random-basis quantum optimization. *Phys. Rev. A* **2011**, *84*, 022326. [CrossRef]
47. Uzdin, R.; Levy, A.; Kosloff, R. Quantum heat machines equivalence, work extraction beyond Markovianity, and strong coupling via heat exchangers. *Entropy* **2016**, *18*, 124. [CrossRef]
48. Abe, T.; Sasaki, T.; Hara, S.; Tsumura, K. Analysis on behaviors of controlled quantum systems via quantum entropy. *IFAC Proc.* **2008**, *41*, 3695–3700. [CrossRef]
49. Sahrai, M.; Arzhang, B.; Seifoory, H.; Navaeipour, P. Coherent control of quantum entropy via quantum interference in a four-level atomic system. *J. Sci. Islam. Repub. Iran* **2013**, *24*, 2.
50. Xing, Y.; Wu, J. Controlling the Shannon entropy of quantum systems. *Sci. World J.* **2013**, *2013*, 381219. [CrossRef] [PubMed]
51. Xing, Y.; Wu, J. Shannon-entropy control of quantum systems. In Proceedings of the World Congress on Engineering and Computer Science 2013, San Francisco, CA, USA, 23–25 October 2013. Available online: https://www.iaeng.org/publication/WCECS2013/WCECS2013_pp862-867.pdf (accessed on 20 December 2023).

52. Xing, Y.; Huang, W.; Zhao, J. Continuous controller design for quantum Shannon entropy. *Intell. Control Autom.* **2016**, *7*, 63–72. [CrossRef]
53. Abbas Khudhair, D.; Fathdal, F.; Raheem, A.-B.F.; Hussain, A.H.A.; Adnan, S.; Kadhim, A.A.; Adhab, A.H. Spatially control of quantum entropy in a three-level medium. *Int. J. Theor. Phys.* **2022**, *61*, 252. [CrossRef]
54. Pechen, A.; Rabitz, H. Unified analysis of terminal-time control in classical and quantum systems. *EPL* **2010**, *91*, 60005. [CrossRef]
55. Landau, L. Das Daempfungsproblem in der Wellenmechanik. *Z. Phys.* **1927**, *45*, 430–464. [CrossRef]
56. Von Neumann, J. *Mathematische Grundlagen der Quantenmechanik*; Springer: Berlin, Germany, 1932.
57. Boscain, U.; Sigalotti, M.; Sugny, D. Introduction to the Pontryagin maximum principle for quantum optimal control. *PRX Quantum* **2021**, *2*, 030203. [CrossRef]
58. Buldaev, A.; Kazmin, I. Operator methods of the maximum principle in problems of optimization of quantum systems. *Mathematics* **2022**, *10*, 507. [CrossRef]
59. Goerz, M.H.; Reich, D.M.; Koch, C.P. Optimal control theory for a unitary operation under dissipative evolution. *New J. Phys.* **2014**, *16*, 055012. [CrossRef]
60. Krotov, V.F.; Morzhin, O.V.; Trushkova, E.A. Discontinuous solutions of the optimal control problems. Iterative optimization method. *Autom. Remote Control* **2013**, *74*, 1948–1968. [CrossRef]
61. Krotov, V.F. *Global Methods in Optimal Control Theory*; Marcel Dekker: New York, NY, USA, 1996.
62. Khaneja, N.; Reiss, T.; Kehlet, C.; Schulte-Herbrüggen, T.; Glaser, S.J. Optimal control of coupled spin dynamics: Design of NMR pulse sequences by gradient ascent algorithms. *J. Magn. Reson.* **2005**, *172*, 296–305. [CrossRef] [PubMed]
63. Müller, M.M.; Said, R.S.; Jelezko, F.; Calarco, T.; Montangero, S. One decade of quantum optimal control in the chopped random basis. *Rep. Prog. Phys.* **2022**, *85*, 076001. [CrossRef] [PubMed]
64. Judson, R.S.; Rabitz, H. Teaching lasers to control molecules. *Phys. Rev. Lett.* **1992**, *68*, 1500. [CrossRef] [PubMed]
65. Brown, J.; Paternostro, M.; Ferraro, A. Optimal quantum control via genetic algorithms for quantum state engineering in driven-resonator mediated networks. *Quantum Sci. Technol.* **2023**, *8*, 025004. [CrossRef]
66. Trushechkin, A. Unified Gorini-Kossakowski-Lindblad-Sudarshan quantum master equation beyond the secular approximation. *Phys. Rev. A* **2021**, *103*, 062226. [CrossRef]
67. Trushechkin, A. Quantum master equations and steady states for the ultrastrong-coupling limit and the strong-decoherence limit. *Phys. Rev. A* **2022**, *106*, 042209. [CrossRef]
68. McCauley, G.; Cruikshank, B.; Bondar, D.I.; Jacobs, K. Accurate Lindblad-form master equation for weakly damped quantum systems across all regimes. *NPJ Quantum Inf.* **2020**, *6*, 74. [CrossRef]
69. Wu, R.; Pechen, A.; Brif, C.; Rabitz, H. Controllability of open quantum systems with Kraus-map dynamics. *J. Phys. A* **2007**, *40*, 5681–5693. [CrossRef]
70. Zhang, W.; Saripalli, R.; Leamer, J.; Glasser, R.; Bondar, D. All-optical input-agnostic polarization transformer via experimental Kraus-map control. *Eur. Phys. J. Plus* **2022**, *137*, 930. [CrossRef]
71. Laforge, F.O.; Kirschner, M.S.; Rabitz, H.A. Shaped incoherent light for control of kinetics: Optimization of up-conversion hues in phosphors. *J. Chem. Phys.* **2018**, *149*, 054201. [CrossRef]
72. Pontryagin, L.S.; Boltyanskii, V.G.; Gamkrelidze, R.V.; Mishchenko, E.F. *The Mathematical Theory of Optimal Processes*; Translated from Russian; Interscience Publishers JohnWiley & Sons, Inc.: New York, NY, USA; London, UK, 1962.
73. Petersen, K.B.; Pedersen, M.S. The Matrix Cookbook; Technical University of Denmark; 2012. Available online: https://www2.imm.dtu.dk/pubdb/pubs/3274-full.html (accessed on 20 December 2023).
74. Polak, E. *Computational Methods in Optimization: A Unified Approach*; Academic Press: New York, NY, USA; London, UK, 1971.
75. Srochko, V.A.; Mamonova, N.V. Iterative procedures for solving optimal control problems based on quasigradient approximations. *Russ. Math.* **2001**, *45*, 52–64.
76. Levitin, E.S.; Polyak, B.T. Constrained minimization methods. *USSR Comput. Math. Math. Phys.* **1966**, *6*, 1–50. [CrossRef]
77. Demyanov, V.F.; Rubinov, A.M. *Approximate Methods in Optimization Problems*; Translated from Russian; American Elsevier Pub. Co.: New York, NY, USA, 1970.
78. Fedorenko, R.P. *Approximate Solution of Optimal Control Problems*; Nauka: Moscow, Russia, 1978. (In Russian)
79. Polyak, B.T. *Introduction to Optimization*; Translated from Russian; Optimization Software Inc., Publ. Division: New York, NY, USA, 1987.
80. Polyak, B.T. Some methods of speeding up the convergence of iteration methods. *USSR Comput. Math. Math. Phys.* **1964**, *5*, 1–17. [CrossRef]
81. Antipin, A.S. Minimization of convex functions on convex sets by means of differential equations. *Differ. Equat.* **1994**, *30*, 1365–1375.
82. Vasil'ev, F.P.; Amochkina, T.V.; Nedić, A. On a regularized version of the two-step gradient projection method. *Moscow Univ. Comput. Math. Cybernet.* **1996**, *1*, 31–37.
83. Sutskever, I.; Martens, J.; Dahl, G.; Hinton, G. On the importance of initialization and momentum in deep learning. *PMLR* **2013**, *28*, 1139–1147. Available online: https://proceedings.mlr.press/v28/sutskever13.html (accessed on 20 December 2023).
84. TensorFlow, Machine Learning Platform: MomentumOptimizer. Available online: https://www.tensorflow.org/api_docs/python/tf/compat/v1/train/MomentumOptimizer (accessed on 20 December 2023).
85. Solgi, R. Genetic Algorithm Package for Python. Available online: https://github.com/rmsolgi/geneticalgorithm, https://pypi.org/project/geneticalgorithm/ (accessed on 20 December 2023).

Disclaimer/Publisher's Note: The statements, opinions and data contained in all publications are solely those of the individual author(s) and contributor(s) and not of MDPI and/or the editor(s). MDPI and/or the editor(s) disclaim responsibility for any injury to people or property resulting from any ideas, methods, instructions or products referred to in the content.

Article

Relationship between Information Scrambling and Quantum Darwinism

Feng Tian [1], Jian Zou [1,*], Hai Li [2], Liping Han [3] and Bin Shao [1]

[1] School of Physics, Beijing Institute of Technology, Beijing 100081, China; tianfeng5g@163.com (F.T.); sbin610@bit.edu.cn (B.S.)
[2] School of Information and Electronic Engineering, Shandong Technology and Business University, Yantai 264005, China; lihai@sdtbu.edu.cn
[3] School of Science, Tianjin University of Technology, Tianjin 300384, China; hanlp9902@bit.edu.cn
* Correspondence: zoujian@bit.edu.cn

Abstract: A quantum system interacting with a multipartite environment can induce redundant encoding of the information of a system into the environment, which is the essence of quantum Darwinism. At the same time, the environment may scramble the initially localized information about the system. Based on a collision model, we mainly investigate the relationship between information scrambling in an environment and the emergence of quantum Darwinism. Our results show that when the mutual information between the system and environmental fragment is a linear increasing function of the fragment size, the tripartite mutual information (TMI) is zero, which can be proved generally beyond the collision model; when the system exhibits Darwinistic behavior, the TMI is positive (i.e., scrambling does not occur); when we see the behavior of an "encoding" environment, the TMI is negative (i.e., scrambling occurs). Additionally, we give a physical explanation for the above results by considering two simple but illustrative examples. Moreover, depending on the nature of system and environment interactions, it is also shown that the single qubit and two-qubit systems behave differently for the emergence of quantum Darwinism, and hence the scrambling, while their relationship is consistent with the above conclusion.

Keywords: quantum Darwinism; information scrambling; collision model; quantum mutual information; tripartite mutual information

Citation: Tian, F.; Zou, J.; Li, H.; Han, L.; Shao, B. Relationship between Information Scrambling and Quantum Darwinism. *Entropy* **2024**, *26*, 19. https://doi.org/10.3390/e26010019

Academic Editors: Paula I. Villar and Fernando C. Lombardo

Received: 27 November 2023
Revised: 16 December 2023
Accepted: 18 December 2023
Published: 24 December 2023

Copyright: © 2023 by the authors. Licensee MDPI, Basel, Switzerland. This article is an open access article distributed under the terms and conditions of the Creative Commons Attribution (CC BY) license (https://creativecommons.org/licenses/by/4.0/).

1. Introduction

Quantum Darwinism is a theoretical framework that allows one to understand the emergence of objectivity out of quantum superpositions [1]. Due to the interaction between system and environment, the latter acquires information about the state of the former with respect to the so-called pointer states [2–5], namely, the eigenstates of the observable, which is coupled with the environment [6–9]. Pointer states are left undisturbed during the interaction with the environment, but if the system is in a coherent superposition of pointer states, it gets entangled with the environment. Many observers can independently access and measure different parts of the environment and independently obtain the same information about the system [10–12]. This redundancy is a characteristic feature of quantum Darwinism [13–15], explaining the emergence of objective reality.

Quantum Darwinism has been extensively studied in various models [16–25]. One of the main issues of these investigations is to understand the fundamental mechanism through which quantum Darwinism emerges. Recently, several works have shown that the emergence of quantum Darwinism is sensitive to the microscopic description of quantum dynamics, such as the nature of interaction and initial conditions [26,27]. It has been shown that the nature of correlations among the environmental constituents, i.e., whether they are quantum or classical, is important for the emergence of quantum Darwinism [27]. The authors in Reference [28] have validated that in quantum Darwinism, when classical objectivity manifests, bipartite quantum correlations between the system and components of its environment are suppressed. Experimental investigations of quantum Darwinism were also reported [29–31]. Generally, quantum Darwinism needs to analyze quantum mutual information between system and environment; hence, it is necessary to keep track not only of the system but also of the environment. The master equation is widely used to obtain the dynamics of an open system by tracing out environmental degrees of freedom, while it cannot treat the full system–environment dynamics. The quantum collision model (CM), which decomposes complicated dynamics in terms of discrete elementary processes, offers an alternative way of describing open quantum system dynamics [32–36]. Additionally, for CM, the correlations between the open quantum system and its environment can be easily traced. In the standard framework of CM, the environment is represented by an ensemble of uncorrelated identical environment constituents termed ancillas, and the system of interest interacts with each ancilla sequentially. Recently, CMs have found an application in the investigation of quantum Darwinism [26,27,37].

How information spreads and becomes distributed over the constituent degrees of freedom is one of the central issues in the study of dynamics in the quantum many-body system. The delocalization of information in a many-body system is referred to as information scrambling, which has attracted more and more attention [38–41]. A general accepted measure of information scrambling is the so-called out-of-time-order correlator (OTOC) [41–44], which is associated with the growth of the square commutator between two initially commuting observables. Besides OTOC, other measures have been proposed, such as the average Pauli weight [45] and the operator entanglement entropy [46]. Beyond these, the tripartite mutual information (TMI) is an important measure of information scrambling [41]. At first, TMI of the evolution operator was used to investigate information scrambling in References [47,48]. Later, instantaneous TMI of a quantum state was also used to study information scrambling in References [49–51]. The method used in this paper is the instantaneous TMI of a quantum state.

In terms of TMI, scrambling in the many-body quantum system has been studied in various models [47,50–52]. Their typical setting of studying information scrambling in the many-body quantum system by the TMI consists of an ancillary system and a many-body system. At the initial time, the information of the ancillary system is locally encoded in the many-body system through entanglement. The many-body system then evolves unitarily, and the locally encoded information might spread over the entire many-body system. Within such a setting, information scrambling in a central spin model has been studied in Reference [53], where information that initially resides in the central spin is first shared with the environment due to the coupling between them. Then, only the environment dynamics are turned on, causing the scrambling of information in the environment. On the other hand, Reference [54] studied a scenario where a nuclear spin (the system) is simultaneously coupled with a large collection of non-interacting bath spins. In this situation, scrambling of information in the bath is due to the interaction between the central spin and bath. This is different from the typical scenario (a many-body system plus an ancillary system) mentioned above, where the scrambling of information is due to the self-dynamics of the many-body system.

In the quantum Darwinistic picture, the environment that acquires redundant information about the system is composed of many subsystems. Therefore, a full characterization of whether the locally encoded information at the beginning is localized or spreads over the entire many-body environment in the time evolution could be helpful for understanding the mechanism of quantum Darwinism. This raises the question of what the relationship is between information scrambling and quantum Darwinism. To this end, we present a unified framework to link these two seemingly unrelated research topics. Specifically, we consider a system interacting with a multipartite environment, which can be considered as a many-body system. Such a model is the basic setting of quantum Darwinism, while it also allows us to analyze information scrambling in the multipartite environment. In this paper, we study scrambling in terms of the TMI among the system and different parts of the environment. There is a reason behind this choice. A widely used measure of quantum Darwinism is the quantum mutual information between the system and a fragment of the environment, which is the key quantifier in capturing redundant information. On the other hand, quantum mutual information is the key ingredient of the TMI in the study on scrambling. It is these two quantifiers that allow us to establish a relationship between the emergence of quantum Darwinism and information scrambling.

We consider a CM as consisting of a system (one or two qubits) and an environment (a collection of ancillas), which can be considered as a quantum-body system. In our model, similar to Reference [54], the scrambling of initially localized system information in the environment is not due to the self-dynamics of a many-body system or environment, but rather the sequential system–ancilla interactions. We mainly consider the pure dephasing and exchange interactions between the system and environment, respectively. We find that when the system exhibits Darwinistic behavior, the TMI is positive; when we see the behavior of an "encoding" environment, the TMI is negative, namely, scrambling occurs; when the mutual information between the system and environmental fragment is a linear function of fragment size (namely, a boundary between the former two cases), the TMI is zero [see Figure 1]. Additionally, we explained the physical mechanism of these results. Moreover, depending on the nature of interactions between the system and environment, our results also show that the single qubit and two-qubit systems behave differently for the emergence of quantum Darwinism, and hence the scrambling, while their relationship is the same as the above conclusion.

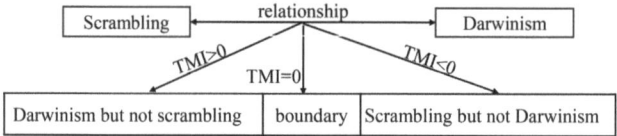

Figure 1. The relationship between information scrambling and quantum Darwinism.

2. Preliminaries

2.1. Model

In this section, we consider a collision model, which consists of a system S and an environment E. The environment consists of a collection of N non-interacting environment ancillas $(E_1, E_2, \ldots, E_N)$, which can be considered as a many-body system. We consider two scenarios where the system S contains a single qubit (Figure 2a) and two qubits (Figure 2b), respectively. For the single qubit system, the Hamiltonian is given by

$$H_S^1 = \frac{1}{2}\sigma_z \tag{1}$$

with Pauli operator σ_z (we set $\hbar = 1$). For the two-qubit system, the Hamiltonian is

$$H_S^2 = H_{S_1} + H_{S_2} + H_{S_1,S_2}, \tag{2}$$

where $H_{S_i} = \frac{1}{2}\sigma_z$ ($i = 1, 2$) is the free Hamiltonian of S_i, and H_{S_1,S_2} is the interaction Hamiltonian between S_1 and S_2. We take H_{S_1,S_2} as

$$H_{S_1,S_2} = \epsilon(\sigma_x \otimes \sigma_x + \sigma_y \otimes \sigma_y) \tag{3}$$

with coupling strength ϵ. We assume that each ancilla of the environment is a qubit, and its Hamiltonian is given by $H_{E_k} = \frac{1}{2}\sigma_z$.

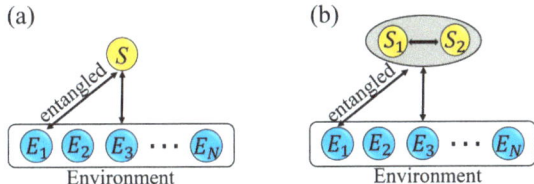

Figure 2. Schematics of the collision model. The system S is coupled to an environment E composed of a collection of environment ancillas ($E_1, E_2, \ldots, E_N$). (**a**) S is a single qubit. (**b**) S is composed of two interacting qubits S_1 and S_2. In both cases, the ancilla E_1 is initially entangled with S, and the remaining ancillas $E_k (k \geq 2)$ are initialized in the same state $|\eta_k\rangle$. The system S interacts sequentially with the ancillas $E_1, E_2, \ldots, E_N$.

Initially, the information of the system is locally encoded in the environment by entangling the system and the first ancilla E_1 in the environment. Specifically, we set the initial state of the system and environmental ancillas to be:

$$|\psi_0\rangle = |\phi_{SE_1}\rangle \bigotimes_{k=2}^{N} |\eta_k\rangle, \qquad (4)$$

where $|\phi_{SE_1}\rangle$ is an entangled state between the system and ancilla E_1, and $|\eta_k\rangle$ is the initial state of the kth ($k \geq 2$) ancilla E_k. The evolution of the whole system is described in terms of pairwise short interactions between S and each environment ancilla: S first collides with E_1; then, S collides with E_2, and so on. In this way, at each step, S collides with a fresh ancilla. A schematic sketch of the collision model is given in Figure 2.

We consider the general Heisenberg interaction between the system and environment ancilla. For the single qubit system, the system–ancilla interaction Hamiltonian is given by

$$H^1_{S,E_k} = \sum_{j=x,y,z} J_j (\sigma^j_S \otimes \sigma^j_{E_k}) \qquad (5)$$

with coupling strength J_j ($j = x, y, z$). For the two-qubit system, the system–ancilla interaction Hamiltonian is given by

$$H^2_{S,E_k} = \sum_{j=x,y,z} J_j (\sigma^j_{S_1} \otimes \sigma^j_{E_k} + \sigma^j_{S_2} \otimes \sigma^j_{E_k}). \qquad (6)$$

For the single qubit system, the interaction between the system S and ancilla E_k in any kth step is realized by the application of the unitary operation

$$U^1_{S,E_k} = \exp\left[-i(H^1_0 + H^1_{SE_k})t\right], \qquad (7)$$

where $H^1_0 = H^1_S + H_{E_k}$, and t stands for the interaction time, i.e., the duration of each collision. Similarly, for the two-qubit system, the interaction between the system S and ancilla E_k in any kth step is realized by

$$U^2_{S,E_k} = \exp\left[-i(H^2_0 + H^2_{SE_k})t\right] \qquad (8)$$

with $H^2_0 = H^2_S + H_{E_k}$. Thus, for the single qubit system, the initial joint system-environment state ρ^1_0, after N steps, evolves into

$$\rho^1_N = U^1_{\{N\}} \rho^1_0 U^{1\dagger}_{\{N\}} \qquad (9)$$

with $U^1_{\{N\}} = U^1_{S,E_N} U^1_{S,E_{N-1}} \cdots U^1_{S,E_2} U^1_{S,E_1}$. Similarly, for the two-qubit system, the initial joint system–environment state ρ^2_0, after N steps, evolves into

$$\rho^2_N = U^2_{\{N\}} \rho^2_0 U^{2\dagger}_{\{N\}} \qquad (10)$$

with $U^2_{\{N\}} = U^2_{S,E_N} U^2_{S,E_{N-1}} \cdots U^2_{S,E_2} U^2_{S,E_1}$.

2.2. Information Scrambling

For the CM we considered in Figure 2, at the initial time, the system information is locally encoded in the environment through $S - E_1$ entanglement, and then this information might be scrambled as the system interacts sequentially with environmental ancillas. Here, we employ the tripartite mutual information as a quantifier of information scrambling [41,49–51]. As shown in Figure 3, we divide the whole environment E into three nonoverlapping subsystems B, C, D, whose sizes (the numbers of the ancillas) are, respectively, given by 1, l, and $N - l - 1$. The tripartite mutual information among three subsystems S, B, and C is defined as

$$I_3(S:B:C) = I(S:B) + I(S:C) - I(S:BC). \qquad (11)$$

Here, $I(S:X)$ is the quantum mutual information between S and X ($X = B, C, BC$), which quantifies the correlations between two subsystems of a composite system. $I(S:X)$ is defined by

$$I(S:X) = H_S + H_X - H_{SX}, \qquad (12)$$

where $H_Y = -\text{Tr}[\rho_Y \ln \rho_Y]$ is the von Neumann entropy for the reduced state ρ_Y of subsystem Y. From an information-theoretic point of view, TMI quantifies how the total (quantum and classical) information is shared among the subsystems A, B, and C. Unlike mutual information, TMI has no definite sign. TMI is negative when $I(S:B) + I(S:C) < I(S:BC)$, which implies that information about S stored in composite BC is larger than the sum of the amounts of information that B and C have individually. In this case, the information about S is nonlocally stored in B and C such that measurements on B and C alone are not able to reconstruct S. When TMI is non-negative at some time, the information at this moment is localized, while at some time when TMI is negative, the information is delocalized. If TMI is non-negative at the beginning and becomes negative as time evolves, the information is gradually delocalized, namely, information scrambling occurs. It is noted that for a given l, there are many partitions of environment when dividing the whole environment E into three nonoverlapping subsystems B, C, D. Therefore, we compute the averaged TMI $\bar{I}_3(S:B:C)$ defined as the averaging over all possible partitions of E for a given l.

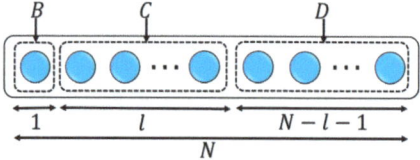

Figure 3. The whole environment E is divided into three nonoverlapping subsystems B, C, D, whose sizes (the numbers of the ancillas) are, respectively, given by 1, l, and $N - l - 1$.

2.3. Quantum Darwinism

In quantum Darwinism, the signature of objectivity is described by the quantum mutual information between the system S and a fragment of the environment E. Any individual environment fragment $F_f \subseteq E$ contains fN individual subsystems if E is compoesd of N individual quantum systems. Here, $f \in [0,1]$ is the fraction of E contained in F_f. According to Equation (12), the mutual information between S and F_f is

$$I(S:F_f) = H_S + H_{F_f} - H_{SF_f}, \qquad (13)$$

where H_S, H_{F_f}, and H_{SF_f} are the von-Neumann entropies of S, F_f, and $S + F_f$, respectively. Since there are many fragments of a given size f, the averaged mutual information $\bar{I}(S:F_f)$ is defined by the average of $I(S:F_f)$ over all possible fragments F_f. The emergence of redundant information is encoded throughout the environment, and therefore, the quantum Darwinism is detected by the existence of the smallest fragment size f_δ such that $\bar{I}(S:F_{f_\delta}) \geq (1-\delta)H_S$, i.e., when the smallest fragment F_{f_δ} contains roughly all (but δ) the information of the system state. Here, the information deficit, $\delta \in (0,1)$, is the information that observers are prepared to forgo. In a plot of $\bar{I}(S:F_f)$ versus fragment size f, such redundancy features are characterized by a rapid rise of $\bar{I}$ at relatively small f, followed by a long "classical plateau" (see line (a) in Figure 4). In the plateau region, as fragment size f increases, F_f provides roughly the same information about the system. $\bar{I}$ increases to $2H_S$ only when the fragment encompasses the whole environment, i.e., $f = 1$. Moreover, the curve ($\bar{I}$ vs. f) can take two other basic shapes [14,15]: the linear profile (see line (b) in Figure 4) corresponds to the behavior of an "independent" environment, where each subenvironment provides unique and independent information about S; the S-shaped profile corresponds to the behavior of an "encoding" environment (see line (c) in Figure 4), where information about S is encoded in multiple subenvironments, and to learn about the system, one requires access to at least half of the environment. For a pure state of the whole system and environment, the curve ($\bar{I}$ vs. f) is always antisymmetric with respect to $\bar{I} = H_S$ at $f = 0.5$ [15].

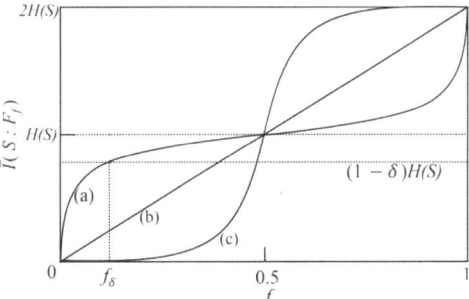

Figure 4. $\bar{I}(S:F_f)$ as a function of f. Line (a) corresponds to the redundancy of Darwinism, where a small fraction f_δ of E contains almost all (but δ) of the information about S. The linear profile (b) shows the behavior of an "independent" environment. Line (c) shows an "encoding" environment or an antiredundancy, i.e., $\bar{I}(S:F_f)$ takes on an S-shaped profile.

3. Results

In this section, we present our numerical results based on the above CM. We are interested in whether or not the system information initially locally encoded in an environment is scrambled as time evolves, whether or not such dynamics can induce the emergence of quantum Darwinism, as well as the relationship between these two phenomena. Different choices of J_j in Equations (5) and (6) create different system–environment interactions. We will consider two types of system–environment interactions, the pure dephasing interaction and exchange interaction, for both the single qubit and two-qubit systems, repectively.

3.1. Dephasing Channel

Now, we consider the pure dephasing channel, i.e., $J_x = J_y = 0$ and $J_z = J$. For the single qubit system, the initial entangled state $|\phi_{SE_1}\rangle$ between the system and ancilla E_1 is supposed to be prepared in

$$|\phi_{SE_1}\rangle = (|-\rangle_S \otimes |+\rangle_{E_1} + |+\rangle_S \otimes |-\rangle_{E_1})/\sqrt{2} \qquad (14)$$

with $|\pm\rangle_{S(E_1)} = (|0\rangle \pm |1\rangle)/\sqrt{2}$. All the ancillas E_k ($k \geq 2$) are initially prepared in the identical state $|\eta_k\rangle = (|0\rangle + |1\rangle)/\sqrt{2}$.

In Figure 5a,b, we show the normalized mutual information $\bar{I}/H_S$ versus time t and environment fraction fN. In Figure 5c,d, we plot the corresponding averaged $\bar{I}_3(S:B:C)$ as a function of t, respectively. It can be seen from Figure 5a,b that as t increases from 0, the redundancy plateau begins to emerge, and it becomes more and more pronounced with the increase of t. Then, this clear signature of objectivity is gradually lost and emerges periodically as t increases further. Interestingly, we find that as time evolves, the dynamics behavior of $\bar{I}/H_S$ is closely related to that of $\bar{I}_3$. From Figure 5c,d, it can be seen that $\bar{I}_3$ is positive, and it experiences successive increasing and decreasing behaviors with the increase of t. Specifically, by comparing Figure 5a,c, or Figure 5b,d, it is clear that as $\bar{I}_3$ increases, the redundancy plateau becomes more and more pronounced and vice versa. To clearly show the relationship between them, in Figure 6, we plot $\bar{I}/H_S$ as a function of fN at some instants of time corresponding to different values of $\bar{I}_3$. From Figure 6, we find that: when $\bar{I}/H_S$ is approximately a linear function of fN, which indicates an "independent" environment, the corresponding $\bar{I}_3 \approx 0$ (black, circles line); when $\bar{I}_3$ increases from 0, the Darwinistic behavior begins to appear, and it becomes more and more pronounced as $\bar{I}_3$ increases further (e.g., $\bar{I}_3 = 0.3746, 0.7740$); when $\bar{I}_3 = 0.9994$, a perfect redundant encoding is observed (blue, diamond line) in the sense that even a single ancilla is sufficient to give all the information about S, i.e., quantum Darwinism emerges perfectly.

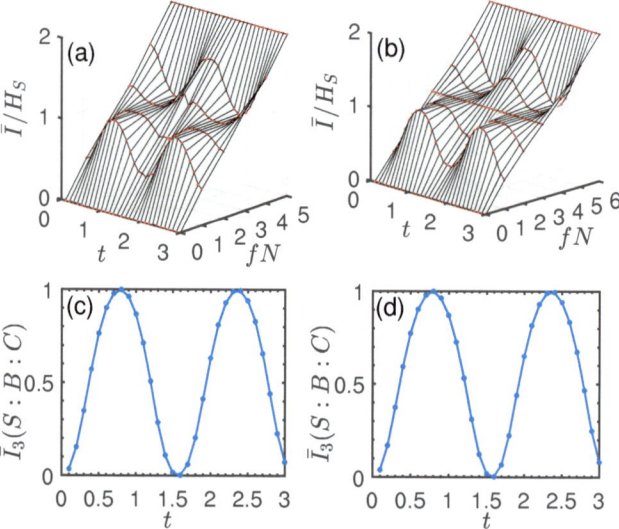

Figure 5. Normalized mutual information $\bar{I}/H_S$ and averaged TMI $\bar{I}_3$ for the pure dephasing system–environment interaction in the case of a single qubit system. Upper panel: $\bar{I}/H_S$ vs. fN and t. Lower panel: $\bar{I}_3(S:B:C)$ as a function of t. (**a,c**) $N = 5$. (**b,d**) $N = 6$. We set $J = 1$ and $l = 2$. $\bar{I}/H_S$ is plotted by averaging over 1000 possible environment fragments for the same f, and $\bar{I}_3$ is plotted by averaging over 1000 possible environment partitions, respectively.

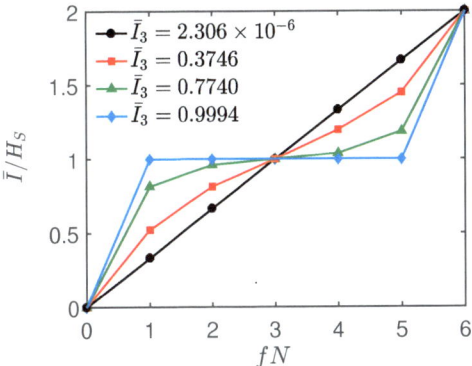

Figure 6. Normalized mutual information $\bar{I}/H_S$ as a function of fN at different t. The red (squares), green (triangles), blue (diamonds), and black (circles) lines correspond to $t = 0.3, 0.5, 0.8$, and 1.59, respectively. All the parameters are the same as those in Figure 5b.

Physically, the above phenomenon can be understood as follows. Consider a general model beyond CM, which is composed of a quantum system S interacting with an environment E consisting of a collection of N ancillas ($E_1, E_2, \ldots, E_N$). Here, we do not have any restrictions on the system–environment dynamics. For this general model, we consider the following two cases separately. First, we consider a situation in which the mutual information $I(S : F_f)$ is a linear function of fragment size f, namely, it exhibits a linear profile like line (b) in Figure 4. In this case, we can easily prove that it is related to zero TMI. Specifically, the linear relationship between $I(S : F_f)$ and f means that $I(S : F_f)$ is proportional to the number of ancillas contained in F_f, i.e.,

$$I(S : F_f) = k \cdot fN,$$

where k is a constant. According to Equation (11), we have

$$\begin{aligned} I_3(S : B : C) &= I(S : B) + I(S : C) - I(S : BC) \\ &= k \cdot 1 + k \cdot l - k \cdot (1 + l) \\ &= 0, \end{aligned}$$

which is independent of the choice of l. Then, we consider the case $I_3(S : B : C) > 0$. Positive TMI indicates that more information about the system S is shared among individual environmental ancillas. As positive TMI increases, more and more system information will be shared. Therefore, the measurement on a small fraction of the environment can have access to almost the same amount of information about the system, leading to the emergence of quantum Darwinism. Assume that in this general model, at the initial time, the joint system–environment state takes the form

$$|\Psi_0\rangle = |\phi_{SE_1}\rangle \otimes |\eta_0^2\rangle \otimes |\eta_0^3\rangle \otimes \ldots |\eta_0^N\rangle. \tag{15}$$

Here, $|\phi_{SE_1}\rangle$ is an entangled state between the system and ancilla E_1, and $|\eta_0^j\rangle$ ($j = 2, \ldots, N$) is the initial state of E_j, respectively. Depending on different system–environment dynamics, the joint system–environment state after the evolution is also different. We consider a situation where the composite system, after time t, evolves into the state,

$$|\Psi_t\rangle = \sum_k \alpha_k |\phi_k\rangle \otimes |\eta_k^1\rangle \otimes |\eta_k^2\rangle \otimes \ldots |\eta_k^N\rangle, \tag{16}$$

where $|\phi_k\rangle$ is the pointer states of the system, and $\{|\eta_k^j\rangle\}$ is the eigenbasis of the ancilla E_j. We first consider the TMI among the S, B, and C. In terms of Equation (16), tracing out all the environmental ancillas, we obtain the reduced density matrix of the system S

$$\rho_S = \sum_k |\alpha_k|^2 |\phi_k\rangle\langle\phi_k|. \tag{17}$$

Similarly, we can obtain

$$\rho_{B(C)} = \sum_k |\alpha_k|^2 |\varphi_k^{B(C)}\rangle\langle\varphi_k^{B(C)}|, \tag{18}$$

$$\rho_{SB(SC)} = \sum_k |\alpha_k|^2 |\phi_k\rangle\langle\phi_k| \otimes |\varphi_k^{B(C)}\rangle\langle\varphi_k^{B(C)}|, \tag{19}$$

where $\varphi_k^B := \otimes_{j \in B} |\eta_k^j\rangle$ and $\varphi_k^C := \otimes_{j \in C} |\eta_k^j\rangle$. Using the definition of Equation (11), we have

$$I_3(S:B:C) = -\sum_k |\alpha_k|^2 \log_2 p_k > 0. \tag{20}$$

Meanwhile, according to Equation (13), the mutual information between S and $B(C)$ can be obtained as

$$I(S:B) = I(S:C) = H(S) = -\sum_k |\alpha_k|^2 \log_2 |\alpha_k|^2, \tag{21}$$

which is independent of l. This means that for the joint system and environment state of Equation (16), one can acquire information about the system through the measurement on any size of the fraction of the environment. In other words, the state of Equation (16) is an example of perfect quantum Darwinism in the sense that even a single environmental ancilla is sufficient to access all the information about S. Therefore, positive TMI is related to the emergence of Darwinism.

For the two-qubit system, the initial entangled state $|\phi_{SE_1}\rangle$ between the system and ancilla E_1 is supposed to be prepared in

$$(|-+\rangle_S \otimes |-\rangle_{E_1} + |+-\rangle_S \otimes |+\rangle_{E_1})/\sqrt{2}, \tag{22}$$

and all the ancillas E_k ($k \geq 2$) are initially prepared in the identical state $|\eta_k\rangle = (|0\rangle + |1\rangle)/\sqrt{2}$. Figure 7a,b shows the normalized mutual information $\bar{I}/H_S$ as a function of t and fN. In this case, its behaviors are qualitatively the same as those of the single qubit system. We also see that a redundancy plateau emerges periodically with the increase of t. In Figure 7c,d, we plot $\bar{I}_3$ as a function of t. From these numerical calculations, we again find that: when $\bar{I}/H_S$ varies linearly with fN, the corresponding $\bar{I}_3 \approx 0$; as $\bar{I}_3$ increases from 0, the Darwinistic behavior gradually emerges, and it becomes more and more pronounced as $\bar{I}_3$ increases further and vice versa.

Until now, we have only considered the environment E consisting of up to $N = 6$ ancillas. From a numerical calculation, in Figure 8, we show that the above results persist for a larger environment $N = 10$ and $N = 12$. Since the global $S + E$ state is pure, the plots ($\bar{I}/H_S$ vs. f) are antisymmetric about $f = 1/2$. In Figure 8, we plot $\bar{I}/H_S$ as a function of fN varying from 0 to $N/2$. It can be seen that the linear dependence of the mutual information $\bar{I}/H_S$ on the environment fraction fN corresponds to the case where $\bar{I}_3 \approx 0$. Additionally, the more pronounced the Darwinistic plateau becomes, the larger the positive $\bar{I}_3$ is and vice versa, showing that the above results are valid for larger environments.

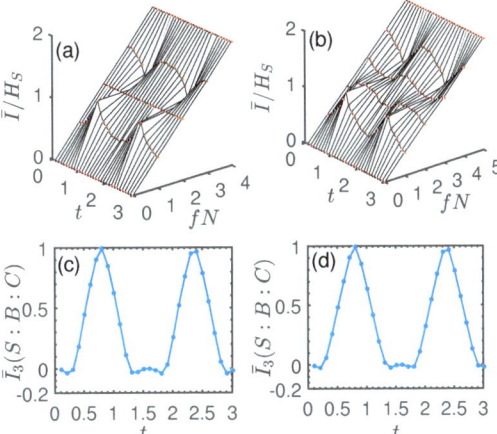

Figure 7. Normalized mutual information $\bar{I}/H_S$ and averaged TMI $\bar{I}_3$ for the pure dephasing system–environment interaction in the case of the two-qubit system. Upper panel: $\bar{I}/H_S$ vs. fN and t. Lower panel: $\bar{I}_3(S:B:C)$ as a function of t. (**a**,**c**) $N=4$. (**b**,**d**) $N=5$. We set $\epsilon=1$, and the remaining parameters are the same as those in Figure 5.

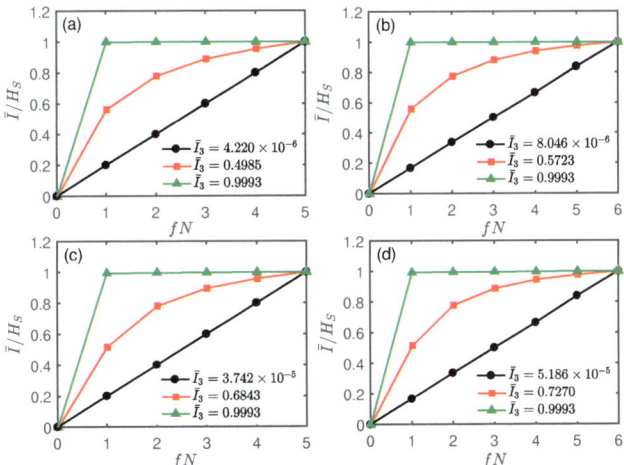

Figure 8. Normalized mutual information $\bar{I}/H_S$ as a function of fN for the pure dephasing system–environment interaction. Upper panel: single qubit system with the same parameters except N as in Figure 5. Lower panel: two-qubit system with the same parameters except N as in Figure 7. The red (squares), green (triangles), and black (circles) lines correspond to $t=0.046$, 0.80, and 1.20, respectively. (**a**,**c**) $N=10$. (**b**,**d**) $N=12$.

3.2. Exchange Interaction

Now, we turn our attention to the exchange interaction, i.e., $J_x = J_y = J$ and $J_z = 0$. First, we consider the single qubit system. The initial entangled state $|\phi_{SE_1}\rangle$ between the system and ancilla E_1 is initially prepared in the same state as Equation (14) in Section 3.1, and all the ancillas E_k ($k \geq 2$) are initially prepared in the identical state $|\eta_k\rangle = |0\rangle$. In Figure 9a,b, we show the normalized mutual information $\bar{I}/H_S$ as a function of fN and t. In this case, there is no perfect redundant encoding compared with that of the pure dephasing interaction. Nonetheless, the key features of quantum Darwinism still persist, i.e., a less manifest redundancy plateau appears at some specific time. In Figure 9c,d, we plot $\bar{I}_3$ as a function of t. It can be seen from Figure 9c,d that although $\bar{I}_3$ is positive, it is much smaller compared with that in Figure 5c,d for all times. This indicates that less information about the system S is shared among individual environment ancillas. Thus, the measurement on a small fraction of the environment can only have access to less information about the system. This is why the redundancy plateau is less manifest in this case. Thus, again, the result is consistent with the conclusion in Section 3.1.

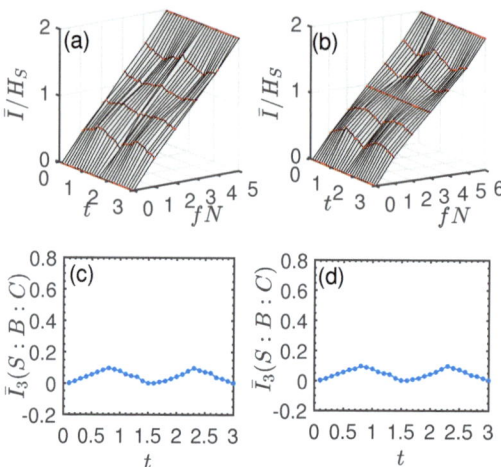

Figure 9. Normalized mutual information $\bar{I}/H_S$ and averaged TMI $\bar{I}_3$ for the exchange system–environment interaction in the case of a single qubit system. Upper panel: $\bar{I}/H_S$ vs. fN and t. Lower panel: $\bar{I}_3(S:B:C)$ as a function of t. (**a**,**c**) $N=5$. (**b**,**d**) $N=6$. The remaining parameters are the same as those in Figure 5.

Now, we consider the two-qubit system. The initial entangled state $|\phi_{SE_1}\rangle$ between the system and ancilla E_1 is prepared in the same state as Equation (22) in Section 3.1, and all the ancillas E_k ($k \geq 2$) are initially prepared in the identical state $|\eta_k\rangle = |0\rangle$. Figure 10a,b shows the normalized mutual information $\bar{I}/H_S$ as a function of t and fN. Different from the single qubit system, it can be seen that there is no redundant encoding of system information and the feature of quantum Darwinism is completely lost. Additionally, at some time, $\bar{I}/H_S$ takes on an S-shaped profile, indicating an "encoding" environment (like line (c) in Figure 4) or an antiredundancy, i.e., information about S is encoded in the multiple environmental ancillas, and to learn about the system, one requires access to at least half of the environmental ancillas. Interestingly, we find that this "encoding" environment behavior is associated with the scrambling of information. From Figure 10c,d, it can be seen that $\bar{I}_3$ becomes negative, indicating the scrambling or delocalization of quantum information.

In Figure 11, we plot the normalized mutual information $\bar{I}/H_S$ as a function of fN at some instants of time corresponding to different values of $\bar{I}_3$. It can be seen from Figure 11 that when $\bar{I}/H_S$ varies linearly with fN, the corresponding $\bar{I}_3 \approx 0$ (cf. the black (circle) line in Figure 11). When $\bar{I}_3 < 0$ (i.e., scrambling occurs), an "encoding" environment appears, where $\bar{I}/H_S$ takes on an S-shaped profile. The larger the absolute value of the negative $\bar{I}_3$, the more pronounced the S-shaped profile of the curve (cf. the red (square) and green (triangle) lines in Figure 11). This is because a negative $\bar{I}_3$ with a larger absolute value means that more information about the system S is nonlocally stored in the joint environment fraction BC and cannot be detected by local measurement just on B or C.

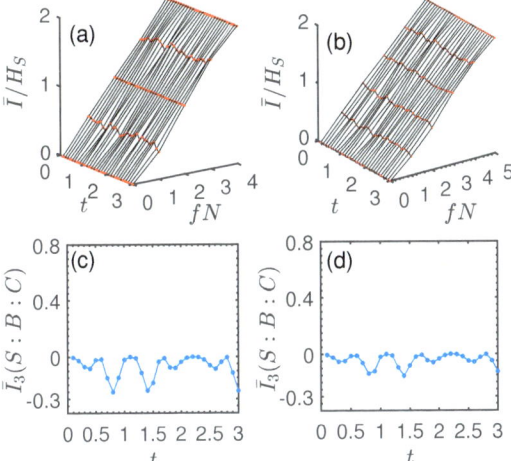

Figure 10. Normalized mutual information $\bar{I}/H_S$ and averaged TMI $\bar{I}_3$ for the exchange system–environment interaction in the case of the two-qubit system. Upper panel: $\bar{I}/H_S$ vs. fN and t. Lower panel: $\bar{I}_3(S:B:C)$ as a function of t. (**a**,**c**) $N=4$. (**b**,**d**) $N=5$. The remaining parameters are the same as those in Figure 7.

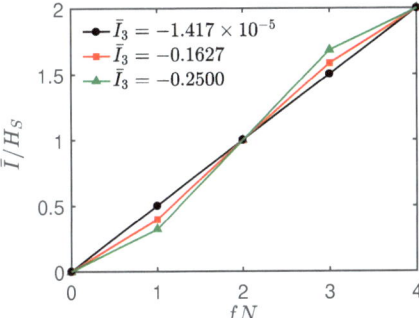

Figure 11. Normalized mutual information $\bar{I}/H_S$ as a function of fN at different t for the exchange system–environment interaction. The black (circles), red (squares), and green (triangles) lines correspond to $t=0.59$, 0.70, and 0.80, respectively. All parameters are the same as those in Figure 10a.

In order to better understand the above phenomenon, we consider again the general model (mentioned in Section 3.1). We suppose that the joint system–environment state is initially the same as Equation (15), and after time t, the whole state of the system and environment evolves into

$$|\Phi_t\rangle = \sum_k \alpha_k |\phi_k\rangle \otimes |\xi_k\rangle, \qquad (23)$$

where $|\xi_k\rangle$ are many-body entangled states of the environment. In this case, due to the presence of entanglement between the environmental ancillas, information about the system is shared among the joint environment ancillas rather than among the individual ancillas. Hence, any measurement on a small fraction of the environment may not obtain enough information about the system. Specifically, we assume that the system S is a qubit and the environment is composed of N qubits for which the evolved state $|\Phi_t\rangle$ is supposed to be

$$|\Phi_t\rangle = |0\rangle \otimes |D_N^{(2)}\rangle + |1\rangle \otimes |D_N^{(1)}\rangle, \qquad (24)$$

where environmental state $|D_N^{(d)}\rangle$ is an N-qubit Dicke state with d excitations, defined as

$$|D_N^{(d)}\rangle = \binom{N}{d}^{-1/2} \sum_i P_i\{|1\rangle^{\otimes d}|0\rangle^{\otimes(N-d)}\}, \tag{25}$$

where $\sum_i P_i\{\cdot\}$ denotes the sum over all possible permutations. For the state of Equation (24), we first calculate the TMI between the S, B, and C and obtain $I_3(S:B:C) < 0$, which is independent of the choice of l. This negative TMI implies that initially localized system information is scrambled in the environment. Then, from Equation (24), in Figure 12, we plot normalized mutual information $\bar{I}/H_S$ as a function of environment fractions fN for different environment sizes $N = 5$, $N = 6$, and $N = 7$. It can be seen that $\bar{I}/H_S$ takes on an S-shaped profile, indicating an "encoding" environment. Therefore, negative TMI is related to an "encoding" environment.

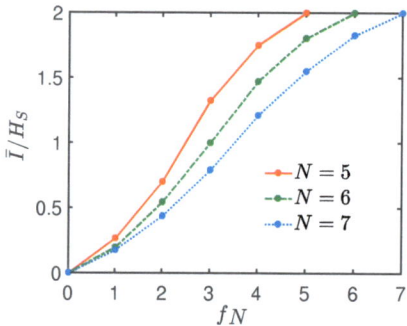

Figure 12. Normalized mutual information $\bar{I}/H_S$ as a function of fN for different N. The plotted curves are the results of averaging over 1000 possible environment fragments with the same f.

Now, we consider the above results for a larger environment with $N = 10$ and $N = 12$. In Figure 13, we plot $\bar{I}$ as a function of fN for different t, where fN varies from 0 to $N/2$. For the single qubit system in Figure 13a,b, we can see the redundant features of Darwinism corresponding to positive TMI, which means that information scrambling does not occur. As the positive $\bar{I}_3$ increases, redundant features of Darwinism become more and more pronounced. For the two-qubit system in Figure 13c,d, we see the behavior of an "encoding" environment, and it is related to the negative TMI. The curve with a more pronounced S-shaped profile coincides with the negative $\bar{I}_3$ with a larger absolute value.

Moreover, for the numerical results in this section, we emphasize the following:

(i) In the case of a single qubit system, the initial system–environment entangled state that we choose in Equation (14) is just one of the Bell states, namely, $(|0\rangle_S \otimes |0\rangle_E - |1\rangle_S \otimes |1\rangle_E)/\sqrt{2}$. The above results are valid when it is replaced by any of the other three orthogonal entangled Bell states.

(ii) In the case of a two-qubit system, from our numerical calculations, we find that our above conclusions are valid when the initial entangled state $|\phi_{SE_1}\rangle$ that we choose (see Equation (22)) is replaced by the GHZ state or W state.

(iii) In the case of a two-qubit system, our results in this section are valid when the interaction Hamiltonian between S_1 and S_2, namely, Equation (3), is replaced by $H_{S_1,S_2} = \epsilon(\sigma_x \otimes \sigma_x + \sigma_y \otimes \sigma_y + \sigma_z \otimes \sigma_z)$.

(iv) As previously mentioned, to compute TMI, we divide the whole environment E into three nonoverlapping subsystems B, C, D, whose sizes (the number of the ancillas) are, respectively, given by 1, l, and $N - l - 1$. Although our above numerical calculation for $\bar{I}_3(S:B:C)$ is in terms of $l = 2$, from numerical calculation, we find that all the above results are independent of l.

(v) For the system–environment interactions besides Equations (5) and (6), we also consider the isotropic Heisenberg interaction (i.e., $J_x = J_y = J_z = 1$) and anisotropic Heisenberg interaction (i.e., $J_x = J_y = 1$ and $J_z = \Delta \in (0,1)$), respectively. For these two interactions, the corresponding results are qualitatively the same as those obtained from the exchange interaction.

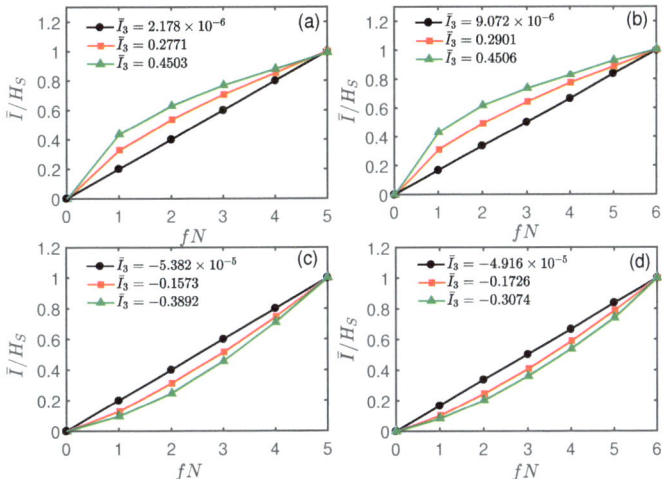

Figure 13. Normalized mutual information $\bar{I}/H_S$ as a function of fN for the exchange system–environment interaction. Upper panel: single qubit system with the same parameters except N as in Figure 9. Lower panel: two-qubit system with the same parameters except N as in Figure 10. The black (circles), red (squares), and green (triangles) lines correspond to $t = 0.26, 0.50$, and 0.80, respectively. (**a**,**c**) $N = 10$. (**b**,**d**) $N = 12$.

For the above results in this paper, the key to the emergence of quantum Darwinism is the particular structure of the joint system–environment state, with a branching structure produced by the decoherence process. Each branch is perfectly correlated with a particular pointer state of the system. The subenvironments are not entangled with each other, only correlated via the system. The state of system is thus redundantly encoded in the environment itself. Different from this, when scrambling takes place, the initially localized information becomes more and more non-local and inaccessible. In this case, it will typically lead to entanglement between the subenvironments, which decreases their information storage capacity. In other words, even though the environment as a whole may still contain a redundant imprint of the state of the system, one can become inaccessible to system information only through local measurements on its small environment fragment. This is why when scrambling occurs, we do not see the Darwinistic behavior but an "encoding" environment. Quantum Darwinism and "encoding" environment behaviors are distinct ways of many-body quantum information spreading [55]. Through the TMI, we can determine whether system information is encoding redundantly in the environment, which contributes to our understanding of quantum Darwinism.

4. Conclusions

How the macroscopic classical world emerges from the framework of quantum mechanics has always been a topic of fundamental interest. Quantum Darwinism provides an explanation of classical objectivity from the quantum formalism, which not only regards the environment as the cause of decoherence, but also as the carrier of information about the system. The environment selects and proliferates the information of the system so that many observers can obtain the same information of the system. While information scrambling is a recent hot topic in the study of quantum many-body systems, in a process of information scrambling, initially localized information gets encoded into the global entanglement among the entire system and hence becomes inaccessible by local measurement. It seems that quantum Darwinism and information scrambling are two different research topics and have no connections. In this paper, from a new perspective and by considering the environment, is a multipartite system where we have explored the relationship between quantum Darwinism and information scrambling in the environment. The key ingredient for both quantum Darwinism and information scrambling is the information: quantum Darwinism is about the proliferation or redundant records of the information of a quantum system in the environment, while information scrambling is about how the initial local information becomes delocalized. The mutual information between the system and a subset environment has been widely used in the study of quantum Darwinism. In order to describe both quantum Darwinism and information scrambling in a unified framework, we have used TMI, which, besides OTOC, is also widely used as a probe of information scrambling. Based on a collision model, we mainly address the relation between the emergence of quantum Darwinism and information scrambling in the environment. We consider a system (one or two qubits) interacting with an ensemble of environment ancillas through two types of system–ancilla interactions, specifically, the pure dephasing interaction and exchange interaction. We find that, for both single qubit and two-qubit systems and both system–ancilla interactions, when the mutual information between the system and environmental fragment exhibits a linear increase with an increasing fragment size, TMI is zero, which can be proved in a general scenario beyond the collision model. In the case of the pure dephasing interaction for both one qubit and two-qubit systems, as TMI increases from zero, the Darwinistic behavior begins to emerge, and the larger the value of TMI, the more pronounced the Darwinistic behavior becomes. In the case of the exchange interaction, we find that the single qubit and two-qubit systems behave differently for the emergence of quantum Darwinism and scrambling. For the single qubit system, although no perfect redundant encoding appeared, the key features of quantum Darwinism still exist. Additionally, this behavior coincides with relatively small but positive TMI. However, for the two-qubit system, the "encoding" environment appears, and there is no redundant encoding of the system information; at the same time, TMI is negative, which means the information scrambling emerges. This can be explained as follows. When the TMI is negative, the system information is predominantly nonlocally shared among the environment ancillas, such that local measurement just on a small fraction of the environment cannot acquire the information of the system, leading to the loss of the quantum Darwinism feature. In contrast, when TMI is positive, more information is shared among individual environment subsystems. Thus, a measurement on a small fraction of the environment can have access to almost the same amount of information about the system, leading to the emergence of quantum Darwinism. Finally, we believe our work might shed some light on the mechanism of quantum Darwinism and stimulate more involved works in this direction.

Author Contributions: Conceptualization, F.T. and J.Z.; methodology, F.T.; validation, F.T. and J.Z.; formal analysis, F.T. and J.Z.; investigation, J.Z. and B.S.; writing—original draft preparation, F.T.; writing—review and editing, J.Z., H.L. and L.H.; supervision, J.Z. and B.S. All authors have read and agreed to the published version of the manuscript.

Funding: This work is supported by the National Natural Science Foundation of China (Grant Nos. 11775019 and 11875086).

Data Availability Statement: Data sharing is not applicable to this article.

Conflicts of Interest: The authors declare no conflict of interest.

References

1. Zurek, W.H. Quantum Darwinism. *Nat. Phys.* **2009**, *5*, 181. [CrossRef]
2. Breuer, H.P.; Petruccione, F. *The Theory of Open Quantum Systems*; Oxford University Press: Oxford, UK, 2002.

3. Zurek, W.H. Pointer basis of quantum apparatus: Into what mixture does the wave packet collapse? *Phys. Rev. D* **1981**, *24*, 1516. [CrossRef]
4. Zurek, W.H. Environment-induced superselection rules. *Phys. Rev. D* **1982**, *26*, 1862. [CrossRef]
5. Zurek, W.H. Decoherence, einselection, and the quantum origins of the classical. *Rev. Mod. Phys.* **2003**, *75*, 715. [CrossRef]
6. Zwolak, M.; Quan, H.T.; Zurek, W.H. Redundant imprinting of information in nonideal environments: Objective reality via a noisy channel. *Phys. Rev. A* **2010**, *81*, 062110. [CrossRef]
7. Zwolak, M.; Zurek, W.H. Redundancy of einselected information in quantum Darwinism: The irrelevance of irrelevant environment bits. *Phys. Rev. A* **2017**, *95*, 030101. [CrossRef]
8. Zwolak, M.; Riedel, C.J.; Zurek, W.H. Amplification, Decoherence and the Acquisition of Information by Spin Environments. *Sci. Rep.* **2016**, *6*, 25277. [CrossRef]
9. Touil, A.; Yan, B.; Girolami, D.; Deffner, S.; Zurek, W.H. Eavesdropping on the Decohering Environment: Quantum Darwinism, Amplification, and the Origin of Objective Classical Reality. *Phys. Rev. Lett.* **2022**, *128*, 010401. [CrossRef]
10. Brandão, F.G.S.L.; Piani, M.; Horodecki, P. Generic emergence of classical features in quantum Darwinism. *Nat. Commun.* **2015**, *6*, 7908. [CrossRef]
11. Riedel, C.J.; Zurek, W.H.; Zwolak, M. The rise and fall of redundancy in decoherence and quantum Darwinism. *New J. Phys.* **2012**, *14*, 083010. [CrossRef]
12. Zurek, W.H. Quantum Theory of the Classical: Einselection, Envariance, Quantum Darwinism and Extantons. *Entropy* **2022**, *24*, 1520. [CrossRef] [PubMed]
13. Ollivier, H.; Poulin, D.; Zurek, W.H. Objective Properties from Subjective Quantum States: Environment as a Witness. *Phys. Rev. Lett.* **2004**, *93*, 220401. [CrossRef] [PubMed]
14. Blume-Kohout, R.; Zurek, W.H. Quantum Darwinism: Entanglement, branches, and the emergent classicality of redundantly stored quantum information. *Phys. Rev. A* **2006**, *73*, 062310. [CrossRef]
15. Blume-Kohout, R.; Zurek, W.H. A Simple Example of "Quantum Darwinism": Redundant Information Storage in Many-Spin Environments. *Found. Phys.* **2005**, *35*, 1857. [CrossRef]
16. Zwolak, M.; Quan, H.T.; Zurek, W.H. Quantum Darwinism in a Mixed Environment. *Phys. Rev. Lett.* **2009**, *103*, 110402. [CrossRef]
17. Pleasance, G.; Garraway, B.M. Application of quantum Darwinism to a structured environment. *Phys. Rev. A* **2017**, *96*, 062105. [CrossRef]
18. Giorgi, G.L.; Galve, F.; Zambrini, R. Quantum Darwinism and non-Markovian dissipative dynamics from quantum phases of the spin-1/2 XX model. *Phys. Rev. A* **2015**, *92*, 022105. [CrossRef]
19. Balaneskovica, N. Random Unitary Evolution Model of Quantum Darwinism with pure decoherence. *Eur. Phys. J. D* **2015**, *69*, 232. [CrossRef]
20. Blume-Kohout, R.; Zurek, W.H. Quantum Darwinism in Quantum Brownian Motion. *Phys. Rev. Lett.* **2008**, *101*, 240405. [CrossRef]
21. Pérez, A. Information encoding of a qubit into a multilevel environment. *Phys. Rev. A* **2010**, *81*, 052326. [CrossRef]
22. Milazzo, N.; Lorenzo, S.; Paternostro, M.; Palma, G.M. Role of information backflow in the emergence of quantum Darwinism. *Phys. Rev. A* **2019**, *100*, 012101. [CrossRef]
23. Oliveira, S.M.; de Paula, A.L.; Drumond, R.C. Quantum darwinism and non-markovianity in a model of quantum harmonic oscillators. *Phys. Rev. A* **2019**, *100*, 052110. [CrossRef]
24. Lampo, A.; Tuziemski, J.; Lewenstein, M.; Korbicz, J.K. Objectivity in the non-Markovian spin-boson model. *Phys. Rev. A* **2017**, *96*, 012120. [CrossRef]
25. Lorenzo, S.; Paternostro, M.; Palma, G.M. Anti-Zeno-based dynamical control of the unfolding of quantum Darwinism. *Phys. Rev. Res.* **2020**, *2*, 013164. [CrossRef]
26. Campbell, S.; Çakmak, B.; Müstecaplıoğlu, Ö.E.; Paternostro, M.; Vacchini, B. Collisional unfolding of quantum Darwinism. *Phys. Rev. A* **2019**, *99*, 042103. [CrossRef]
27. Ryan, E.; Paternostro, M.; Campbell, S. Quantum darwinism in a structured spin environment. *Phys. Lett. A* **2021**, *416*, 127675. [CrossRef]
28. Girolami, D.; Touil, A.; Yan, B.; Deffner, S.; Zurek, W.H. Redundantly amplified information suppresses quantum correlations in many-body systems. *Phys. Rev. Lett.* **2022**, *129*, 010401. [CrossRef]
29. Ciampini, M.A.; Pinna, G.; Mataloni, P.; Paternostro, M. Experimental signature of quantum Darwinism in photonic cluster states. *Phys. Rev. A* **2018**, *98*, 020101. [CrossRef]
30. Chen, M.C.; Zhong, H.S.; Li, Y.; Wu, D.; Wang, X.L.; Li, L.; Liu, N.L.; Lu, C.Y.; Pan, J.W. Emergence of classical objectivity of quantum Darwinism in a photonic quantum simulator. *Sci. Bull.* **2019**, *64*, 580–585. [CrossRef]
31. Unden, T.K.; Louzon, D.; Zwolak, M.; Zurek, W.H.; Jelezko, F. Revealing the Emergence of Classicality Using Nitrogen-Vacancy Centers. *Phys. Rev. Lett.* **2019**, *123*, 140402. [CrossRef]
32. Scarani, V.; Ziman, M.; Štelmachovič, P.; Gisin, N.; Bužek, V. Thermalizing Quantum Machines: Dissipation and Entanglement. *Phys. Rev. Lett.* **2002**, *88*, 097905. [CrossRef] [PubMed]
33. Ciccarello, F.; Lorenzo, S.; Giovannetti, V.; Palma, G.M. Quantum collision models: Open system dynamics from repeated interactions. *Phys. Rep.* **2022**, *954*, 1. [CrossRef]
34. Ziman, M.; Štelmachovič, P.; Bužek, V. Description of quantum dynamics of open systems based on collision-like models. *Open Syst. Inf. Dyn.* **2005**, *12*, 81. [CrossRef]

35. Ziman, M.; Bužek, V. All (qubit) decoherences: Complete characterization and physical implementation. *Phys. Rev. A* **2005**, *72*, 022110. [CrossRef]
36. Li, Y.W.; Li, L. Hierarchical-environment-assisted non-Markovian and its effect on thermodynamic properties. *EPJ Quantum Technol.* **2021**, *8*, 9. [CrossRef]
37. Çakmak, B.; Müstecaplıoğlu, Ö.E.; Paternostro, M.; Vacchini, B.; Campbell, S. Quantum Darwinism in a Composite System: Objectivity versus Classicality. *Entropy* **2021**, *23*, 995. [CrossRef] [PubMed]
38. Mezei, M.; Stanford, D. On entanglement spreading in chaotic systems. *J. High Energy Phys.* **2017**, *2017*, 65. [CrossRef]
39. Seshadri, A.; Madhok, V.; Lakshminarayan, A. Tripartite mutual information, entanglement, and scrambling in permutation symmetric systems with an application to quantum chaos. *Phys. Rev. E* **2018**, *98*, 052205. [CrossRef]
40. Roberts, D.A.; Yoshida, B. Chaos and complexity by design. *J. High Energy Phys.* **2017**, *2017*, 121. [CrossRef]
41. Hosur, P.; Qi, X.L.; Roberts, D.A.; Yoshida, B. Chaos in quantum channels. *J. High Energy Phys.* **2016**, *2016*, 4. [CrossRef]
42. Swingle, B. Unscrambling the physics of out-of-time-order correlators. *Nat. Phys.* **2018**, *14*, 988. [CrossRef]
43. Maldacena, J.; Shenker, S.H.; Stanford, D. A bound on chaos. *J. High Energy Phys.* **2016**, *2016*, 106 [CrossRef]
44. Hartman, T.; Maldacena, J. Time evolution of entanglement entropy from black hole interiors. *J. High Energy Phys.* **2013**, *2013*, 14. [CrossRef]
45. Zhou, T.; Chen, X. Operator dynamics in a Brownian quantum circuit. *Phys. Rev. E* **2019**, *99*, 052212. [CrossRef] [PubMed]
46. Zanardi, P. Entanglement of quantum evolutions. *Phys. Rev. A* **2001**, *63*, 040304. [CrossRef]
47. Schnacck, O.; Bölter, N.; Paeckel, S.; Manmana, S.R.; Kehrein, S.; Schmitt, M. Tripartite information, scrambling, and the role of Hilbert space partitioning in quantum lattice models. *Phys. Rev. B* **2019**, *100*, 224302. [CrossRef]
48. Bölter, N.; Kehrein, S. Scrambling and many-body localization in the XXZ chain. *Phys. Rev. B* **2022**, *105*, 104202. [CrossRef]
49. Iyoda, E.; Sagawa, T. Scrambling of quantum information in quantum many-body systems. *Phys. Rev. A* **2018**, *97*, 042330. [CrossRef]
50. Sun, Z.H.; Cui, J.; Fan, H. Quantum information scrambling in the presence of weak and strong thermalization. *Phys. Rev. A* **2021**, *104*, 022405. [CrossRef]
51. Li, Y.; Li, X.L.; Jin, J. Information scrambling in a collision model. *Phys. Rev. A* **2020**, *101*, 042324. [CrossRef]
52. Wanisch, D.; Fritzsche, S. Delocalization of quantum information in long-range interacting systems. *Phys. Rev. A* **2021**, *104*, 042409. [CrossRef]
53. Niknam, M.; Santos, L.F.; Cory, D.G. Sensitivity of quantum information to environment perturbations measured with a nonlocal out-of-time-order correlation function. *Phys. Rev. Res.* **2020**, *2*, 013200. [CrossRef]
54. Yan, B.; Sinitsyn, N.A. Recovery of Damaged Information and the Out-of-Time-Ordered Correlators. *Phys. Rev. Lett.* **2020**, *125*, 040605. [CrossRef] [PubMed]
55. Ferté, B.; Cao, X. Quantum Darwinism-encoding transitions on expanding trees. *arXiv* **2023**, arXiv:2312.04284.

Disclaimer/Publisher's Note: The statements, opinions and data contained in all publications are solely those of the individual author(s) and contributor(s) and not of MDPI and/or the editor(s). MDPI and/or the editor(s) disclaim responsibility for any injury to people or property resulting from any ideas, methods, instructions or products referred to in the content.

Article

A Wigner Quasiprobability Distribution of Work

Federico Cerisola [1,2,3,4,*], Franco Mayo [1,2] and Augusto J. Roncaglia [1,2]

1. Departamento de Física, Facultad de Ciencias Exactas y Naturales, Universidad de Buenos Aires, Buenos Aires C1428EGA, Argentina; fmayo@df.uba.ar (F.M.); augusto@df.uba.ar (A.J.R.)
2. Instituto de Física de Buenos Aires (IFIBA), CONICET—Universidad de Buenos Aires, Buenos Aires C1121A6B, Argentina
3. Department of Physics and Astronomy, University of Exeter, Stocker Road, Exeter EX4 4QL, UK
4. Department of Engineering Science, University of Oxford, Parks Road, Oxford OX1 3PJ, UK
* Correspondence: federico.cerisola@eng.ox.ac.uk

Abstract: In this article, we introduce a quasiprobability distribution of work that is based on the Wigner function. This proposal rests on the idea that the work conducted on an isolated system can be coherently measured by coupling the system to a quantum measurement apparatus. In this way, a quasiprobability distribution of work can be defined in terms of the Wigner function of the apparatus. This quasidistribution contains the information of the work statistics and also holds a clear operational definition that can be directly measured in a real experiment. Moreover, it is shown that the presence of quantum coherence in the energy eigenbasis is related with the appearance of features related to non-classicality in the Wigner function such as negativity and interference fringes. On the other hand, from this quasiprobability distribution, it is straightforward to obtain the standard two-point measurement probability distribution of work and also the difference in average energy for initial states with coherences.

Keywords: quantum thermodynamics; work statistics; quantum coherence

Citation: Cerisola, F.; Mayo, F.; Roncaglia, A.J. A Wigner Quasiprobability Distribution of Work. *Entropy* **2023**, *25*, 1439. https://doi.org/10.3390/e25101439

Academic Editors: Gioacchino Massimo Palma and Lawrence Horwitz

Received: 18 September 2023
Revised: 6 October 2023
Accepted: 8 October 2023
Published: 11 October 2023

Copyright: © 2023 by the authors. Licensee MDPI, Basel, Switzerland. This article is an open access article distributed under the terms and conditions of the Creative Commons Attribution (CC BY) license (https://creativecommons.org/licenses/by/4.0/).

1. Introduction

The notion of work is one of the most basic and fundamental concepts in physics, particularly in thermodynamics. During the last decades, several attempts have been made to obtain the work statistics for non-equilibrium thermodynamic transformations in the quantum regime. These definitions were motivated by the idea of extending classical fluctuation theorems [1–5] to quantum operations. In order to describe the thermodynamics of general non-equilibrium quantum processes, it is necessary to provide a general definition of work valid for any quantum system and process. However, this task presents serious difficulties. This is due to the fact that many concepts belonging to the classical definition of work cannot be directly translated to quantum mechanics. For example, the basic definition of the work that a force performs on a particle along a trajectory cannot be used in quantum mechanics because of the lack of a ubiquitous meaning of trajectories in the theory, although recently a definition of quantum work was made by considering Bohmian trajectories [6]. A great advancement came in the area with the definition of the two-point measurement protocol (TPM) to define work in driven isolated quantum systems [3,4,7–9]. This definition is based on the simple observation that, for an isolated system, work is a random variable associated to the difference in energy along the process. Thus, in order to determine this random value, one should make an energy measurement at the beginning and another at the end of the process. This definition is not only straightforward in an operational sense, but it also recovers the results of the fluctuation theorems for quantum systems [3–5,7,8,10] and was verified experimentally in different platforms [11–16].

However, there is a caveat with the TPM when one considers initial states that have coherences in the energy basis. This is because the first energy measurement destroys these coherences, and therefore the TPM scheme is insensitive to quantum coherence between different energy subspaces. This leads to undesirable consequences, for instance, related with the fact that the average work performed in the process is different from the change in the average energy of the system. Moreover, it has been shown that it is impossible to define a probability distribution of work that satisfies at the same time the fluctuation theorems and whose mean value of work equals the average energy change for states with coherences [17]. This has led the community to consider different approaches to generalize the work distribution [6,18–20], including some proposals for quasiprobability distributions of work [21–32].

In this article, we propose a distribution based on the Wigner function. This definition relies on the fact that the work probability distribution can also be coherently measured by coupling the system to a quantum apparatus and making a single measurement over the apparatus, i.e., a single-measurement protocol (SM) [13,33,34]. In this way, the final state of the apparatus contains the information about the work distribution and one can define a quasiprobability distribution [35]. This approach provides a clear operational definition with an immediate experimental implementation. In addition, the Wigner function is represented using coordinates that have an intuitive interpretation in terms of time and energy associated with the work. Moreover, it can be shown that the presence of quantum coherence is related with the appearance of features related to non-classicality in the Wigner function, such as negativity and interference fringes. On the other hand, for coherence-free states, this definition agrees with the standard two-point measurement probability distribution of work.

The paper is organized as follows. In Section 2, we briefly discuss the two-point measurement scheme and the single-measurement protocol. In Section 3, we introduce the quasiprobability distribution of work based on the Wigner function, showing how it works for initial states of the system with and without coherence. In Section 4, we discuss experimental implementations, and we end with discussions and conclusion in Section 5.

2. Work Statistics

We are interested in the work distribution for isolated quantum systems that are subjected to an external driving. In this way, the external work can be associated to the energy change of the system. The typical scenario consists of a system $\mathcal{S}$ that starts in a given initial state, ρ_S, and is subjected to an external driving, represented by a unitary evolution $\mathcal{U}$. The driving is such that it changes the Hamiltonian from an initial H to a final one $\tilde{H}$, such that

$$H = \sum_n E_n \Pi_n, \qquad \tilde{H} = \sum_m \tilde{E}_m \tilde{\Pi}_m, \qquad (1)$$

where Π_n ($\tilde{\Pi}_m$) are the projectors on each energy subspace of the initial (final) Hamiltonian. In this case, what we know is that the average change of energy in the system is

$$\Delta E = \text{tr}\left[\tilde{H}\mathcal{U}\rho_S\mathcal{U}^\dagger\right] - \text{tr}[H\rho_S], \qquad (2)$$

where $\mathcal{U}\rho_S\mathcal{U}^\dagger$ is the final state after the driving. Clearly, it would be desirable that the average work obtained from the corresponding probability distribution equals this average energy change. This requisite is equivalent to asking that the first law of thermodynamics for mean values is satisfied for an isolated system. However, it can be shown that, if one imposes that the statistics of work is consistent with the standard fluctuation theorems, the distribution of work should be defined by the two-point measurement protocol [3,4,7,8]. In this case, although the resulting work average coincides with the mean energy difference for initial stationary states (i.e., diagonal in the initial energy eigenbasis), it is different for initial states with coherences.

2.1. The Two-Point Measurement Protocol

The two-point measurement protocol allows us to define a probability distribution of work consistent with fluctuation theorems. In order to do so, one should define a stochastic work value for each realization of the given driving protocol. This is conducted in terms of the difference of two energy values that are obtained by making two projective energy measurements: one at the beginning, and the other one at the end of the driving. In this way, the corresponding probability distribution can be written as

$$P_{\text{TPM}}(w) = \sum_{n,m} p_n p_{m|n} \, \delta\big(w - (\tilde{E}_m - E_n)\big), \tag{3}$$

where p_n is the probability of obtain E_n in the first energy measurement, and $p_{m|n}$ is the conditional probability of obtaining $\tilde{E}_m$ at the end given that E_n was obtained at the beginning. Therefore, if the initial state is already diagonal in the energy eigenbasis, the first measurement does not modify the state, and it is straightforward to verify that the mean value of work equals the average energy difference. Indeed, from (3), we have that, in general, the mean value of work is

$$\begin{aligned}
\langle w \rangle &= \int dw \, P_{\text{TPM}}(w) \, w = \sum_{n,m} p_n p_{m|n} (\tilde{E}_m - E_n) \\
&= \sum_{n,m} \text{tr}\big[\tilde{\Pi}_m \, \mathcal{U} \, \Pi_n \, \rho_S \, \Pi_n \, \mathcal{U}^\dagger\big] (\tilde{E}_m - E_n) \\
&= \text{tr}\big[\tilde{H} \, \mathcal{U} \, \bar{\rho}_S \, \mathcal{U}^\dagger\big] - \text{tr}[H \bar{\rho}_S],
\end{aligned} \tag{4}$$

where $\bar{\rho}_S = \sum_n \Pi_n \rho_S \Pi_n$ is the dephased initial state. This state is obtained by removing all the coherences between different energy subspaces of the initial Hamiltonian, and it is equivalent to the state resulting the following asymptotic temporal average

$$\bar{\rho}_S = \lim_{T \to \infty} \frac{1}{T} \int_{-T/2}^{T/2} d\tau \, e^{-\frac{i}{\hbar}H\tau} \rho_S e^{\frac{i}{\hbar}H\tau}. \tag{5}$$

Therefore, unless the initial state is diagonal in the basis of the initial Hamiltonian, the work average given by the TPM is different from the difference of average energy of the system. In fact, if the initial state is diagonal in this basis, then ρ_S can be interpreted as a 'classical' probability distribution over the different energies. In that case, the first measurement is not invasive, in the sense that it only 'reveals' the value of the energy in each realization of the experiment. On the other hand, for an initial state with coherences, the initial energy is not well defined and this interpretation is not straightforward.

2.2. The Single-Measurement Protocol

Another method for assessing the work probability distribution was introduced in [33]. The method is based on the idea that the work measurement can be described in terms of a generalized measurement (POVM). That is, by coupling the system to an ancilla, which is finally subjected to a 'single measurement' (SM). In this way, it can be shown that one can obtain the same probability distribution provided the ancilla is properly initialized.

Let us now describe briefly the general method that is summarized in the circuit of Figure 1. Initially, the system is in the state ρ_S and there is an auxiliary system (ancilla) $\mathcal{A}$ whose state is described terms of a continuous degree of freedom. In the ancilla's space, one can consider two canonically conjugated operators, $\mathcal{W}_\mathcal{A}$ and $\mathcal{T}_\mathcal{A}$, such that $[\mathcal{W}_\mathcal{A}, \mathcal{T}_\mathcal{A}] = i\hbar$. Thus, the evolution contains two coherent interactions between $\mathcal{S}$ and $\mathcal{A}$: one before, $e^{iH \otimes \mathcal{T}_\mathcal{A}/\hbar}$, and another, $e^{-i\tilde{H} \otimes \mathcal{T}_\mathcal{A}/\hbar}$, at the end of the driving. Each interaction can be viewed either as a coherent translation in the variable w of the ancilla in an amount that depends on the energy of the system or, conversely, as a coherent time-translation (free evolution) of the system whose time interval is proportional to the variable τ of the ancilla. Therefore, one can immediately associate the variables w and τ to energy (work) and time, respectively.

This analogy between the variables of the ancilla with work and time will become clearer after analyzing some examples of our proposed distribution.

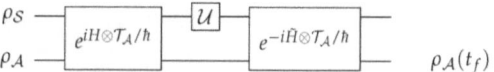

Figure 1. Circuit that describes the single-measurement protocol from which the work probability distribution can be obtained.

Following the protocol of the circuit, if the initial state of $\mathcal{A}$ is $\rho_\mathcal{A}$, then at time t_f after both interactions with the system, its reduced state is

$$\rho_\mathcal{A}(t_f) = \sum_{n,n',m} \mathrm{tr}\left[\tilde{\Pi}_m \mathcal{U} \Pi_n \rho_S \Pi_{n'} \mathcal{U}^\dagger\right] \times e^{-i w_{nm} T_\mathcal{A}/\hbar} \rho_\mathcal{A} e^{i w_{n'm} T_\mathcal{A}/\hbar}, \tag{6}$$

where $w_{nm} = \tilde{E}_m - E_n$ are the different work values. The SM protocol finishes by performing a projective measurement of the observable $\mathcal{W}_\mathcal{A}$. In this case, for highly localized initial pure states of $\mathcal{A}$, the resulting probability distribution is equivalent to the work distribution of the TPM protocol [33]. Notably, within this formulation, one can associate work to an observable that is acting over the ancillary system. Of course, work is not an observable acting on the system's space [4].

It is important to stress at this point that the entangling interaction between system and apparatus establishes a coherent record of the different values of work. Therefore, the reduced state of the ancilla contains information not only about the probability distribution given by the TPM, but also about the initial state of the system. At the end, the type of measurement that is conducted over the ancilla determines which information is extracted from the protocol. It is also interesting to note that this type of interaction appears in a very related task: the work extraction from a quantum system. This can be modeled by adding an interaction between the system and an auxiliary system that acts as a battery in which work is stored [36–38]. In general, the battery can be thought of as a continuous variable system, an ideal weight, with a Hamiltonian like the operator $\mathcal{W}_\mathcal{A}$. The work extraction process consists on some unitary evolution on the joint system (where the driving on the system is included) that can change the system Hamiltonian from H to $\tilde{H}$. The extracted work, in this way, is stored in the battery. There are a few conditions that should be imposed in this framework in order to ensure that the weight does not provide any thermodynamical resource to the work extraction process [37], one of them is of course energy conservation. It has been shown in [37] that the unitary operations that satisfy these conditions are of the form $e^{i H \otimes T_\mathcal{A}/\hbar}(\mathcal{U} \otimes \mathbb{I}_\mathcal{A}) e^{-i \tilde{H} \otimes T_\mathcal{A}/\hbar}$ where $\mathcal{U}$ is the driving of the system. Therefore, it is straightforward to see that these are the same operations (up to a sign) used in the SM protocol for measuring work. Thus, there is also a clear operational interpretation of the state of the ancilla as the state of a battery where work is stored.

3. The Wigner Distribution of Work

In the following, we will define a generalized work distribution. The general idea is inspired by the SM protocol. As we just mentioned, the state of the ancilla after the interaction not only holds information about work, but also about the coherences present in the initial state. In order to extract such information, we will evaluate their Wigner function [39,40], P_W. The Wigner function is a quasiprobability distribution that is used to represent quantum states in phase space. This is a real-valued function that, unlike their classical counterparts, can be negative for generic quantum states. This property has been widely used as an indicator of quantumness in different contexts, for instance in the study of the quantum-classical transition [41,42].

In our case, we will define it for the final state of the ancilla and in terms of the conjugate variables w and τ as

$$P_W(w,\tau) = \frac{1}{2\pi\hbar} \int_{-\infty}^{\infty} dy \left\langle w + \frac{y}{2} \middle| \rho_A(t_f) \middle| w - \frac{y}{2} \right\rangle e^{-i\tau y/\hbar}$$

$$= \frac{1}{2\pi\hbar} \sum_{n,n',m} \mathrm{tr}\left[\tilde{\Pi}_m \mathcal{U}\Pi_n \rho_S \Pi_{n'} \mathcal{U}^\dagger\right] \int_{-\infty}^{\infty} dy \left\langle w + \frac{y}{2} - w_{nm} \middle| \rho_A \middle| w - \frac{y}{2} - w_{n'm} \right\rangle e^{-i\tau y/\hbar} \quad (7)$$

This expression is valid for a generic initial state of the ancilla. In order to evaluate it, we will assume that the initial state of the ancilla is a coherent Gaussian state. This assumption not only will allow us to easily perform analytical calculations, but is also an appropriate choice for the description of typical experimental situations. Moreover, Gaussian states are classical, in the sense that they have a positive Wigner function. This guarantees that any negativity appearing in the Wigner function of the ancilla comes exclusively from their interaction with the system. Thus, we consider $\rho_A = |0,\sigma\rangle\langle 0,\sigma|$ as a coherent Gaussian state with zero mean and variance σ^2 in $\mathcal{W}_A$ (and hence zero mean and variance $\frac{\hbar^2}{2\sigma^2}$ in $\mathcal{T}_A$). After replacing this in Equation (7) (see Appendix A) and using that $\Pi_n \rho_S e^{i\tau E_n/\hbar} = \Pi_n e^{i\tau H/\hbar} \rho_S$, we obtain an expression for the quasidistribution of work for a generic process

$$P_W(w,\tau) = \sum_{n,n',m} \mathrm{tr}\left[\tilde{\Pi}_m \mathcal{U}\Pi_n \rho_S(-\tau) \Pi_{n'} \mathcal{U}^\dagger\right]$$

$$\times \mathcal{N}\left(w \middle| \frac{w_{nm} + w_{n'm}}{2}, \sigma\right) \mathcal{N}\left(\tau \middle| 0, \frac{\hbar}{\sqrt{2}\sigma}\right), \quad (8)$$

where $\rho_S(-\tau)$ is the state obtained after performing a free evolution of the initial state of the system for a time $-\tau$, and $\mathcal{N}(w|\mu,\sigma)$ is a normal probability density in w with mean μ and variance σ^2 (analogously for τ). The fact that the evolved state of the system appears in the distribution is a consequence of quantum coherence. If the initial state has quantum coherence in the energy basis then it is not a steady state, and it will evolve with its free Hamiltonian; therefore, it becomes important the amount of time τ that passes between the preparation of the state and the beginning of the work measurement protocol. From this expression, we can appreciate again the operational interpretation of the variables w and τ that characterize the state of the ancilla.

In the following, we will introduce some notation that will be useful to simplify forthcoming expressions. First, let us recall that the distribution $P_{\mathrm{TPM}}(w)$ does not take into account any coherence between the different energy subspaces of H in the initial state ρ_S. Therefore, we can associate this probability distribution to the dephased state $\bar{\rho}_S$. It would then be convenient to define the probability distribution $P_\mathcal{N}(w|\sigma)$ that is the convolution of $P_{\mathrm{TPM}}(w)$ with a normal distribution with zero mean and variance σ^2

$$P_\mathcal{N}(w|\sigma) = \int_{-\infty}^{\infty} du\, P_{\mathrm{TPM}}(w-u)\mathcal{N}(u|0,\sigma) \quad (9)$$

$$= \sum_{n,m} \mathrm{tr}\left[\tilde{\Pi}_m \mathcal{U}\Pi_n \rho_S \Pi_n \mathcal{U}^\dagger\right] \mathcal{N}(w|w_{nm},\sigma).$$

Notice that $P_\mathcal{N}(w|\sigma)$ is simply the TPM distribution, Equation (3), with the Dirac deltas replaced by a normal distribution with the corresponding mean values of work and variance σ^2. Thus, for a highly localized normal distribution, it satisfies $P_\mathcal{N}(w|\sigma) \xrightarrow[\sigma \to 0]{} P_{\mathrm{TPM}}(w)$.

In order to illustrate the effect of initial coherences, let us consider Equation (8), and split it into diagonal ($n = n'$) and non-diagonal ($n \neq n'$) contributions

$$P_W(w,\tau) = P_\mathcal{N}(w|\sigma)\mathcal{N}\left(\tau \middle| 0, \frac{\hbar}{\sqrt{2}\sigma}\right) + P_W^{(c)}(w,\tau). \quad (10)$$

The non-diagonal one corresponds to the contribution of the so-called initial coherences and it is easy to see that

$$P_W^{(c)}(w,\tau) = \sum_{n\neq n',m} \text{tr}\left[\tilde{\Pi}_m \mathcal{U} \Pi_n \rho_S(-\tau) \Pi_{n'} \mathcal{U}^\dagger\right] \qquad (11)$$
$$\times \mathcal{N}\left(w \left| \frac{w_{nm}+w_{n'm}}{2}, \sigma\right.\right) \mathcal{N}\left(\tau \left| 0, \frac{\hbar}{\sqrt{2}\sigma}\right.\right).$$

3.1. Quasidistribution for Initial Dephased States

When the initial state of the system is diagonal in the energy basis ($\Pi_n \rho_S \Pi_{n'} = 0$ for $n \neq n'$), then $P_W^{(c)}(w,\tau) = 0$ and the Wigner function is just

$$P_W(w,\tau) = P_\mathcal{N}(w|\sigma) \mathcal{N}\left(\tau \left| 0, \frac{\hbar}{\sqrt{2}\sigma}\right.\right), \qquad (12)$$

that is, it is proportional to the convoluted TPM distribution for every value of τ. Moreover, if we calculate the marginal $P_W(w)$,

$$P_W(w) = \int_{-\infty}^\infty d\tau\, P_W(w,\tau) = P_\mathcal{N}(w|\sigma), \qquad (13)$$

we recover the probability distribution that would be obtained if one measures the observable $\mathcal{W}_A$. This expression reflects a characteristic property of the Wigner function: The partial integration provides the probability distribution corresponding to the other variable. Therefore, for initial states without coherences in the initial energy eigenbasis, $P_W(w)$ is exactly $P_\mathcal{N}(w|\sigma)$. If, in addition, $\sigma \ll (w_{nm} - w_{n'm'}), \forall n, n', m, m'$, then we recover the probability distribution of work given by the TPM protocol.

In Figure 2b, upper panel, we show the distribution $P_W(w,\tau)$ for a two-level system $\mathcal{S}$ without initial coherences. In the lower panel, we show the marginal of the distribution in w, $P_W(w)$, and compare it with the discrete probabilities associated to the TPM. Notice that the area under each Gaussian in the marginal is equal to the corresponding probability in the TPM protocol. We can further see that it effectively reproduces the ideal TPM distribution. On the other hand, in Figure 2a, we show the distribution of work obtained for the same system but using an ancilla that has an initial state with a standard deviation five times smaller. One can easily note that this case is much closer to the ideal projective measurement regime. In this case, the Wigner function is invariant under translations in τ, as expected, since the initial state of the system commutes with the initial Hamiltonian. In Figure 2c, we show the distribution for a standard deviation even bigger than the one in Figure 2b. As we can see, while the position of the peaks matches the correct work values, there is a significant overlap between the different Gaussians.

3.2. Effects of Quantum Coherences

Let us now consider a system with initial coherences. From Equation (10), we can notice that, in this case, the Wigner function also has Gaussian peaks on each work value w_{nm}, just as it happens for the dephased state. However, there are some additional terms centered around the average of two work values with different initial energy, $(w_{nm} + w_{n'm})/2$. These terms are the ones that hold the non-trivial dependence on the variable τ and, as we will see, they can be negative. This can be easily seen from the following argument. If we look at Equation (10), we can see that $\int_{-\infty}^\infty d\tau\, dw\, P_W(w,\tau) = 1$, and, in addition, also the integral over the phase space of the first term is equal to one, as it is the Wigner function of the initial dephased state. Therefore, the integral of the second term must be zero. In order for it to be so, some of the terms in the sum must be negative. In these terms, besides the global Gaussian modulation, the variable τ appears as a time evolution of the state.

The fact that time appears explicitly only for initial states with coherences has a clear interpretation. If the initial state ρ_S is diagonal, then it is a steady state of the initial Hamiltonian, and the state is the same for every instant in time before the driving is applied. On the other hand, if ρ_S has coherences, the state evolves due to the free evolution induced by the initial Hamiltonian. This time, of course, is irrelevant at the moment of performing the first projecting energy measurement for the TPM distribution. However, it appears in our approach due to the fact that we are performing coherent operations between system and ancilla. Notably, one can also observe that the mean energy difference Equation (2) is not invariant under initial time translations for states with coherences. Thus, given a reference state ρ_S, the calculated distribution contains, in principle, information about every initial state that is unitarily connected with ρ_S by the initial Hamiltonian. Nevertheless, the amplitude of the Wigner function decays exponentially to zero when $\tau \to \pm\infty$ due to the Gaussian modulation, putting in practice some cut-off to the maximum time for which such information can be obtained. At the same time, given the complementary nature of the variables work w and time τ, when localizing the Gaussian in w, we are delocalizing it in τ. We will come back to this issue when we consider the marginals of the distribution.

We have already shown that, if the initial state does not have coherence in the energy basis, the resulting quasidistribution of work function is positive, because the diagonal terms in Equation (10) are all positive. Therefore, if the distribution $P_W(w,\tau)$ has some negativities, it is a signature of the presence of coherences in the initial state. This can be clearly seen in the upper panel of Figure 3, where the quasiprobability distribution of work $P_W(w,\tau)$ of a two-level system is plotted. The Hamiltonians, drivings, and ancilla parameters are identical to those of Figure 2. Moreover, in both cases, the initial state of the system has the same probability distribution in the energy basis. The only difference between Figures 2 and 3 is that, in the latter, the initial state has coherence between the two energy levels. Comparing both figures, we can notice that we effectively have the same Gaussian distributions over the same work values. The key difference lies in the fact that, for the initial coherent state, the quasidistribution displays additional oscillations that become negative. This interference fringes indicate the presence of non-classicality in the Wigner function and in the initial state of the system.

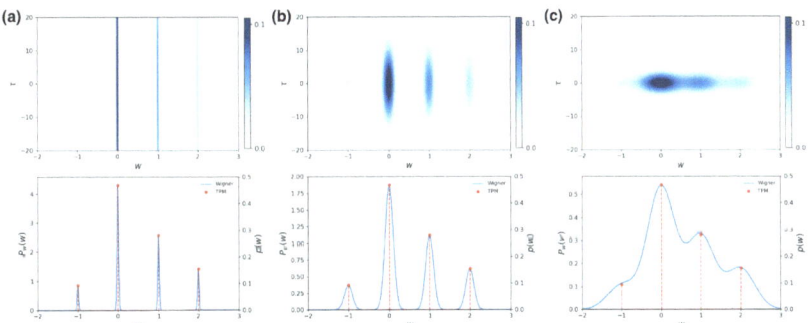

Figure 2. Wigner function of work for a two-level system using a Gaussian ancilla. The initial Hamiltonian is $H = E\sigma_+\sigma_-$, with $\sigma_\pm$ the Pauli creation and annihilation operators. The unitary driving is given by $\mathcal{U} = (\sqrt{2}\mathbb{I} + i\sigma_x + i\sigma_z)/2$ and the final Hamiltonian is $\hat{H} = 2E\sigma_+\sigma_-$. The initial state of the system is $\rho_S = (\mathbb{I} + \sigma_z/4)/2$ and the variance of the initial Gaussian packets of the ancilla are (**a**) $\sigma = 0.02E$, (**b**) $\sigma = 0.1E$, and (**c**) $\sigma = 0.35E$. The upper panel shows the distribution $P_W(w,\tau)$ of Equation (10) based on the Wigner function. In the lower panel, we show the marginal of w, given by Equation (13), along with the discrete probabilities $p(w)$ corresponding to the usual TPM distribution.

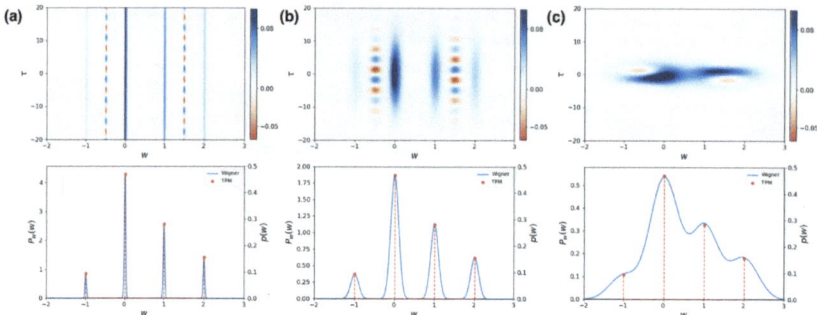

Figure 3. Wigner function for work of a two-level system using a Gaussian ancilla with with variance (a) $\sigma = 0.02E$, (b) $\sigma = 0.1E$ and (c) $\sigma = 0.35E$. The parameters used are the same as in Figure 2, except that now the initial density matrix has non-diagonal elements, $\rho_S = (\mathbb{I} + \sigma_x/2 + \sigma_y/2 + \sigma_z/4)/2$. The upper panel shows the distribution $P_W(w, \tau)$ Equation (10) based on the Wigner function. We notice now, because of of the initial coherences, the appearance of negative values in the distribution. The lower panel shows the marginal of w, given by Equation (13), and it is compared to the work values and respective discrete probabilities $p(w)$ that appear in the usual TPM distribution.

In the lower panel of Figure 3, we show the marginal probability distribution for w. Comparing them with Figure 2, we can notice that the marginal distributions are equivalent. In Figure 3, we also show the work distribution for the same system but using an initial state of the ancilla $\mathcal{A}$ with different standard deviations. In the case of the smaller standard deviation (corresponding to an ideal projective measurement), we can see that the marginal probability recovers that of the TPM distribution. Notably, in the case of a bigger standard deviation, the interference between different Gaussian peaks modifies the distribution of w, and there are corrections due to coherences, as shown in Equation (A2). In this limit, the marginal distribution may not even coincide with that of the corresponding dephased state. This behavior is similar to what happen when one makes a weak measurement [43]. To understand when it is possible to observe these differences, lets note that, when the marginal for w is calculated from Equation (10), the diagonal terms give exactly the convoluted distribution $P_\mathcal{N}(w|\sigma)$. For the non-diagonal contributions, we have time-averages of the form

$$\int_{-\infty}^{\infty} d\tau\, \rho_S(-\tau) \mathcal{N}\left(\tau \middle| 0, \frac{\hbar}{\sqrt{2}\sigma}\right). \tag{14}$$

This operation is similar to a dephasing map on the energy basis, but there is a significant difference since this average is weighted with a normal distribution with variance $\hbar^2/(2\sigma^2)$ centered in the origin. The bigger the variance of the Gaussian (and therefore the smaller σ), more values of τ enter in the time-average. Therefore, in the limit of small σ, we expect the non-diagonal terms to average to zero. Hence, one can show that, if $\sigma \ll (E_n - E_{n'})/2$, $\forall n, n'$, independent of the initial state,

$$\begin{aligned} P_W(w) &= \int_{-\infty}^{\infty} d\tau\, P_W(w, \tau) \\ &\approx P_\mathcal{N}(w|\sigma). \end{aligned} \tag{15}$$

Thus, the marginal of the quasiprobability distribution reproduces the TPM distribution.

3.3. Calculation of Mean Values

Given the formalism associated to the Wigner function [40], one can easily obtain average values from this quasidistribution. In fact, using the Wigner–Weyl representation [40] of an operator $\mathcal{A}$ acting on the ancillary space,

$$A(w,\tau) = \int_{-\infty}^{\infty} dy \left\langle w + \frac{y}{2} \middle| \mathcal{A} \middle| w - \frac{y}{2} \right\rangle e^{-i\tau y/\hbar}, \tag{16}$$

their mean value is just

$$\text{tr}\left[\mathcal{A}\, \rho_A(t_f)\right] = \int_{-\infty}^{\infty} d\tau \int_{-\infty}^{\infty} dw\, P_W(w,\tau)\, A(w,\tau). \tag{17}$$

For instance, the mean value of work is just the mean value of the operator $\mathcal{W}_A$, and it is obtained by integrating the function w over the phase space

$$\langle w \rangle \equiv \text{tr}\left[\mathcal{W}_A\, \rho_A(t_f)\right]$$
$$= \int_{-\infty}^{\infty} d\tau \int_{-\infty}^{\infty} dw\, P_W(w,\tau)\, w. \tag{18}$$

The other typical average that is calculated in the context of fluctuation theorems, where the system is initially in thermal equilibrium at inverse temperature β, is $\langle e^{-\beta w} \rangle$. This is easily conducted by integration of the function $e^{-\beta w}$. In all cases, the calculated mean values depend on the initial state of the ancilla. As it can be easily proven, for any observable of the type $f(\mathcal{W}_A)$, in the limit of $\sigma \to 0$, their averages converge to the values associated with the TPM distribution.

3.4. Energy Difference in the Presence of Coherences

Finally, we will show another interesting property of the work quasidistribution we have defined. As previously discussed, unless the initial state of the system is diagonal in the energy eigenbasis, the difference in mean energy and the mean value of work (Equations (2) and (4)) do not coincide. Thus, the TPM distribution does not provide any information about the initial coherences. Notably, as we will show, this information is also contained in the quasiprobability distribution.

In order to do so, let us consider the average in phase space of the function $g_{\tau_0}(w,\tau) = w\, \delta(\tau - \tau_0)$ (see Appendix A). Using the Wigner–Weyl transform [40], it corresponds to the expectation value of the operator $\hat{G}_{p_0} = (\mathcal{W}_A|\tau_0\rangle\langle\tau_0| + |\tau_0\rangle\langle\tau_0|\mathcal{W}_A)/2$ measured over the ancilla. It can be easily shown that this average, which is equivalent to the integral of the function w weighed by the Wigner function along an horizontal line at τ_0, is proportional to

$$\int_{-\infty}^{\infty} d\tau \int_{-\infty}^{\infty} dw\, P_W(w,\tau) g_{\tau_0}(w,\tau) \propto \Delta E_{\tau_0}, \tag{19}$$

where $\Delta E_{\tau_0} = \text{tr}[\tilde{H}\mathcal{U}\rho_S(-\tau_0)\mathcal{U}^\dagger] - \text{tr}[H\rho_S(-\tau_0)]$ is the mean energy difference for a situation where the driving $\mathcal{U}$ is turned on at time $-\tau_0$, and the proportionality constant is just equal to the Gaussian modulation at τ_0, $\mathcal{N}\left(\tau_0 \middle| 0, \frac{\hbar}{\sqrt{2}\sigma}\right)$ (see Appendix A). Therefore, when $\tau_0 = 0$, this is just proportional to the 'initial' energy difference ΔE in Equation (2). As we have shown, from this quasiprobability distribution, we can calculate not only the energy difference corresponding to the actual initial state, but also for the set of states $\rho_S(\tau)$, $\tau \in \mathbb{R}$. This set can be viewed as different 'initial times' at which the driving is turned on starting from a reference state ρ_S at time zero. This is so because this set of initial states is connected with ρ_S by a free Hamiltonian evolution.

Interestingly, for a Gaussian initial state of the ancilla, one obtains the correct value ΔE independent of their initial variance σ. However, since there is a Gaussian modulation centered around $\tau_0 = 0$ (the proportionality constant), the error in its determination increases as one localizes the initial state of the ancilla in the variable w. However, if one reduces the value of σ, the estimation of $P_{\text{TPM}}(w)$ becomes worse. Hence, one can again appreciate in this case the complementary nature of the variables w and τ.

4. Possible Experimental Implementations

For any proposal of a generalized work distribution to be of practical interest, it should be experimentally accessible and measurable. Here, we discuss how the Wigner work distribution can be measured. The measurement of the quasiprobability distribution that we propose requires two fundamental ingredients: (i) coherent control of two degrees of freedom of system and ancilla in order to implement the interactions of the SM protocol; (ii) being able to measure the Wigner function of the ancilla. In particular, implementing the SM requires the ability of performing translations of the ancilla conditioned on the energy of the degree of freedom on which the work is performed. There is a great variety of systems where this sort of interaction can be implemented, and an experimental realization of the SM protocol has been realized using cold atoms [13]. However, it is not clear how one can implement the measurement of the Wigner function in such platform. Nevertheless, there are systems where both requirements are, in principle, satisfied, and in what follows we will briefly describe two of them.

The first example is given by superconducting qubits coupled to a cavity, e.g., circuit quantum electrodynamics (cQED) [44]. Here, the qubit circuit can be coupled to a waveguide that acts as a microwave cavity where coherent states or states with a well defined number of photons can be stored [45]. For instance, in Ref. [46], they generate coherent displacements of the state of the cavity depending on the state of the qubit. This interaction is exactly what is needed for implementing the protocol where the qubit acts as the system and the cavity as the ancilla. On the other hand, in a different coupling regime between qubit and cavity, this same scheme has been used to measure the Wigner function of the state of the field in the cavity [47].

The second possible platform are trapped ions. In this case, ions are trapped in an electric potential such that the motion degrees of freedom of the ion are subjected to an effective harmonic oscillator potential [48]. At the same time, using the interactions between the electronic degree of freedom and the position of the ion, it is possible to generate coherent, squeezed, and Fock states of the oscillator [48]. In particular, in different experiments [49,50], it has been shown that one can apply forces on the ion depending on its electronic state, and in this way, displacements in phase space depending on the qubit state can be coherently implemented. Again, this is the interaction needed to perform the protocol. The Wigner function of the motion degree of freedom of trapped ions has been successfully measured [51].

5. Conclusions

In this work, we introduced a generalization of the probability distribution of work based on the Wigner function. The starting point is the single-measurement protocol proposed in [33], where an ancilla is coupled to the system whose work one wants to measure in order to keep a coherent record of all possible work values. Following this idea, we define the Wigner function of the final state of the ancilla. This quasiprobability distribution contains all the information regarding both work and coherence in the initial state of the system. In fact, initial quantum coherence in the system results in negativities in the quasiprobability distribution of work, a clear signature of non-classicality. In this case, we can also recover the mean value of energy, which is different from the average work for states with coherences. Moreover, we show that, from this quasiprobability distribution, one can easily recover the standard TPM distribution simply by integrating over the time variable. In addition, we show that, given that the average work and other quantities of interest can be obtained as the mean value of an operator acting on the ancillary space, it is easy to calculate mean values using the formalism of the Wigner function. The quasiprobability distribution here defined has certain similarities with the one proposed in [22,23,52]. The way in which the distribution is defined there is also inspired by the SM scheme [53] and requires the preparation of a coherent superposition of the ancilla between two momentum eigenstates, $|p\rangle + |-p\rangle$, together with the implementation of an interaction analogous to that of the SM. At the end of the protocol, the relative phase between these states is measured and a quasiprobability distribution that contains information about work and coherence is obtained [53]. In contrast, our proposal has a clear operational interpretation and direct experimental application, as it is simply the Wigner function of the final state of the measurement apparatus. Moreover, our protocol not only contains all the information of Refs. [22,23,52], but for coherent initial states, it has additional information on the dependence of the time variable, τ. From a practical point of view, our protocol does not need ideal (non-physical) states and it is easy to adapt to any initial state of the ancilla. Here, we have just developed the case of Gaussian states given that they are easy to treat analytically and are typically appropriate to model experimental conditions. However, this whole analysis can be repeated for any initial state. We hope that this approach to the work distribution can shed some light to elucidate the effects of quantum coherences in thermodynamic transformations.

Author Contributions: Conceptualization, F.C., F.M. and A.J.R.; Investigation, F.C., F.M. and A.J.R.; Writing—original draft, F.C., F.M. and A.J.R. All authors contributed to all parts of the paper. All authors have read and agreed to the published version of the manuscript.

Funding: This work was partially supported by CONICET, UBACyT, and ANPCyT. FC acknowledges support from grant number FQXi-IAF19-01 from the Foundational Questions Institute Fund, a donor-advised fund of the Silicon Valley Community Foundation.

Institutional Review Board Statement: Not applicable.

Data Availability Statement: Data sharing not applicable.

Acknowledgments: We thank C. Brukner and M. Saraceno for interesting discussions.

Conflicts of Interest: The authors declare no conflict of interest.

Appendix A. Quasiprobability Distribution for Gaussian States

We start with the general expression in Equation (7) for the Wigner function and replace the initial state of the ancilla with a coherent (squeezed) Gaussian state $\rho_A = |0, \sigma\rangle\langle 0, \sigma|$ centered in the origin of coordinates of the phase space and with variance σ^2 in $\mathcal{W}_A$ and $\frac{\hbar^2}{2\sigma^2}$ in $\mathcal{T}_A$. Then, we obtain

$$P_W(w, \tau) = \frac{1}{2\pi\hbar} \sum_{n,n',m} \text{tr}\left[\tilde{\Pi}_m \mathcal{U} \Pi_n \rho_S \Pi_{n'} \mathcal{U}^\dagger\right] \times$$

$$\times \frac{1}{\sqrt{2\pi}\sigma} \int_{-\infty}^{\infty} dy \exp\left[-\frac{(w + \frac{y}{2} - w_{nm})^2}{4\sigma^2}\right] \exp\left[-\frac{(w - \frac{y}{2} - w_{n'm})^2}{4\sigma^2}\right] e^{-i\tau y/\hbar}$$

$$= \frac{1}{2\pi\hbar} \sum_{n,n',m} \text{tr}\left[\tilde{\Pi}_m \mathcal{U} \Pi_n \rho_S \Pi_{n'} \mathcal{U}^\dagger\right] \exp\left[-\frac{\left(w - \frac{w_{nm} + w_{n'm}}{2}\right)^2}{2\sigma^2}\right] \times$$

$$\times \frac{1}{\sqrt{2\pi}\sigma} \int_{-\infty}^{\infty} dy \exp\left[-\frac{(y - (w_{nm} - w_{n'm}))^2}{2\sigma^2}\right] e^{-i\tau y/\hbar}$$

$$= \sum_{n,n',m} \text{tr}\left[\tilde{\Pi}_m \mathcal{U} \Pi_n \rho_S \Pi_{n'} \mathcal{U}^\dagger\right] \mathcal{N}\left(w \left| \frac{w_{nm} + w_{n'm}}{2}, \sigma\right.\right) \mathcal{N}\left(\tau \left| 0, \frac{\hbar}{\sqrt{2}\sigma}\right.\right) e^{i\tau(E_n - E_{n'})/\hbar}, \quad \text{(A1)}$$

Now, let us calculate the marginal of Equation (A1) in order to see that it gives us the convoluted probability distribution of work of Equation (9). First, we split the function into diagonal and non-diagonal terms as in Equation (10), and then integrate each of them:

$$\begin{aligned}
P_W(w) &= \int_{-\infty}^{\infty} d\tau\, P_W(w,\tau) \\
&= P_{\mathcal{N}}(w|\sigma) \int_{-\infty}^{\infty} d\tau\, \mathcal{N}\left(\tau\,\Big|\,0, \frac{\hbar}{\sqrt{2}\sigma}\right) \\
&\quad + \sum_{n\neq n',m} \text{tr}\left[\tilde{\Pi}_m \mathcal{U}\Pi_n \rho_S \Pi_{n'} \mathcal{U}^\dagger\right] \mathcal{N}\left(w\,\Big|\,\frac{w_{nm}+w_{n'm}}{2},\sigma\right) \int_{-\infty}^{\infty} d\tau\, e^{i\tau(E_n - E_{n'})/\hbar} \mathcal{N}\left(\tau\,\Big|\,0,\frac{\hbar}{\sqrt{2}\sigma}\right) \\
&= P_{\mathcal{N}}(w|\sigma) + \sum_{n\neq n',m} \text{tr}\left[\tilde{\Pi}_m \mathcal{U}\Pi_n \rho_S \Pi_{n'}\mathcal{U}^\dagger\right] \mathcal{N}\left(w\,\Big|\,\frac{w_{nm}+w_{n'm}}{2},\sigma\right) e^{-\frac{(E_n-E_{n'})^2}{4\sigma^2}}. \quad \text{(A2)}
\end{aligned}$$

Thus, if $\sigma \ll (E_n - E_{n'})/2$, meaning that the dispersion is much smaller than all the energy gaps of the initial Hamiltonian, then the non-diagonal terms become exponentially small and we have

$$P_W(w) = \int_{-\infty}^{\infty} d\tau\, P_W(w,\tau) \approx P_{\mathcal{N}}(w|\sigma), \qquad \text{if } \sigma \ll \frac{E_n - E_{n'}}{2} \,\forall n, n'. \quad \text{(A3)}$$

Finally, let us consider the mean value of the $g_{\tau_0}(w,\tau) = w\delta(\tau - \tau_0)$; this is equivalent to an average (weighted by the Wigner function) of work variable w at a given fixed time $\tau = \tau_0$:

$$\begin{aligned}
\int_{-\infty}^{\infty} d\tau \int_{-\infty}^{\infty} dw\, P_W(w,\tau) g_{\tau_0}(w,\tau) &= \int_{-\infty}^{\infty} dw\, P_W(w,\tau_0) w \\
&= \mathcal{N}\left(\tau_0\,\Big|\,0,\frac{\hbar}{\sqrt{2}\sigma}\right) \sum_{n,n',m} \text{tr}\left[\tilde{\Pi}_m \mathcal{U}\Pi_n \rho_S(-\tau_0)\Pi_{n'}\mathcal{U}^\dagger\right] \int_{-\infty}^{\infty} dw\, \mathcal{N}\left(w\,\Big|\,\frac{w_{nm}+w_{n'm}}{2},\sigma\right) w \\
&= \mathcal{N}\left(\tau_0\,\Big|\,0,\frac{\hbar}{\sqrt{2}\sigma}\right) \sum_{n,n',m} \text{tr}\left[\tilde{\Pi}_m \mathcal{U}\Pi_n \rho_S(-\tau_0)\Pi_{n'}\mathcal{U}^\dagger\right] \left(\frac{w_{nm}+w_{n'm}}{2}\right) \\
&= \mathcal{N}\left(\tau_0\,\Big|\,0,\frac{\hbar}{\sqrt{2}\sigma}\right) \sum_{n,n',m} \text{tr}\left[\tilde{\Pi}_m \mathcal{U}\Pi_n \rho_S(-\tau_0)\Pi_{n'}\mathcal{U}^\dagger\right] \frac{1}{2}(2\tilde{E}_m - E_n - E_{n'}) \\
&= \mathcal{N}\left(\tau_0\,\Big|\,0,\frac{\hbar}{\sqrt{2}\sigma}\right) \left(\text{tr}\left[\tilde{H}\mathcal{U}\rho_S(-\tau_0)\mathcal{U}^\dagger\right] - \text{tr}[H\rho_S(-\tau_0)]\right) \\
&\equiv \mathcal{N}\left(\tau_0\,\Big|\,0,\frac{\hbar}{\sqrt{2}\sigma}\right) \Delta E_{\tau_0}
\end{aligned}$$

That is, this integral is proportional to the mean energy difference for an initial state, which may have coherences, when the driving is turned on at time $-\tau_0$. Thus, for $\tau_0 = 0$, it is the usual mean energy difference ΔE of (2).

References

1. Jarzynski, C. Nonequilibrium Equality for Free Energy Differences. *Phys. Rev. Lett.* **1997**, *78*, 2690–2693. [CrossRef]
2. Crooks, G.E. Entropy production fluctuation theorem and the nonequilibrium work relation for free energy differences. *Phys. Rev. E* **1999**, *60*, 2721–2726. [CrossRef]
3. Talkner, P.; Hänggi, P. The Tasaki–Crooks quantum fluctuation theorem. *J. Phys. Math. Theor.* **2007**, *40*, F569–F571. [CrossRef]
4. Talkner, P.; Lutz, E.; Hänggi, P. Fluctuation theorems: Work is not an observable. *Phys. Rev. E* **2007**, *75*, 050102. [CrossRef]
5. Campisi, M.; Hänggi, P.; Talkner, P. Colloquium: Quantum fluctuation relations: Foundations and applications. *Rev. Mod. Phys.* **2011**, *83*, 771–791. [CrossRef]
6. Sampaio, R.; Suomela, S.; Ala-Nissila, T.; Anders, J.; Philbin, T. Quantum work in the Bohmian framework. *Phys. Rev. A* **2018**, *97*, 012131. [CrossRef]
7. Tasaki, H. Jarzynski Relations for Quantum Systems and Some Applications. *arXiv* **2000**, arXiv:cond-mat/0009244.
8. Kurchan, J. A Quantum Fluctuation Theorem. *arXiv* **2000**, arXiv:cond-mat/0007360.
9. Mukamel, S. Quantum Extension of the Jarzynski Relation: Analogy with Stochastic Dephasing. *Phys. Rev. Lett.* **2003**, *90*, 170604. [CrossRef]

10. Esposito, M.; Harbola, U.; Mukamel, S. Nonequilibrium fluctuations, fluctuation theorems, and counting statistics in quantum systems. *Rev. Mod. Phys.* **2009**, *81*, 1665–1702. [CrossRef]
11. Batalhão, T.B.; Souza, A.M.; Mazzola, L.; Auccaise, R.; Sarthour, R.S.; Oliveira, I.S.; Goold, J.; De Chiara, G.; Paternostro, M.; Serra, R.M. Experimental Reconstruction of Work Distribution and Study of Fluctuation Relations in a Closed Quantum System. *Phys. Rev. Lett.* **2014**, *113*, 140601. [CrossRef] [PubMed]
12. An, S.; Zhang, J.N.; Um, M.; Lv, D.; Lu, Y.; Zhang, J.; Yin, Z.Q.; Quan, H.T.; Kim, K. Experimental test of the quantum Jarzynski equality with a trapped-ion system. *Nat. Phys.* **2015**, *11*, 193–199. [CrossRef]
13. Cerisola, F.; Margalit, Y.; Machluf, S.; Roncaglia, A.J.; Paz, J.P.; Folman, R. Using a quantum work meter to test non-equilibrium fluctuation theorems. *Nat. Commun.* **2017**, *8*, 1241. [CrossRef] [PubMed]
14. Smith, A.; Lu, Y.; An, S.; Zhang, X.; Zhang, J.N.; Gong, Z.; Quan, H.T.; Jarzynski, C.; Kim, K. Verification of the quantum nonequilibrium work relation in the presence of decoherence. *New J. Phys.* **2018**, *20*, 013008. [CrossRef]
15. Hernández-Gómez, S.; Gherardini, S.; Poggiali, F.; Cataliotti, F.S.; Trombettoni, A.; Cappellaro, P.; Fabbri, N. Experimental test of exchange fluctuation relations in an open quantum system. *Phys. Rev. Res.* **2020**, *2*, 023327. [CrossRef]
16. Solfanelli, A.; Santini, A.; Campisi, M. Experimental verification of fluctuation relations with a quantum computer. *PRX Quantum* **2021**, *2*, 030353. [CrossRef]
17. Perarnau-Llobet, M.; Bäumer, E.; Hovhannisyan, K.V.; Huber, M.; Acin, A. No-Go Theorem for the Characterization of Work Fluctuations in Coherent Quantum Systems. *Phys. Rev. Lett.* **2017**, *118*, 070601. [CrossRef]
18. Lostaglio, M. Quantum Fluctuation Theorems, Contextuality, and Work Quasiprobabilities. *Phys. Rev. Lett.* **2018**, *120*, 040602. [CrossRef]
19. Xu, B.M.; Zou, J.; Guo, L.S.; Kong, X.M. Effects of quantum coherence on work statistics. *Phys. Rev. A* **2018**, *97*, 052122. [CrossRef]
20. Sagawa, T., Second law-like inequalities with quantum relative entropy: An introduciton. In *Lectures on Quantum Computing, Thermodynamics and Statistical Physics*; Word Scientific Publishing Co.: Singapore, 2012; pp. 125–190. [CrossRef]
21. Allahverdyan, A.E. Nonequilibrium quantum fluctuations of work. *Phys. Rev. E* **2014**, *90*, 032137. [CrossRef]
22. Solinas, P.; Gasparinetti, S. Full distribution of work done on a quantum system for arbitrary initial states. *Phys. Rev. E* **2015**, *92*, 042150. [CrossRef]
23. Solinas, P.; Gasparinetti, S. Probing quantum interference effects in the work distribution. *Phys. Rev. A* **2016**, *94*, 052103. [CrossRef]
24. Wiseman, H.M. Weak values, quantum trajectories, and the cavity-QED experiment on wave-particle correlation. *Phys. Rev. A* **2002**, *65*, 032111. [CrossRef]
25. Hall, M.J.W. Prior information: How to circumvent the standard joint-measurement uncertainty relation. *Phys. Rev. A* **2004**, *69*, 052113. [CrossRef]
26. Miller, H.J.D.; Anders, J. Time-reversal symmetric work distributions for closed quantum dynamics in the histories framework. *New J. Phys.* **2017**, *19*, 062001. [CrossRef]
27. Łobejko, M. Work and Fluctuations: Coherent vs. Incoherent Ergotropy Extraction. *Quantum* **2022**, *6*, 762. [CrossRef]
28. Francica, G. Class of quasiprobability distributions of work with initial quantum coherence. *Phys. Rev. E* **2022**, *105*, 014101. [CrossRef] [PubMed]
29. Francica, G. Most general class of quasiprobability distributions of work. *Phys. Rev. E* **2022**, *106*, 054129. [CrossRef]
30. Lostaglio, M.; Belenchia, A.; Levy, A.; Hernández-Gómez, S.; Fabbri, N.; Gherardini, S. Kirkwood-Dirac quasiprobability approach to quantum fluctuations: Theoretical and experimental perspectives. *arXiv* **2022**, arXiv:2206.11783.
31. Hernández-Gómez, S.; Gherardini, S.; Belenchia, A.; Lostaglio, M.; Levy, A.; Fabbri, N. Projective measurements can probe non-classical work extraction and time-correlations. *arXiv* **2023**, arXiv:2207.12960.
32. Santini, A.; Solfanelli, A.; Gherardini, S.; Collura, M. Work statistics, quantum signatures and enhanced work extraction in quadratic fermionic models. *arXiv* **2023**, arXiv:2302.13759.
33. Roncaglia, A.J.; Cerisola, F.; Paz, J.P. Work Measurement as a Generalized Quantum Measurement. *Phys. Rev. Lett.* **2014**, *113*, 250601. [CrossRef] [PubMed]
34. De Chiara, G.; Roncaglia, A.J.; Paz, J.P. Measuring work and heat in ultracold quantum gases. *New J. Phys.* **2015**, *17*, 035004. [CrossRef]
35. Cerisola, F. Trabajo y Correlaciones en Mecánica Cuántica. Ph.D. Thesis, University of Buenos Aires, Buenos Aires, Argentina, 2020.
36. Skrzypczyk, P.; Short, A.J.; Popescu, S. Work extraction and thermodynamics for individual quantum systems. *Nat. Commun.* **2014**, *5*, 4185. [CrossRef]
37. Alhambra, Á.M.; Masanes, L.; Oppenheim, J.; Perry, C. Fluctuating work: From quantum thermodynamical identities to a second law equality. *Phys. Rev. X* **2016**, *6*, 041017. [CrossRef]
38. Richens, J.G.; Masanes, L. Work extraction from quantum systems with bounded fluctuations in work. *Nat. Commun.* **2016**, *7*, 13511. [CrossRef]
39. Wigner, E. On the Quantum Correction For Thermodynamic Equilibrium. *Phys. Rev.* **1932**, *40*, 749–759. [CrossRef]
40. Hillery, M.; O'Connell, R.; Scully, M.; Wigner, E. Distribution functions in physics: Fundamentals. *Phys. Rep.* **1984**, *106*, 121–167. [CrossRef]

41. Kenfack, A.; Życzkowski, K. Negativity of the Wigner function as an indicator of non-classicality. *J. Opt. B Quantum Semiclassical Opt.* **2004**, *6*, 396. [CrossRef]
42. Tan, K.C.; Choi, S.; Jeong, H. Negativity of quasiprobability distributions as a measure of nonclassicality. *Phys. Rev. Lett.* **2020**, *124*, 110404. [CrossRef]
43. Aharonov, Y.; Albert, D.Z.; Vaidman, L. How the result of a measurement of a component of the spin of a spin-1/2 particle can turn out to be 100. *Phys. Rev. Lett.* **1988**, *60*, 1351–1354. [CrossRef] [PubMed]
44. Makhlin, Y.; Schön, G.; Shnirman, A. Quantum-state engineering with Josephson-junction devices. *Rev. Mod. Phys.* **2001**, *73*, 357–400. [CrossRef]
45. Paik, H.; Schuster, D.I.; Bishop, L.S.; Kirchmair, G.; Catelani, G.; Sears, A.P.; Johnson, B.R.; Reagor, M.J.; Frunzio, L.; Glazman, L.I.; et al. Observation of High Coherence in Josephson Junction Qubits Measured in a Three-Dimensional Circuit QED Architecture. *Phys. Rev. Lett.* **2011**, *107*, 240501. [CrossRef]
46. Naghiloo, M.; Alonso, J.J.; Romito, A.; Lutz, E.; Murch, K.W. Information Gain and Loss for a Quantum Maxwell's Demon. *Phys. Rev. Lett.* **2018**, *121*, 030604. [CrossRef] [PubMed]
47. Sun, L.; Petrenko, A.; Leghtas, Z.; Vlastakis, B.; Kirchmair, G.; Sliwa, K.M.; Narla, A.; Hatridge, M.; Shankar, S.; Blumoff, J.; et al. Tracking photon jumps with repeated quantum non-demolition parity measurements. *Nature* **2014**, *511*, 444–448. [CrossRef]
48. Leibfried, D.; Blatt, R.; Monroe, C.; Wineland, D. Quantum dynamics of single trapped ions. *Rev. Mod. Phys.* **2003**, *75*, 281–324. [CrossRef]
49. Haljan, P.C.; Brickman, K.A.; Deslauriers, L.; Lee, P.J.; Monroe, C. Spin-Dependent Forces on Trapped Ions for Phase-Stable Quantum Gates and Entangled States of Spin and Motion. *Phys. Rev. Lett.* **2005**, *94*, 153602. [CrossRef]
50. von Lindenfels, D.; Gräb, O.; Schmiegelow, C.T.; Kaushal, V.; Schulz, J.; Mitchison, M.T.; Goold, J.; Schmidt-Kaler, F.; Poschinger, U.G. Spin Heat Engine Coupled to a Harmonic-Oscillator Flywheel. *Phys. Rev. Lett.* **2019**, *123*, 080602. [CrossRef]
51. Leibfried, D.; Meekhof, D.M.; King, B.E.; Monroe, C.; Itano, W.M.; Wineland, D.J. Experimental Determination of the Motional Quantum State of a Trapped Atom. *Phys. Rev. Lett.* **1996**, *77*, 4281–4285. [CrossRef]
52. Solinas, P.; Miller, H.J.D.; Anders, J. Measurement-dependent corrections to work distributions arising from quantum coherences. *Phys. Rev. A* **2017**, *96*, 052115. [CrossRef]
53. De Chiara, G.; Solinas, P.; Cerisola, F.; Roncaglia, A.J. Ancilla-Assisted Measurement of Quantum Work. In *Thermodynamics in the Quantum Regime: Fundamental Aspects and New Directions*; Binder, F., Correa, L.A., Gogolin, C., Anders, J., Adesso, G., Eds.; Springer International Publishing: Cham, Switzerland, 2018; pp. 337–362. [CrossRef]

Disclaimer/Publisher's Note: The statements, opinions and data contained in all publications are solely those of the individual author(s) and contributor(s) and not of MDPI and/or the editor(s). MDPI and/or the editor(s) disclaim responsibility for any injury to people or property resulting from any ideas, methods, instructions or products referred to in the content.

Article

Adiabatic Shortcuts Completion in Quantum Field Theory: Annihilation of Created Particles

Nicolás F. Del Grosso [1,2], Fernando C. Lombardo [1,2,*], Francisco D. Mazzitelli [3] and Paula I. Villar [1,2,*]

1. Departamento de Física, Facultad de Ciencias Exactas y Naturales, Universidad de Buenos Aires, Buenos Aires 1428, Argentina; ngrosso@df.uba.ar
2. Instituto de Física de Buenos Aires (IFIBA), CONICET—Universidad de Buenos Aires, Buenos Aires 1428, Argentina
3. Centro Atómico Bariloche and Instituto Balseiro, Comisión Nacional de Energía Atómica, Bariloche 8400, Argentina; fdmazzi@cab.cnea.gov.ar
* Correspondence: lombardo@df.uba.ar (F.C.L.); paula@df.uba.ar (P.I.V.)

Abstract: Shortcuts to adiabaticity (STA) are relevant in the context of quantum systems, particularly regarding their control when they are subjected to time-dependent external conditions. In this paper, we investigate the completion of a nonadiabatic evolution into a shortcut to adiabaticity for a quantum field confined within a one-dimensional cavity containing two movable mirrors. Expanding upon our prior research, we characterize the field's state using two Moore functions that enables us to apply reverse engineering techniques in constructing the STA. Regardless of the initial evolution, we achieve a smooth extension of the Moore functions that implements the STA. This extension facilitates the computation of the mirrors' trajectories based on the aforementioned functions. Additionally, we draw attention to the existence of a comparable problem within nonrelativistic quantum mechanics.

Keywords: shortcuts to adiabaticity; optomechanical cavity; quantum thermodynamics

Citation: Del Grosso, N.F.; Lombardo, F.C.; Mazzitelli, F.D.; Villar, P.I. Adiabatic Shortcuts Completion in Quantum Field Theory: Annihilation of Created Particles. *Entropy* **2023**, *25*, 1249. https://doi.org/10.3390/e25091249

Academic Editor: Sebastian Deffner

Received: 18 July 2023
Revised: 16 August 2023
Accepted: 19 August 2023
Published: 23 August 2023

Copyright: © 2023 by the authors. Licensee MDPI, Basel, Switzerland. This article is an open access article distributed under the terms and conditions of the Creative Commons Attribution (CC BY) license (https:// creativecommons.org/licenses/by/ 4.0/).

1. Introduction

Quantum thermodynamics constitutes a burgeoning research field that explores the interplay between thermodynamic principles and quantum systems [1]. By merging the foundational tenets of quantum mechanics with classical thermodynamics, it seeks to unravel the intricacies of thermal phenomena manifested at the microscopic scale. Within this insight, quantum thermal machines have emerged as pioneering devices capable of harnessing quantum effects to execute thermodynamic operations such as work extraction from heat reservoirs and refrigeration. Operating in the quantum regime, these machines exploit the distinctive attributes of quantum coherence and entanglement, thus surpassing the limitations imposed by their classical counterparts. However, to this end, it is crucial to isolate these systems from the interaction with their surroundings in order to maintain quantum correlation or even cool atoms to absolute zero.

Quantum open systems investigate the dynamics and interactions of quantum systems under the influence of an environment, accounting the reasons for which it is often challenging to isolate or completely control the quantum system [2]. These interactions introduce complexities that can lead to undesired effects, such as decoherence, dissipation, and errors. A comprehensive understanding and characterizing of the open dynamics of a system is essential for controlling it effectively and minimizing sources of errors. This knowledge allows for the design of strategies to manipulate and engineer quantum systems while mitigating the impact of unwanted interactions. In quantum thermodynamics, where precision and accuracy are of utmost importance, controlling and reducing errors is critical to achieving reliable and efficient operations.

Quantum machines play a crucial role in quantum thermodynamics by enabling the manipulation and control of quantum states and energy exchanges at the microscopic level [3–5]. They serve as experimental platforms for studying fundamental aspects of quantum thermodynamics. Most of the research in this area has been conducted on qubits [3] or harmonic oscillators [4] subjected to different thermodynamic cycles. In [5], a thermal machine using a quantum field subjected to an Otto cycle, implemented with a superconducting circuit (consisting of a transmission line terminated by a superconducting quantum interference device), has been considered. The performance of this machine has been studied when acting as both a heat engine and a refrigerator. It has been shown that in a nonadiabatic regime, the efficiency of the quantum cycle is affected by the dynamical Casimir effect (DCE) [6–10], which induces a kind of quantum friction that diminishes the efficiency. Superconducting qubits, the building blocks of circuit QED systems, provide long coherence times and high-fidelity operations and therefore offer a versatile platform for implementing thermodynamic protocols and studying quantum heat engines and refrigerators [11].

In some cases of discrete stroke quantum machines, such as a quantum harmonic oscillator or a quantum field undergoing an Otto cycle, it has been shown that the efficiency of the resulting machine is maximum for adiabatic (i.e., infinitely slow) driving [5,12]. The problem is that these conditions imply a slow evolution that can be impractical and inefficient in terms of time. Furthermore, it can lead to the loss of efficiency of a heat engine. In this scenario, a shortcut to adiabaticity (STA) appears as a promising technique to overcome the efficiency loss associated with finite-time operations and achieve results comparable to adiabatic processes. Adiabatic shortcuts is a technique that allows a system to evolve rapidly between two adiabatic states without violating the adiabatic constraints. In general, this means that for an adiabatic shortcut, no new excitations will be generated in the final state; however, it is worth noting that some STA methods, such as transitionless quantum driving [13,14], also ensure that nonadiabatic excitations are suppressed even at intermediate times. Other methods for implementing shortcuts to adiabaticity include the use of invariants [15], fast forward techniques [16], optimal protocols [17], fast quasiadiabatic (FAQUAD), etc. [18].

STA has been considered from a theoretical and/or an experimental point of view for different physical systems: trapped ions [19], cold atoms [20], ultracold Fermi gases [21], Bose–Einstein condensates in atom chips [22], spin systems [23], etc. STA has been also proposed to relieve the trade-off of efficiency and power [24–26], both in single-particle quantum heat engines (QHEs) [27] and in many-particle QHEs [28–30].

In a previous work [31], we explored the possibility of applying STA in quantum field theory. Particularly, we showed how to implement an STA for a massless scalar field inside a cavity with a moving wall in $(1+1)$ dimensions. The approach is based on the already known solution to the problem by exploiting the conformal symmetry. The shortcuts take place whenever the solution matches the adiabatic Wentzel–Kramers–Brillouin (WKB) solution [32] and there is no DCE. In [33], we generalized the results of a quantum scalar field in a one-dimensional optomechanical cavity to two moving mirrors. We showed that given the trajectories for the left and right mirrors, it is possible to find an STA ruled by the effective trajectories of the mirrors. When implemented in finite time, these trajectories result in the same state as if the original ones had been evolved adiabatically. This protocol has the advantage that it can be easily implemented experimentally using either an optomechanical cavity or superconducting circuits, as it does not require additional exotic potentials. Moreover, the effective trajectory can be computed from the original one quite simply, paving the way for more efficient quantum field thermal machines.

In this context, herein, we find a general approach to complete an STA in the optomechanical cavity. By completing an STA, we refer to the following scenario: let us consider the system initially in its ground state; then, subject it to a time-dependent, nonadiabatic transformation. As a result of this transformation, the system transitions to an excited state. The arising question is if it is feasible to carry out a subsequent transformation in a manner that leads the system back to its ground state. If such a possibility exists, we refer to it as an

STA completion. This completion provides an additional tool for the control of quantum systems.

The paper is organized as follows. In Section 2, we review the main results for the quantization of a massless scalar field in a cavity with two moving boundaries. The excitation of the system can be described in terms of the so-called Moore's functions [6,34], which are the main tools to construct the STA for the field. Before discussing the STA completion for this system, and as a warm-up, in Section 3, we describe a simple analogy using a quantum harmonic oscillator with time-dependent frequency. We see that there is a simple way to unfold the evolution and construct an STA by an inverse engineering method based on a smooth continuation of the so-called Ermakov function [35]. The striking similarity between the Ermakov and Moore functions is used in Section 4 to construct the STA completion in the optomechanical cavity. We show that there is a general procedure to build up STA completions and that in some particular cases, the protocol is extremely simple and shows time-inversion invariance. Section 5 contains the conclusions of our work.

2. The Optomechanical Cavity

The system we consider is a scalar field, $\Phi(x,t)$, inside a one-dimensional cavity delimited by a moving mirror at each end whose positions are given by $L(t)$ and $R(t)$, respectively (see Figure 1). The evolution of the field is determined by the wave equation inside the cavity

$$(\partial_x^2 - \partial_t^2)\Phi(x,t) = 0, \tag{1}$$

and Dirichlet boundary conditions on each mirror

$$\Phi(L(t),t) = \Phi(R(t),t) = 0. \tag{2}$$

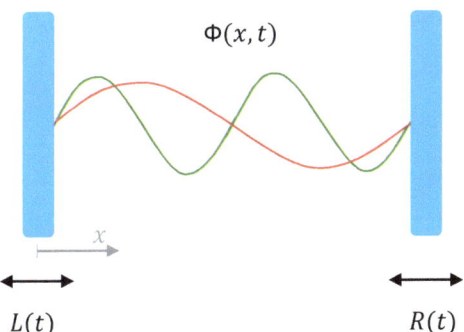

Figure 1. Schematics of the one-dimensional cavity with a scalar quantum field $\Phi(x,t)$ inside and two moving mirrors with trajectories $L(t)$ and $R(t)$: The red and green curves illustrate two of the infinite modes of the field in the cavity.

Here, and in the rest of the paper, we use units where $c = \hbar = k_B = 1$.

It is known that the time evolution of the field is solved by expanding the field in modes

$$\Phi(x,t) = \sum_{k=1}^{\infty}\left[a_k\psi_k(x,t) + a_k^{\dagger}\psi_k^*(x,t)\right], \tag{3}$$

where the modes are given by [34]

$$\psi_k(x,t) = \frac{i}{\sqrt{4\pi k}}[e^{-ik\pi G(t+x)} + e^{ik\pi F(t-x)}]. \tag{4}$$

Here, $F(z)$ and $G(z)$ are functions determined by the so-called Moore's equations

$$G(t+L(t)) - F(t-L(t)) = 0 \quad (5)$$
$$G(t+R(t)) - F(t-R(t)) = 2. \quad (6)$$

The functions $F(z)$ and $G(z)$ implement the conformal transformation

$$\bar{t} + \bar{x} = G(t+x) \quad \bar{t} - \bar{x} = F(t-x) \quad (7)$$

such that, in the new coordinates, the left and right mirrors are static at $\bar{x}_L = 0$ and $\bar{x}_R = 1$. In the particular case in which the left mirror is static at $x = 0$, we have $G(t) = F(t)$ and a single nontrivial Moore equation.

The description of the dynamics of the quantum field in the presence of moving mirrors is therefore reduced to solving the Moore's equations. Once this is achieved, the renormalized energy density of the field can be obtained from [34]

$$\langle T_{tt}(x,t)\rangle_{\text{ren}} = f_G(t+x) + f_F(t-x), \quad (8)$$

where

$$\begin{aligned} f_G &= -\frac{1}{24\pi}\left[\frac{G'''}{G'} - \frac{3}{2}\left(\frac{G''}{G'}\right)^2\right] + \frac{(G')^2}{2}\left[-\frac{\pi}{24} + Z(Td_0)\right] \\ f_F &= -\frac{1}{24\pi}\left[\frac{F'''}{F'} - \frac{3}{2}\left(\frac{F''}{F'}\right)^2\right] + \frac{(F')^2}{2}\left[-\frac{\pi}{24} + Z(Td_0)\right], \end{aligned} \quad (9)$$

and $d_0 = |R(0) - L(0)|$ is the initial length of the cavity. The above result is valid when the state of the field is initially in a thermal state at temperature T and $Z(Td_0)$ is related to the initial mean energy

$$Z(Td_0) = \sum_{n=1}^{\infty} \frac{n\pi}{\exp\left(\frac{n\pi}{Td_0}\right) - 1}. \quad (10)$$

The expression for the renormalized energy–momentum tensor above can be obtained using the standard approach based on point-splitting regularization (see for instance [36]). It can also be derived using the conformal anomaly associated with the conformal transformation Equation (7) [37]. In the rest of the paper, we consider the $T = 0$ case, in which the field is initially in the vacuum state.

It is important to note that for a static cavity with $L(t) = 0$, $R(t) = d_0$, the general solution for $F(z)$ and $G(z)$ is

$$F(z) = G(z) = \frac{(z-z_0)}{d_0} + p(t), \quad (11)$$

where z_0 is a constant, and $p(t)$ is a $2d_0$-periodic function. However, a closer look at Equation (8) shows that the renormalized energy density reduces to the vacuum energy (the static Casimir energy density) if and only if $p(t) = 0$, in other words, if the Moore functions are linear. Thus, it is the initial state of the cavity that determines this function, and the phenomenon of particle creation appears when $F(z)$ and $G(z)$ are nonlinear functions. The periodic function contains all the information of the excited state of the field.

2.1. STA for the Field

It is particularly challenging to find an STA for the quantum field in the cavity, since the only parameters that we can control and that affect the time evolution of the field are the positions of the left and right walls, $L(t)$ and $R(t)$, respectively. However, in previous papers [31,33], we have shown that this is indeed possible and can be achieved as follows. First, we find the adiabatic Moore functions

$$F_{\rm ad}(t) = \int dt \frac{1}{R_{\rm ref}(t) - L_{\rm ref}(t)} + \frac{1}{2}\frac{R_{\rm ref}(t) + L_{\rm ref}(t)}{R_{\rm ref}(t) - L_{\rm ref}(t)} \tag{12}$$

$$G_{\rm ad}(t) = \int dt \frac{1}{R_{\rm ref}(t) - L_{\rm ref}(t)} - \frac{1}{2}\frac{R_{\rm ref}(t) + L_{\rm ref}(t)}{R_{\rm ref}(t) - L_{\rm ref}(t)}. \tag{13}$$

which correspond to the infinitely slow evolution of the field for reference trajectories $L_{\rm ref}(t)$ and $R_{\rm ref}(t)$. Then, we look for effective trajectories $L_{\rm eff}(t)$ and $R_{\rm eff}(t)$ such that they give rise to the adiabatic Moore functions previously found

$$G_{\rm ad}(t + L_{\rm eff}(t)) - F_{\rm ad}(t - L_{\rm eff}(t)) = 0 \tag{14}$$

$$G_{\rm ad}(t + R_{\rm eff}(t)) - F_{\rm ad}(t - R_{\rm eff}(t)) = 2. \tag{15}$$

The effective trajectories obtained produce an evolution of the field in finite time that at the end replicates the adiabatic one for the reference trajectories; hence, they constitute an STA.

This protocol has the potential to dramatically improve the efficiency of a thermodynamical cycle, but the energy cost of the shortcut should be taken into account [24–26,32]. Although there is no universal consensus on exactly how to measure this cost, one possible metric in standard quantum mechanics is given by

$$\langle \delta W \rangle = \frac{1}{\tau} \int_0^\tau [\langle H_{\rm eff}(t) \rangle - \langle H_{\rm ref}(t) \rangle] dt, \tag{16}$$

where $H_{\rm eff}$ is the Hamiltonian that implements a shortcut to the adiabatic evolution for a reference Hamiltonian $H_{\rm ref}$, and protocols have a duration τ. However, in quantum field theory, the reference and effective protocols have different durations, $\tau_{\rm ref}$ and $\tau_{\rm eff}$, respectively, and so the previous measure should be adapted. The simplest possible generalization of the energy cost to QFT would be

$$\langle \delta W \rangle = \frac{1}{\tau_{\rm eff}} \int_0^{\tau_{\rm eff}} \int_{L_{\rm eff}(t)}^{R_{\rm eff}(t)} \langle T_{tt}^{\rm eff} \rangle_{\rm ren} dx dt - \frac{1}{\tau_{\rm ref}} \int_0^{\tau_{\rm ref}} \int_{L_{\rm ref}(t)}^{R_{\rm ref}(t)} \langle T_{tt}^{\rm ref} \rangle_{\rm ren} dx dt, \tag{17}$$

but further investigations are needed in order to understand whether this is in fact a faithful measure.

We discuss STA completions for this system in Section 4. Before doing that, we analyze the same problem in a simpler context that serves to illustrate the reverse engineering method used to build the STA completions.

3. A Simple Analogue: The Ermakov Equation for the Harmonic Oscillator with Time-Dependent Frequency

In this Section, we describe a simple analogue of the STA in QFT using a quantum harmonic oscillator. As we see, the excitation of the harmonic oscillator caused by the time dependence of the frequency can be described by the so-called Ermakov function. This function exhibits a behavior similar to that of the Moore functions for the scalar field in the optomechanical cavity.

The dynamics of a harmonic oscillator with a time-dependent frequency are given by

$$\ddot{q} + \omega^2(t)q = 0. \tag{18}$$

The position operator $\hat{q}$ can be written in terms of annihilation and creation operators ($\hat{a}$ and $\hat{a}^\dagger$) as

$$\hat{q}(t) = q(t)\hat{a} + q^*(t)\hat{a}^\dagger, \tag{19}$$

where $q(t)$ is a solution of Equation (18) with Wronskian given by

$$\dot{q}q^* - \dot{q}^*q = i. \qquad (20)$$

The Wronskian condition implies that the solutions can be written in terms of a real function $W(t)$ as

$$q(t) = \frac{1}{\sqrt{2W(t)}} e^{i \int^t W(t') dt'} \qquad (21)$$

that satisfies the equation

$$\omega^2 = W^2 + \frac{1}{2}\left(\frac{\ddot{W}}{W} - \frac{3}{2}\left(\frac{\dot{W}}{W}\right)^2 \right), \qquad (22)$$

which is equivalent to Equation (18).

For a slowly varying function $\omega(t)$, we have $W \simeq \omega$, and Equation (21) gives the usual lowest-order WKB solution. Equation (22) can be used to obtain the higher-order corrections by solving it recursively using an expansion in the number of derivatives of ω. Alternatively, one can use an inverse engineering approach and think of Equation (21) as the exact solution of the problem with a frequency ω^2 given by Equation (22).

Assuming that the frequency tends to constants values $\omega^{\text{in,out}}$ for $t \to \pm\infty$, and that the oscillator is in the ground state $|0_{\text{in}}\rangle$ for $t \to -\infty$, in the case of a nonadiabatic evolution the oscillator will be excited for $t \to +\infty$, that is $|\langle 0_{\text{out}}|0_{\text{in}}\rangle| \neq 1$. The in and out basis are the solutions of Equation (21) that satisfy

$$q^{\text{in,out}}(t) \xrightarrow[t \to \pm\infty]{} \frac{1}{\sqrt{2\omega^{\text{in,out}}}} e^{-i\omega^{\text{in,out}} t}. \qquad (23)$$

The Bogoliubov transformation that connects the in and out basis and in and out Fock spaces, when nontrivial, is an indication of the excitation of the system due to the external time dependence:

$$\begin{aligned} q^{\text{out}} &= \alpha\, q^{\text{in}} + \beta\, q^{\text{in}*} \\ \hat{a}^{\text{out}} &= \alpha^* \hat{a}^{\text{in}} - \beta^* \hat{a}^{\text{in}\dagger}. \end{aligned} \qquad (24)$$

As is well known, and described in more detail below, the in vacuum can be written as a squeezed state in terms of the out states.

A frequency $\omega(t)$ that leads to an evolution that does not produce an excitation of the harmonic oscillator constitutes an STA. It is important to remark that the evolution at intermediate times is in general nonadiabatic, but the system returns to the initial state when the effective frequency becomes constant at $t \to +\infty$. The system is excited at intermediate times and subsequently returns to its ground state.

3.1. The Lewis-Riesenfeld Approach and the Ermakov Equation

In the Lewis–Riesenfeld approach, the solutions to Equation (18) are written in terms of the so-called Ermakov function $\rho = 1/\sqrt{W}$ that satisfies the Ermakov equation

$$\ddot{\rho} + \omega^2(t)\rho - \frac{1}{\rho^3} = 0, \qquad (25)$$

which is equivalent to Equation (22).

It is possible to show that within a temporal interval where $\omega = \omega_0$ is constant, the general solution of the Ermakov equation reads [38,39]

$$\rho^2(t) = \frac{1}{\omega_0}\left[\cosh\delta - \sinh\delta \sin(2\omega_0 t + \varphi)\right], \qquad (26)$$

where δ and φ are arbitrary constants. For $\delta = 0$, we have the usual solution for the harmonic oscillator with frequency ω_0. If the frequency is ω_0 for $t < 0$, time-dependent in the interval $0 < t < \tau$, and then stops at ω_1, for $t < 0$ we will have $\rho^2 = 1/\omega_0$, and at the end of the motion, for $t > \tau$, ρ^2 is given by Equation (26) with $\omega_0 \to \omega_1$. The values of δ and φ will depend on the whole temporal evolution of the frequency. In general, the oscillator will end up in a squeezed state. In ref. [40], it has been shown that for the particular case $\omega_0 = \omega_1$, different evolutions $\omega(t)$ may lead to the same Ermakov function for $t > \tau$ and therefore to the same excited state. If the evolution is such that ρ is also constant for $t > \tau$, then we have an unexciting evolution.

One can use reverse engineering to find effective unexciting evolutions $\omega_{\text{eff}}^2(t)$ by an adequate choice of $\rho_{\text{ref}}^2 = 1/\omega_{\text{ref}}(t)$; indeed, assuming that ρ_{ref}^2 is constant both for $t < 0$ and $t > \tau$, and plugging this "reference" Ermakov function into Equation (25), one can obtain the effective evolution as

$$\omega_{\text{eff}}^2 = \frac{1}{\rho_{\text{ref}}^4} - \frac{\ddot{\rho}_{\text{ref}}}{\rho_{\text{ref}}}. \tag{27}$$

Note that for this effective evolution the function $q(t)$ evolves as the adiabatic solution for the reference frequency ω_{ref} in a finite time. The model may admit, or not, situations where $\omega_{\text{eff}}^2(t) < 0$; so, one should choose $\rho_{\text{ref}}(t)$ appropriately in models where this is physically unacceptable.

3.2. De-Excitation of the Harmonic Oscillator

Now, we address the following question: assume that the frequency is equal to ω_- for $t < 0$ and evolves from ω_- to ω_+ during the interval $0 < t < \tau_1$ in such a way that the evolution "generates particles", that is, that the oscillator is in an excited state for $t > \tau_1$. We denote by $\omega_I(t)$ the function that interpolates between ω_- and ω_+. Is it possible to find a subsequent time evolution of the frequency, $\omega_{II}(t)$, in an interval $t_1 < t < t_2$ such that the final frequency is ω_{++} and that the final state of the oscillator is $|0_{++}\rangle$? In other words, we are looking for a complementary time-dependent function $\omega_{II}(t)$ such that the joint effective protocol

$$\omega_{\text{eff}}(t) = \begin{cases} \omega_- & t < 0 \\ \omega_I(t) & 0 < t < \tau_1 \\ \omega_+ & \tau_1 < t < t_1 \\ \omega_{II}(t) & t_1 < t < t_1 + \tau_2 \\ \omega_{++} & t_1 + \tau_2 < t \end{cases}, \tag{28}$$

converts an initially nonadiabatic evolution into an STA.

After the initial evolution (described by $\omega_I(t)$), one can prove that the initial vacuum state becomes a squeezed state. That is, for $\tau_1 < t < t_1$, we have [41]

$$|0_-\rangle = c_0 \sum_{n \geq 0} \left(-\frac{\beta^*}{\alpha}\right)^n \frac{\sqrt{(2n)!}}{2^n n!} |2n_+\rangle, \tag{29}$$

where α and β are the coefficients of the Bogoliubov transformation. The mean occupation number of the + states reads

$$\langle 0_-|a_+^\dagger a_+|0_-\rangle = |\beta|^2. \tag{30}$$

In order to unfold the evolution and generate the corresponding antisqueezing, one could choose an adequate Ermakov function as follows: for $t < \tau_1$, $\rho(t)$ is determined by $\omega_I(t)$. It is an oscillating function for $\tau_1 < t < t_1$. We can now consider a smooth continuation of this function that starts at t_1 and becomes constant $\rho^2(t) = 1/\omega_{++}$ for $t > t_1 + \tau_2$. From the complete Ermakov function, one can determine the evolution of the "de-exciting frequency" $\omega_{\text{II}}(t)$ in the interval $t_1 < t < t_1 + \tau_2$ that interpolates between ω_+ and ω_{++}. The combination of the two evolutions implements the STA.

When $\omega_- = \omega_{++}$, the second evolution $\omega_{\text{II}}(t)$ can be chosen to be the time reversal of the first evolution $\omega_{\text{I}}(t)$. This symmetric trajectory always exists and can be constructed as follows. After the first evolution, for $t > \tau_1$, the square of the Ermakov function is given by Equation (26). At any time t_n that corresponds to a maximum or minimum of this periodic function, one can extend the Ermakov function symmetrically, that is $\rho_{\text{II}}(t) = \rho_{\text{I}}(2t_n - t)$ for $t > t_n$. From the Ermakov Equation (25), one can easily show that the whole evolution of the frequency is time-symmetric around t_n.

It is worth remarking that the temporal reverse of an adiabatic shortcut is also an adiabatic shortcut, as can be easily checked using the Ermakov equation. We show a similar property for the moving mirrors in the next section.

Finally, we point out that for a general state, the unfolding would not be possible, i.e., given an arbitrary state with the same $|\beta|^2$, an unitary evolution will not lead to a vacuum state. Note that a given value of $|\beta|^2$ only gives the mean occupation number but does not have the information about the full quantum state of the oscillator.

4. Completing an STA in the Optomechanical Cavity

In this section, we come back to the STA in the optomechanical cavity. We present different alternatives for completing a given trajectory into an STA for the quantum field. That is, we assume the cavity was initially in a vacuum state (zero temperature) at position R_- and has suffered a perturbation that moved the right wall according to the trajectory $R_{\text{I}}(t)$ with an associated Moore function $F(z)$. Our goal will be to find a second trajectory $R_{\text{II}}(t)$ such that the joint trajectory $R_{\text{eff}}(t)$ has a Moore function that is linear at early and late times, which will result in an adiabatic evolution of the field. We present different strategies to achieve this based on the same ideas described in the previous section for the quantum harmonic oscillator.

Before proceeding, we need to establish a magnitude to decide whether an STA has been achieved and measure how far we are from one. Hence, we define the adiabaticity coefficient

$$Q(t) := \frac{E(t)}{E_{\text{ad}}(t)}, \tag{31}$$

where $E(t)$ is the total energy in the cavity

$$E(t) = \int_{L(t)}^{R(t)} dx \langle T_{00}(x,t) \rangle_{\text{ren}}, \tag{32}$$

while the adiabatic energy is given by

$$E_{\text{ad}}(t) = -\frac{\pi}{24d}, \tag{33}$$

where $d = |R(t) - L(t)|$ is the length of the cavity. We are assuming that the field is initially in the vacuum state.

Once the effective trajectories that complete a shortcut and the associated Moore functions are obtained, the energy and adiabaticity coefficients can then be calculated using Equations (8) and (32).

4.1. Reverse Engineering

We now present a first method for completing a trajectory to be an adiabatic shortcut. The method is theoretically quite simple and works for an arbitrary final position. Its practical implementation is not so simple, and it involves three steps: the computation of the Moore functions associated with the initial evolution, the extension of them smoothly into linear functions, and the computation of the effective trajectory using inverse engineering.

As mentioned, our goal here is to complete an adiabatic shortcut for a cavity that was initially in a vacuum state and was perturbed by an arbitrary trajectory of one of the mirrors that left it in an excited state. Because of the initial condition, we know that the associated Moore function was a linear function before the motion started and, since an STA is achieved by having a Moore function that is linear at early and late times, we can theoretically complete an adiabatic shortcut by simply extending the Moore function continuously to a linear function at late times and computing the associated effective trajectory. Notice that this method for completing an STA is completely analogous to the one previously presented for the harmonic oscillator.

In order to illustrate this strategy, let us consider an initial trajectory for the mirrors given by

$$L_I(t) = 0$$
$$R_I(t) = \begin{cases} R_- & t < 0 \\ f(t) & 0 < t < \tau \\ R_+ & \tau < t \end{cases}, \quad (34)$$

where $f(t)$ is given by the following polynomial

$$f(t) = R_-(1 - \epsilon \delta(t/\tau))$$
$$\delta(x) = 35x^4 - 84x^5 + 70x^6 - 20x^7, \quad (35)$$

which verifies $\delta(0) = 0$ and $\delta(1) = 1$. The choice of the polynomial that defines $\delta(t)$ ensures that $R_I(t)$ and its first three derivatives are continuous. This is needed to avoid spurious divergences in the energy–momentum tensor (see Equation (9)).

Using this trajectory, we can compute the associated Moore function F_I, which is oscillating at late times due to the creation of photons, and extend it continuously into a linear function F_{eff} with any desired slope (which in turn determines the final position of the boundary). Then, we can recover the corresponding trajectory of the mirror $R_{\text{eff}}(t)$ for the extended function by solving Equation (6) (Figure 2).

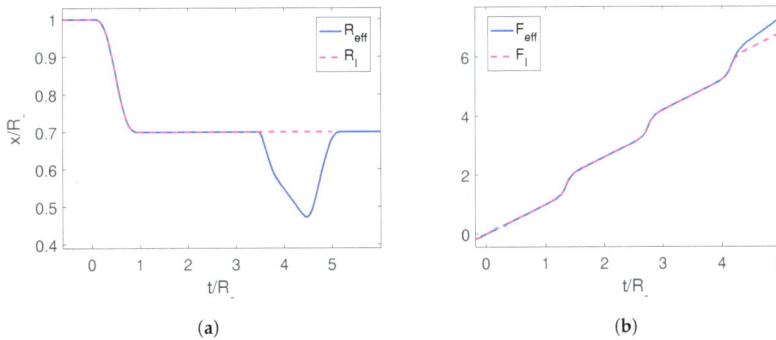

Figure 2. (a) Initial trajectory for the right mirror in magenta and its completion to an adiabatic shortcut in blue. (b) Moore functions for the corresponding trajectories. The parameters employed for the first trajectory are $R_- = 1$, $\epsilon = 0.3$ and $\tau = 1$.

The resulting trajectory is a well-defined continuous function that starts at the final position of the initial perturbation and ends at the position set by us through the slope of the linear function at late times. The speed can be seen to be below the speed of light. We can also check that the end-to-end trajectory constitutes a shortcut, since the adiabaticity parameter Q_{eff} starts and ends at 1 (Figure 3).

The disadvantage of this method is that in order to erase the photons generated by the initial motion, one needs to compute the Moore function of the field, then extend it smoothly, and subsequently compute the trajectory that completes an adiabatic shortcut by solving Moore's equation.

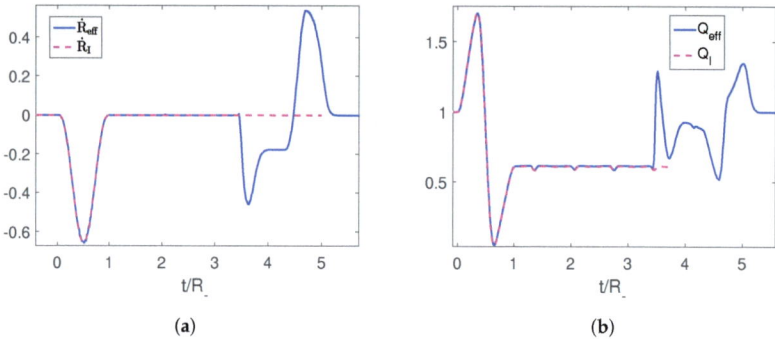

Figure 3. (a) Velocity for the initial trajectory of the mirror and for its completion to an adiabatic shortcut in blue. (b) Adiabaticity parameters for the corresponding trajectories. The parameters employed for the first trajectory are $R_- = 1$, $\epsilon = 0.3$ and $\tau = 1$.

4.2. Short Pulses

In this section, we explore how to complete an STA using the idea of the previous section, now applied to an arbitrary short pulse. In other words, we show how to erase the photons generated by any brief motion of one of the cavity mirrors and reset the cavity to its initial state.

We again consider an initial trajectory given by Equations (34) and (35) such that $\tau \leq R_\pm$. It can be seen that in this case, the derivative of the Moore function, $F'_I(z)$, has a simple structure. This is given by an initial constant, $1/R_-$, followed by a pulse that starts at $z = R_-$ and ends at $z = \tau + R_+$. From there, and up to $z = 2R_+$, the function again takes the value of the initial constant. This structure is repeated periodically with period $2R_+$ (Figure 4).

In order to show this, we consider the derivative of the Moore Equation (6)

$$F'_I[t + R(t)][1 + \dot{R}_I(t)] - F'_I[t - R_I(t)][1 - \dot{R}_I(t)] = 0, \quad \forall t. \tag{36}$$

Using the initial condition $R_I(t < 0) = R_-$ and $F'_I(z < 0) = 1/R_-$, we have

$$F'_I[t + R_I(t)] = \frac{1}{R_-} \frac{[1 - \dot{R}_I(t)]}{[1 + \dot{R}_I(t)]} = \frac{1}{R_-} \quad \text{if } t < 0 \tag{37}$$

from which we conclude that the derivative of the Moore function is constant up to $z < R_-$

$$F'_I[z] = \frac{1}{R_-} \quad \text{if } z < R_-. \tag{38}$$

Additionally, we can use Equation (37) to show that if $0 < t < \tau < R_\pm$, then $t - R(t) < 0$, and we have

$$F'_I[t + R_I(t)] = \frac{1}{R_-} \frac{[1 - \dot{R}_I(t)]}{[1 + \dot{R}_I(t)]} \quad \text{if } 0 < t < \tau. \tag{39}$$

This equation sets the shape of F'_I for $R_- < z = t + R_I(t) < \tau + R_+$. Lastly, once the trajectory has stopped, we have

$$F'_I[t + R_+] = F'_I[t - R_+] \quad \text{if } \tau < t, \tag{40}$$

which relates the F'_I at late times to its value at an earlier time. Indeed, we can rewrite

$$F'_I[z] = F'_I[z - 2R_+] \quad \text{if } \tau + R_+ < t + R_+ = z, \tag{41}$$

and this sets the $2R_+$ periodicity at late times. We can use this to find the value of the derivative for $z < R_- + 2R_+$ using Equation (38) to find

$$F'_I[z] = \frac{1}{R_-} \quad \text{if } \tau + R_+ < z < R_- + 2R_+, \tag{42}$$

which establishes that the derivative of the Moore function is constant in that interval.

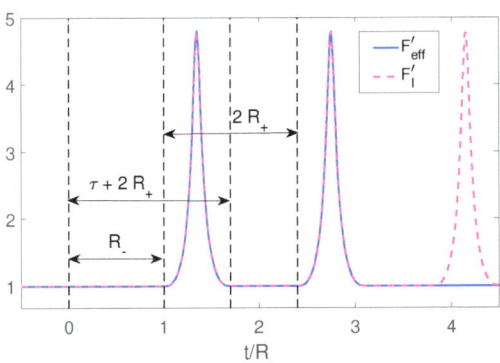

Figure 4. Structure of the derivative of the Moore function for a brief pulse trajectory in dashed magenta and the completion to an adiabatic shortcut in solid blue line.

From this structure, it is very easy to find an extension of the Moore function that is linear at late times, taking advantage of the fact that the derivative of the Moore function is constant in the interval $\tau + R_+ < z < R_- + 2R_+$. The natural extension is to complete the derivative of the Moore function, assuming that is constant for $t > R_- + 2R_+$. Therefore, $F_{\text{eff}}(z) = z/R_-$ for $z > R_- + 2R_+$. The extended Moore function will give rise to a trajectory $R_{\text{eff}}(t)$ that will coincide with $R_I(t)$ initially; then, the trajectory will be constant, and finally, there will be an erasing trajectory $R_{II}(t)$ which will have to come back to R_- to satisfy the final slope of the Moore function.

This erasing trajectory will satisfy Equation (36)

$$\frac{1}{R_-}\left(\frac{1 + \dot{R}_{II}(t)}{1 - \dot{R}_{II}(t)}\right) = F'_{\text{eff}}[t - R_{II}(t)], \quad \text{if } t > R_- + R_+ \tag{43}$$

where we used the extension condition $F'_{\text{eff}}(t + R_{\text{eff}}(t)) = 1/R_-$ for $z = t + R_{\text{eff}}(t) > R_- + 2R_+$. Of course, this equation should be coupled with the condition that the erasing trajectory begins where R_I ended, i.e.,

$$R_{II}(t = R_+ + R_-) = R_+. \tag{44}$$

The previous equation can be solved exactly in terms of the initial trajectory by taking

$$R_{II}(t) = R_+ - [R_I(t - t_1) - R_-], \text{if } t_1 < t < t_1 + \tau \tag{45}$$

where $t_1 = R_- + R_+$.

This can be seen by replacing in Equation (43)

$$\frac{1}{R_-}\left(\frac{1 - \dot{R}_I(t - (R_- + R_+))}{1 + \dot{R}_I(t - (R_- + R_+))}\right) = F'_{\text{eff}}[t - (R_- + R_+) + R_I(t - (R_- + R_+))], \tag{46}$$

which is satisfied because it is simply Equation (37) evaluated at $0 < t' = t - (R_- + R_+) < \tau$.

It is worth mentioning a couple of generalizations that can be derived from the previous result. The first one is that since the initial Moore function F_I is $2R_+$-periodic, it can be extended to a linear function at $z_n = R_- + 2nR_+$ for any natural n, which corresponds to applying the erasing trajectory $R_{II}(t)$ at $t_n = R_- + (2n-1)R_+$. Therefore, so far, we have shown that any given initial trajectory $R_I(t)$ with a small enough duration ($\tau < R_\pm$) can be completed to an adiabatic shortcut by following the protocol

$$R_{\text{eff}}(t) = \begin{cases} R_- & t < 0 \\ R_I(t) & 0 < t < \tau \\ R_+ & \tau < t < t_n \\ R_{II}(t) & t_n < t < t_n + \tau \\ R_- & t_n + \tau < t \end{cases}, \quad (47)$$

with $R_{II}(t)$ set by Equation (45).

A second generalization that can also illustrate the mechanism behind shortcut completion for short pulses is allowing the erasing trajectory to be executed by the left mirror. In this case, since initially the left mirror is at rest, we have $L(0 < t < \tau) = 0$. Equations (5) and (6) then imply that $F_I = G_I$ and

$$F_I(t + R_I(t)) - F_I(t - R_I(t)) = 2. \quad (48)$$

Therefore, the derivative of F satisfies Equation (37), and we can thus extend it smoothly in the same manner as before.

However, now the erasing trajectory has a static right mirror, $R_{II}(t) = R_+$, and a nontrivial trajectory for the left mirror, $L_{II}(t)$, to be determined by the Moore equations

$$G_{\text{eff}}(t + L_{II}(t)) - F_{\text{eff}}(t - L_{II}(t)) = 0 \quad (49)$$
$$G_{\text{eff}}(t + R_-) - F_{\text{eff}}(t - R_-) = 2. \quad (50)$$

The second equation implies $G_{\text{eff}}(t) = 2 + F_{\text{eff}}(t - 2R_-)$ which, by replacing in the first one and taking the time derivative, leads to

$$F'_{\text{eff}}(t + L_{II}(t) - 2R_-) = \frac{1}{R_-} \frac{[1 - \dot{L}_{II}(t)]}{[1 + \dot{L}_{II}(t)]}. \quad (51)$$

This equation determines the erasing trajectory for the left mirror and can be solved by taking

$$L_{II}(t) = R(t - t_n) - R_-, \quad t_n < t < t_n + \tau \quad (52)$$

where $t_n = 2nR_+ + R_-$ for any natural number n. This can be seen simply by replacing it in the previous equation and comparing again with Equation (37).

In order to illustrate these results, we consider an initial trajectory given by Equation (34) with

$$f(t) = R_- \cos\left[A \sin^2(\omega t)\right], \quad (53)$$

where A and ω are fixed parameters. As in the previous example, this choice ensures the continuity of $R_I(t)$ and its first derivatives. In Figure 5, we can see that by following the protocol given by Equation (47), i.e., executing the trajectory R_{II} at the precise time $t_1 = 2R_-$, the derivative Moore function remains constant, thereby erasing the photons generated by the pulse R_I and producing an adiabatic shortcut.

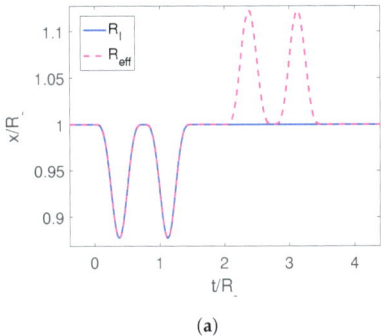
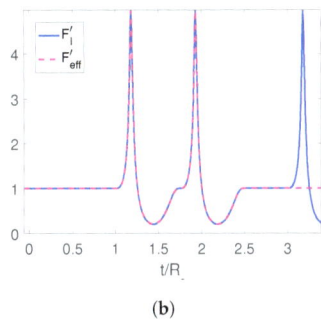

Figure 5. (**a**) Trajectory for the initial velocity of the mirror in blue and for its completion to an adiabatic shortcut in magenta. (**b**) Derivative of the Moore function for the corresponding trajectories. The parameters employed for the first trajectory are $R_- = R_+ = 1$, $A = 0.5$ and $\omega = 2\pi/\tau = 2\pi/1.5$.

To have a better understanding of how this is actually achieved, we can look at the energy density inside the cavity in Figure 6. There, we can observe that a pulse of energy is emitted, it then reflects off the left mirror and comes back to the right mirror precisely when the erasing trajectory R_{II} begins and reabsorbs it by moving in the opposite direction of the photons propagation. This mechanism can also be seen in the case of two moving mirrors. Once again, we see that the destructive interference necessary to erase the initially generated photons requires that during the second trajectory, the mirror should move in a direction opposite to the propagation of the pulse at the precise time when the pulse reaches that boundary (see Equation (52)). Therefore, one would expect that the adiabaticity parameter and thus the final state of the cavity would be highly dependent on the time t_n. This is indeed the case, as can be seen in Figure 6, where although we have $Q = 1$ for $t = t_1 = 2R_-$, $t_2 = 4R_-$, at intermediate times, Q deviates greatly from adiabaticity.

The advantage of this method over the previously presented is clear: for short pulses, one does not need to compute the Moore function; it is enough to apply the erasing trajectory for the right mirror R_{II} (or L_{II} for the left mirror) at any of the precise times t_n. This will annihilate the particles and execute an STA that will return the state of the cavity to the ground state with the same initial length.

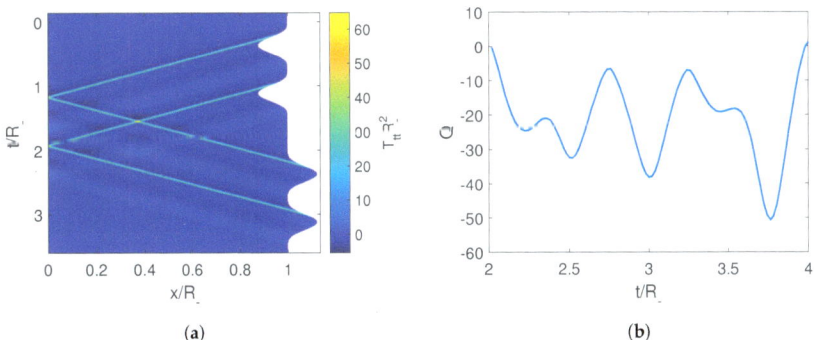

Figure 6. (**a**) Energy density of the field inside the cavity for the effective trajectory composed of an initial trajectory followed by an erasing trajectory at t_1. (**b**) Adiabaticity parameter obtained when by implementing the erasing trajectory Equation (45) at different times t_n. Note that $Q = 1$ for $t = t_1$ and $t = t_2$. The parameters employed for the first trajectory are $R_- = R_+ = 1$, $A = 0.5$, $\omega = 2\pi/1.5$, $t_1 = 2R_-$.

4.3. Time Inversion

In this section, we present a third method for completing STA by taking advantage of a time-inversion symmetry of the system. As a byproduct of this approach, we show that the time reversal of an STA is also an STA, which is a result that may be useful when considering thermodynamic cycles.

First, we note that although the physical theory does not have time-inversion invariance due to the moving boundary condition, the theory in the conformal variables Equation (7) is symmetric with respect to the time inversion $\tilde{t} \to -\tilde{t}$. It is simple to see that this symmetry in the conformal variables generates a symmetry in the physical theory given by the transformation $t \to -t$, i.e., a time reversal, and $F(z) \to -G(-z), G(z) \to -F(-z)$. We call this conformal time-reversal symmetry.

From this, we can establish two results. One is that even when a trajectory $L(t), R(t)$ is not an STA for the field, if there are constants $\tau, \tilde{z}, C_F, C_G$ such that the Moore functions satisfy

$$F(z) = -G(-z + \tilde{z}) + C_F \quad z \gg \tau \tag{54}$$
$$G(z) = -F(-z + \tilde{z}) + C_G \quad z \gg \tau, \tag{55}$$

then there are times at which implementing $L(t), R(t)$ followed by its temporal reverse $L(-t), R(-t)$, generates an STA. This is because, under these conditions, it is possible to smoothly continue the Moore functions of the trajectories with those associated with their temporal reverse. Since the former are initially linear, the latter must also be linear at late times. Therefore, a smooth continuation of the first Moore function into the other generates an STA formed by the trajectory followed by its temporal reversal.

In the case that $L(t) = 0$, we have $F(z) = G(z)$, and the previous condition can be expressed more simply as

$$F'(z) = F'(-z + \tilde{z}) \quad z \gg \tau, \tag{56}$$

meaning that the derivative of the Moore function at late times should be an even function for some suitable choice of the origin. In fact, since the Moore function is R_+-periodic if $\tilde{z}$ satisfies this condition, then $\tilde{z}_n = \tilde{z} + 2nR_+$ also does, and so to complete a shortcut the time-reversed trajectory has to be implemented at certain discrete times t_n.

This method works only when the complete shortcut starts and ends at the same position. However, it works for any initial evolution, i.e., it does not need to be a short pulse. It can have any duration as long as the Moore function verifies the condition given by Equation (54).

In order to check this result numerically, we consider the initial trajectories R_I given by Equations (34) and (35), and then we apply it to the cavity followed by the time-reversed trajectory at times t_n, carefully chosen so that the two Moore functions coincide to form a longer trajectory R_{eff}. As we can see in Figure 7, acting on the time-reverse trajectory at discrete times, we manage to take the adiabaticity parameter Q from 0.6 back to 1, signaling a successful adiabatic shortcut.

A second result that can be obtained from the conformal time-reversal symmetry is that if the trajectories $L(t), R(t)$ constitute an STA, then the time-reversed trajectories are also an STA. To see that this is the case, note that the Moore functions of the original STA are linear both at early and late times

$$F(z \to \pm\infty) = \frac{z}{R_\pm - L_\pm} + \frac{1}{2}\frac{R_\pm + L_\pm}{R_\pm - L_\pm} \tag{57}$$

$$G(z \to \pm\infty) = \frac{z}{R_\pm - L_\pm} - \frac{1}{2}\frac{R_\pm + L_\pm}{R_\pm - L_\pm}. \tag{58}$$

Therefore, applying a conformal time-reversal transformation, we can conclude that the time-reversed trajectories $L(-t), R(-t)$ have reverse Moore functions given by

$$F_{\text{rev}}(z \to \pm\infty) = \frac{z}{R_{\mp} - L_{\mp}} + \frac{1}{2}\frac{R_{\mp} + L_{\mp}}{R_{\mp} - L_{\mp}} \qquad (59)$$

$$G_{\text{rev}}(z \to \pm\infty) = \frac{z}{R_{\mp} - L_{\mp}} - \frac{1}{2}\frac{R_{\mp} + L_{\mp}}{R_{\mp} - L_{\mp}}, \qquad (60)$$

and so these trajectories are also adiabatic shortcuts but with initial and final positions exchanged. This property is particularly useful when considering thermodynamic cycles, in which we need not only an STA for the position of the boundaries to change from L_-, R_- to L_+, R_+ but also during a second stroke that returns the mirrors to their original positions.

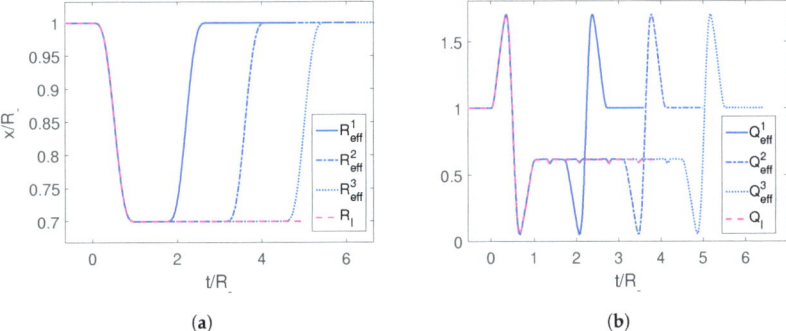

Figure 7. (a) First trajectory for the mirror in magenta and the effective trajectories R^n_{eff} using its temporal reverse at different times t_n in blue. (b) Adiabaticity parameters for the corresponding trajectories. The parameters employed for the first trajectory are $R_- = 1$, $\epsilon = 0.3$ and $\tau = 1$.

We can illustrate this numerically considering a reference trajectory given by Equations (34) and (35) and use them to calculate the corresponding effective STA following Section 2.1, $R_{\text{eff}}(t)$, which starts at t_i and ends at t_f. We then compute the temporal reverse $R_{\text{rev}}(t) = R_{\text{eff}}(t_f - t)$ and the adiabaticity coefficient Q_{rev}. We can clearly see from Figure 8 that Q_{rev} starts equal to 1, oscillates, and goes back to unity when the motion stops signaling that the evolution was indeed an adiabatic shortcut.

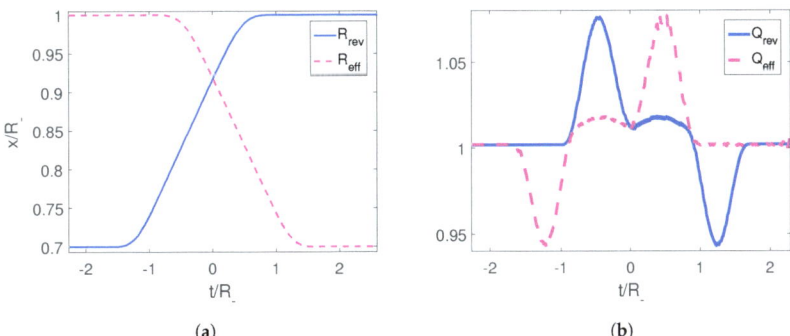

Figure 8. (a) Trajectories for the right mirror for an adiabatic shortcut in magenta and its temporal reverse in blue. (b) Adiabaticity parameters for the adiabatic shortcut and its temporal reverse. The parameters employed for the effective shortcut trajectory aew $R_- = 1$, $\epsilon = 0.3$ and $\tau = 1$.

As already mentioned, this property is also valid for a harmonic oscillator with time-dependent frequency: the temporal reverse of an STA is also an STA.

5. Discussion

The fast manipulation of micromachines can result in excessive losses in the form of quantum friction, which reduces its efficiency. Thus, finding an STA is of paramount importance in quantum thermodynamics since, even after considering the energy cost associated, STA-enhanced thermal machines have the potential for greatly enhanced efficiency.. It also allows for the design of rapid quantum gates, which makes it useful for quantum information processing as well. Here, we propose an additional application of an STA for state preparation. Once an initial state has been prepared, if a perturbation of the parameters of the system change suddenly, it will greatly reduce the fidelity of the desired state; however, knowing how they have been changed, we can complete this time variation into an STA which, by definition, will restore the state of the system to the initial one. This method can then allow for longer times between the preparation of a state and its processing [42].

In this paper, we addressed the problem of how to transform an initially nonadiabatic evolution into an STA, for a quantum field confined within a one-dimensional cavity featuring two moving mirrors. As outlined in our previous studies, the state of the field is determined by two Moore functions, F and G. In a time interval where the cavity remains static, these functions can be expressed as the sum of a linear and a periodic function. The presence of a nonzero periodic component indicates an excited state of the field. By characterizing the field's state using F and G, we can employ reverse engineering techniques to construct an STA. Regardless of the initial evolution, it is possible to smoothly extend the Moore functions to asymptotically linear functions, from which we can compute the mirrors trajectories.

We identified an inspiring similarity between the Moore functions that describe the state of the field in the one-dimensional cavity and the Ermakov function ρ that provides a formal solution for the quantum harmonic oscillator with a time-dependent frequency. Within a time interval of constant frequency, ρ^2 can be expressed as the sum of a constant and a periodic function. When the periodic component is nonzero, it describes an oscillator in a squeezed state. The STA completion can be constructed by extending the Ermakov function in such a manner that ρ tends to a constant as time approaches infinity.

We presented three different methods for completing an STA for the optomechanical cavity. All of them are based on the fact that the Moore functions should be linear before and after the perturbation in order for a trajectory of the mirrors to comprise an adiabatic shortcut.

The first method, presented in Section 4.1, consists of three steps: (1) computing the Moore functions, (2) extending them smoothly into a linear functions, and then (3) using them to compute the trajectory using inverse engineering. This method is conceptually simple, it can be used for any perturbation and executes an STA for any final position of the mirror; however, the three steps require implementing precise measurements or demand challenging computations.

The second method, described in Section 4.2, addresses these problems and solves them in the case that the perturbation consists of a short pulse. In that case, the second trajectory, which erases the excitations and restores the state of the system, can be given explicitly in terms of the initial perturbation. This allows us to avoid the computation and inversion of the Moore function entirely. The only restrictions are that perturbation time must be short and the final length of the cavity should be the same as the initial one.

The third method, presented in Section 4.3, takes advantage of a symmetry of the system to show that the time-reversed trajectory of the perturbation can be used to complete an adiabatic shortcut if the derivative of the Moore function of the perturbation is even for late times. In this case, the perturbation can have an arbitrary duration, but it is not necessary to compute its Moore function to check the parity. Nonetheless, when comparing this method with the first one, we avoid extending it smoothly and inverting it to compute the trajectory, which constitutes a great simplification.

With regard to the possible experimental implementation, it should be noted that the dynamical Casimir effect has been effectively demonstrated in superconducting circuits more than a decade ago [43]. In that case, a SQUID was used instead of a mechanical moving mirror, which simulated a time-dependent boundary condition for the field. The protocols presented here could be implemented in this type of setup by making use of two SQUIDs at both ends of a superconducting cavity in a way similar to [44]. Additionally, it was recently proposed to measure the dynamical Casimir effect due to true mechanical motion by using film bulk acoustic resonators in the GHz spectral range [45], which could also lead to an implementation of the schemes proposed here.

Further investigations should be conducted in order to establish how the different methods discussed here to complete an STA can be adapted for other systems, as well as how much they could improve the time delay between state preparation and processing under experimental conditions.

Author Contributions: N.F.D.G., F.C.L., F.D.M. and P.I.V. have contributed equally to the work. All authors have read and agreed to the published version of the manuscript.

Funding: This research was funded by Agencia Nacional de Promoción Científica y Tecnológica (ANPCyT), Consejo Nacional de Investigaciones Científicas y Técnicas (CONICET), Universidad de Buenos Aires (UBA) and Universidad Nacional de Cuyo (UNCuyo).

Institutional Review Board Statement: Not applicable.

Informed Consent Statement: Not applicable.

Data Availability Statement: Not applicable.

Acknowledgments: P.I.V. acknowledges the ICTP-Trieste Associate Program.

Conflicts of Interest: The authors declare no conflict of interest.

References

1. Binder, F.; Correa, L.A.; Gogolin, C.; Anders, J.; Adesso, G. (Eds.) *Thermodynamics in the Quantum Regime*; Springer: Berlin/Heidelberg, Germany, 2019.
2. Zurek, W.H. Environment-assisted invariance, entanglement, and probabilities in quantum physics. *Rev. Mod. Phys.* **2003**, *75*, 715. [CrossRef]
3. Karimi, B.; Pekola, J.P. Otto refrigerator based on a superconducting qubit: Classical and quantum performance. *Phys. Rev. B* **2016**, *94*, 184503. [CrossRef]
4. Kosloff, R.; Rezek, Y. The quantum harmonic Otto cycle. *Entropy* **2017**, *19*, 136. [CrossRef]
5. Del Grosso, N.F.; Lombardo, F.C.; Mazzitelli, F.D.; Villar, P.I. Quantum Otto cycle in a superconducting cavity in the nonadiabatic regime. *Phys. Rev. A* **2022**, *105*, 022202. [CrossRef]
6. Moore, G.T.J. Quantum theory of the electromagnetic field in a variable-length one-dimensional cavity. *Math. Phys.* **1970**, *11*, 2679. [CrossRef]
7. Dodonov, V.V. Current status of the dynamical Casimir effect. *Phys. Scr.* **2010**, *82*, 038105. [CrossRef]
8. Dalvit, D.A.R.; Neto, P.A.M.; Mazzitelli, F.D. Fluctuations, dissipation and the dynamical casimir effect. *Lect. Notes Phys.* **2011**, *834*, 419.
9. Nation, P.D.; Johansson, J.R.; Blencowe, M.P.; Nori, F. Colloquium: Stimulating uncertainty: Amplifying the quantum vacuum with superconducting circuits. *Rev. Mod. Phys.* **2012**, *84*, 1. [CrossRef]
10. Dodonov, V.V. Fifty years of the dynamical Casimir effect. *Physics* **2020**, *2*, 67–104. [CrossRef]
11. Blais, A.; Grimsmo, A.L.; Girvin, S.M.; Wallraff, A. Circuit quantum electrodynamics. *Rev. Mod. Phys.* **2021**, *93*, 025005. [CrossRef]
12. Gluza, M.; Sabino, J.; Ng, N.H.Y.; Vitagliano, G.; Pezzutto, M.; Omar, Y.; Mazets, I.; Huber, M.; Schmiedmayer, J.; Eisert, J. Quantum field thermal machines. *PRX Quantum* **2021**, *2*, 030310. [CrossRef]
13. Berry, M.V.J. Transitionless quantum driving. *Phys. A* **2009**, *42*, 365303. [CrossRef]
14. del Campo, A. Shortcuts to adiabaticity by counterdiabatic driving. *Phys. Rev. Lett.* **2013**, *111*, 100502. [CrossRef] [PubMed]
15. Chen, X.; Ruschhaupt, A.; Schmidt, S.; del Campo, A.; Guéry-Odelin, D.; Muga, J.G. Fast optimal frictionless atom cooling in harmonic traps: Shortcut to adiabaticity. *Phys. Rev. Lett.* **2010**, *104*, 063002. [CrossRef] [PubMed]
16. Masuda, S.; Nakamura, K. Fast-forward problem in quantum mechanics. *Phys. Rev. A* **2008**, *78*, 062108. [CrossRef]

17. Torrontegui, E.; Chen, X.; Modugno, M.; Schmidt, S.; Ruschhaupt, A.; Muga, J.G. Fast transport of Bose–Einstein condensates. *New J. Phys.* **2012**, *14*, 013031. [CrossRef]
18. Guéry-Odelin, D.; Ruschhaupt, A.; Kiely, A.; Torrontegui, E.; Martínez-Garaot, S.; Muga, J.G. Shortcuts to adiabaticity: Concepts, methods, and applications. *Rev. Mod. Phys.* **2019**, *91*, 045001. [CrossRef]
19. Palmero, M.; Bowler, R.; Gaebler, J.P.; Leibfried, D.; Muga, J.G. Fast transport of mixed-species ion chains within a Paul trap. *Phys. Rev. A* **2014**, *90*, 053408. [CrossRef]
20. Torrontegui, E.; Chen, X.; Modugno, M.; Ruschhaupt, A.; Guéry-Odelin, D.; Muga, J.G. Fast transitionless expansion of cold atoms in optical Gaussian-beam traps. *Phys. Rev. A* **2012**, *85*, 033605. [CrossRef]
21. Dowdall, T.; Benseny, A.; Busch, T.; Ruschhaupt, A. Fast and robust quantum control based on Pauli blocking. *Phys. Rev. A* **2017**, *96*, 043601. [CrossRef]
22. Amri, S.; Corgier, R.; Sugny, D.; Rasel, E.M.; Gaaloul, N.; Charron, E. Optimal control of the transport of Bose-Einstein condensates with atom chips. *Sci. Rep.* **2019**, *9*, 5346. [CrossRef] [PubMed]
23. Cakmak, B.; Mustecaplioulu, O.E. Spin quantum heat engines with shortcuts to adiabaticity. *Phys. Rev. E* **2019**, *99*, 032108. [CrossRef]
24. Abah, O.; Abah, M. Paternostro Shortcut-to-adiabaticity Otto engine: A twist to finite-time thermodynamics. *Phys. Rev. E* **2019**, *99*, 022110. [CrossRef] [PubMed]
25. Abah, O.; Lutz, E. Performance of shortcut-to-adiabaticity quantum engines. *Phys. Rev. E* **2018**, *98*, 032121. [CrossRef]
26. Abah, O.; Lutz, E. Energy efficient quantum machines. *Europhys. Lett.* **2017**, *118*, 40005. [CrossRef]
27. del Campo, A.; Goold, J.; Paternostro, M. More bang for your buck: Super-adiabatic quantum engines. *Sci. Rep.* **2014**, *4*, 6208.
28. Beau, M.; Jaramillo, J.; del Campo, A. Scaling-Up Quantum Heat Engines Efficiently via Shortcuts to Adiabaticity. *Entropy* **2016**, *18*, 168. [CrossRef]
29. Keller, T.; Fogarty, T.; Li, J.; Busch, T. Feshbach engine in the Thomas-Fermi regime. *Phys. Rev. Res.* **2020**, *2*, 033335. [CrossRef]
30. Li, J.; Fogarty, T.; Campbell, S.; Chen, X.; Busch, T. An efficient nonlinear Feshbach engine. *New J. Phys.* **2018**, *20*, 015005. [CrossRef]
31. Del Grosso, N.F.; Lombardo, F.C.; Mazzitelli, F.D.; Villar, P.I. Shortcut to adiabaticity in a cavity with a moving mirror. *Phys. Rev. A* **2022**, *105*, 052217. [CrossRef]
32. Calzetta, E. Not-quite-free shortcuts to adiabaticity. *Phys. Rev. A* **2018**, *98*, 032107. [CrossRef]
33. Del Grosso, N.F.; Lombardo, F.C.; Mazzitelli, F.D.; Villar, P.I. Fast adiabatic control of an optomechanical cavity. *Entropy* **2023**, *25*, 18. [CrossRef] [PubMed]
34. Dalvit, D.A.; Mazzitelli, F.D. Creation of photons in an oscillating cavity with two moving mirrors. *Phys. Rev. A* **1999**, *59*, 3049. [CrossRef]
35. Lewis, H.R., Jr. Classical and quantum systems with time-dependent harmonic-oscillator-type Hamiltonians. *Phys. Rev. Lett.* **1967**, *18*, 510. [CrossRef]
36. Davies, P.C.W.; Fulling, S.A. Radiation from a moving mirror in two-dimensional space-time: Conformal anomaly. *Proc. Roy. Soc. Lond. A* **1976**, *348*, 393.
37. Birrell, N.D.; Davies, P.C.W. *Quantum Fields in Curved Space*; Cambridge University Press: Cambridge, UK, 1984.
38. Lewis, H.R., Jr.; Riesenfeld, W.B. An exact quantum theory of the time-dependent harmonic oscillator and of a charged particle in a time-dependent electromagnetic field. *J. Math. Phys.* **1969**, *10*, 1458. [CrossRef]
39. Gjaja, I.; Bhattacharjee, A. Asymptotics of reflectionless potentials. *Phys. Rev. Lett.* **1992**, *68*, 2413. [CrossRef]
40. Coelho, S.S.; Queiroz, L.; Alves, D.T. Squeezing equivalence of quantum harmonic oscillators under different frequency jumps. *arXiv* **2023**, arXiv:2306.05577.
41. Hu, B.L.; Kang, G.; Matacz, A. Squeezed vacua and the quantum statistics of cosmological particle creation. *Int. J. Mod. Phys. A* **1994**, *9*, 991. [CrossRef]
42. Theis, L.S.; Motzoi, F.; Machnes, S.; Wilhelm, F.K. Counteracting systems of diabaticities using DRAG controls: The status after 10 years. *Europhys. Lett.* **2018**, *123*, 60001. [CrossRef]
43. Wilson, C.M.; Johansson, G.; Pourkabirian, A.; Simoen, M.; Johansson, J.R.; Duty, T.; Nori, F.; Delsing, P. Observation of the dynamical Casimir effect in a superconducting circuit. *Nature* **2011**, *479*, 376–379. [CrossRef] [PubMed]
44. Svensson, I.M.; Pierre, M.; Simoen, M.; Wustmann, W.; Krantz, P.; Bengtsson, A.; Johansson, G.; Bylander, J.; Shumeiko, V.; Delsing, P. Microwave photon generation in a doubly tunable superconducting resonator. *J. Phys. Conf. Ser.* **2011**, *969*, 012146. [CrossRef]
45. Sanz, M.; Wieczorek, W.; Gröblacher, S.; Solano, E. Electro-mechanical Casimir effect. *Quantum* **2018**, *2*, 91. [CrossRef]

Disclaimer/Publisher's Note: The statements, opinions and data contained in all publications are solely those of the individual author(s) and contributor(s) and not of MDPI and/or the editor(s). MDPI and/or the editor(s) disclaim responsibility for any injury to people or property resulting from any ideas, methods, instructions or products referred to in the content.

MDPI AG
Grosspeteranlage 5
4052 Basel
Switzerland
Tel.: +41 61 683 77 34

Entropy Editorial Office
E-mail: entropy@mdpi.com
www.mdpi.com/journal/entropy

Disclaimer/Publisher's Note: The title and front matter of this reprint are at the discretion of the Guest Editors. The publisher is not responsible for their content or any associated concerns. The statements, opinions and data contained in all individual articles are solely those of the individual Editors and contributors and not of MDPI. MDPI disclaims responsibility for any injury to people or property resulting from any ideas, methods, instructions or products referred to in the content.